Stellar Evolution and Nucleosynthesis

This astrophysics textbook, focusing on stellar evolution and nucleosynthesis, is aimed at advanced undergraduates with a background in maths, physics and astronomy. The book begins with a consideration of the properties of the Sun and other stars on the main sequence, then moves on to describe how gravitational contraction is the key driver of stellar evolution, before examining the physics of nuclear fusion. Subsequent chapters explain stellar evolution off the main sequence and helium-burning stars, followed by chapters exploring late stages of stellar evolution and stellar end-points. The book concludes with a chapter on star formation that demonstrates the cyclical nature of stellar evolution. Produced by academics drawing on decades of Open University experience in supported open learning, the book is completely self-contained with numerous worked examples and exercises (with full solutions provided), and illustrated in full-colour throughout. Designed to be worked through sequentially by a self-guided student, it also includes clearly identified key facts and equations as well as informative chapter summaries and an Appendix of useful data.

Sean G. Ryan is Professor of Astrophysics and Dean of the School of Physics, Astronomy and Mathematics at the University of Hertfordshire, where he leads a research programme into the formation and chemical enrichment of the Galaxy and the evolution of its first stars. Prior to this he worked for The Open University, contributing to the development of courses in physics and astronomy.

Andrew J. Norton is Senior Lecturer and Awards Director in the Department of Physical Science, The Open University, where his research focuses on time domain astrophysics of binary stars. He has authored teaching materials in physics and astronomy across the undergraduate curriculum, including twelve Open University study texts.

Stellar Evolution and Nucleosynthesis

Authors:

Sean G. Ryan

Andrew J. Norton

The Open University

CAMBRIDGE
UNIVERSITY PRESS

CAMBRIDGE UNIVERSITY PRESS

Cambridge, New York, Melbourne, Madrid, Cape Town, Singapore, São Paulo, Delhi, Dubai, Tokyo

Cambridge University Press
The Edinburgh Building, Cambridge CB2 8RU, UK

In association with THE OPEN UNIVERSITY

The Open University, Walton Hall, Milton Keynes MK7 6AA, UK

Published in the United States of America by Cambridge University Press, New York.

www.cambridge.org
Information on this title: www.cambridge.org/9780521133203

First published 2010.

Edited and designed by The Open University.

Typeset by The Open University.

Printed and bound in the UK by Henry Ling Limited, at the Dorset Press, Dorchester DT1 1HD.

This book forms part of an Open University course S382 *Astrophysics*. Details of this and other Open University courses can be obtained from the Student Registration and Enquiry Service, The Open University, PO Box 197, Milton Keynes MK7 6BJ, United Kingdom: tel. +44 (0)845 300 60 90, email general-enquiries@open.ac.uk

http://www.open.ac.uk

British Library Cataloguing in Publication Data available on request.

Library of Congress Cataloguing in Publication Data available on request.

ISBN   978-0-521-19609-3 Hardback
ISBN   978-0-521-13320-3 Paperback

Additional resources for this publication at www.cambridge.org/9780521133203

Cambridge University Press has no responsibility for the persistence or accuracy of URLs for external or third-party internet websites referred to in this publication, and does not guarantee that any content on such websites is, or will remain, accurate or appropriate.

1.2

STELLAR EVOLUTION AND NUCLEOSYNTHESIS

Contents

Introduction

In this book you will study the processes that lead to the formation of stars, the energy sources that fuel them, what they do during their lifetimes, and what happens when their fuel runs out. It is assumed that you already have a general, qualitative idea of some events in the life cycles of stars. This book is intended to take you beyond a *description* of events to an understanding of the physical processes that bring them about. That is, the book will not only tell you *what* happens, but will provide you with the physical tools to understand *why*, and to use the physics to work out what will happen as a star ages.

A key point is that stellar evolution is a cyclic process: stars are born, live their lives, and then die, but in doing so they seed the interstellar medium with nuclear-processed material that forms the building blocks of a subsequent generation of stars. We can, therefore, chose to begin an exploration of this cycle at any point, and follow the process until we get back to where we started.

This book is divided into eight chapters, each of which provides a fairly self-contained discussion of a particular phase in the life cycle of a star, or a particular physical process. We begin in Chapter 1 by looking at the general properties of stars on the main sequence of the Hertzsprung–Russell diagram, then in Chapter 2 explore the key driver of stellar evolution, namely the process of gravitational collapse. In Chapter 3 we explore the physics of nuclear fusion in detail, concentrating on the fusion reactions that power a star on the main sequence. Chapter 4 examines how stars evolve from the main sequence to the red-giant branch whilst the phase of helium burning is described in Chapter 5. The next two Chapters, 6 and 7, consider advanced stages of stellar evolution and the fate of stars including white dwarfs, neutron stars and black holes. The last part of the book, in Chapter 8, considers how stars form out of the interstellar material which has been enriched by previous generations of stars. This brings us full circle back to the main sequence again, where we started.

The book is designed to be worked through in sequence; some aspects of later chapters build on the knowledge gained in earlier chapters. So, whilst you could dip in at any point, you will find if you do so that you are often referred back to concepts developed elsewhere in the book. Our intention is, that if the book is studied sequentially, it provides a self-contained, self-study course in stellar astrophysics.

A special comment should be made about the exercises in this book. You may be tempted to regard them as optional extras that are only there to help you refresh your memory about certain concepts when you re-read the text in preparation for an exam. *Do not fall into this trap!* The exercises are part of the *learning*. Several of the important concepts are developed through the exercises and nowhere else. Therefore, you should attempt each of them when you come to it. You will find full solutions for every exercise at the end of this book, but do try to complete the exercise yourself first before looking at the answer. A table of physical constants is also given at the end of the book; use these values as appropriate in your calculations.

For most calculations presented here, use of a scientific calculator is essential. In some cases, you will be able to work out order of magnitude estimates without the use of a calculator, and such estimates are invariably useful to check whether an expression is correct. In some calculations you may find that use of a computer spreadsheet, or graphing calculator, provides a convenient means of visualizing a particular function. If you have access to such tools, please feel free to use them.

Finally, a note about the genesis of this book. It was originally written for an astrophysics course which used, in addition, a book called *The Physics of Stars* by A. C. Phillips. We commend that book to readers, and acknowledge the influence that it had on the development of our approach to teaching this subject.

Chapter 1 Main-sequence stars

Introduction

In this chapter we present a set of topics concerning the physics of stars on the main sequence. These topics serve to illustrate some of the general principles of stellar astrophysics and allow you to begin a quantitative exploration of the structure and behaviour of stars. We review the Hertzsprung–Russell diagram, introduce some parameters of the Sun as a typical star, and consider the general equations of stellar structure. We then present a summary of the nuclear fusion reactions that occur in main-sequence stars and consider the amount of energy released by the fusion of hydrogen. At this stage, we will not go into the detailed quantum physics which underlies the nuclear fusion reactions of hydrogen burning; that is left for Chapter 3. We begin by considering the question of just what we mean by 'a star'.

The starting point for the formation of a star is a cloud of cold gas, composed primarily of hydrogen and helium, with traces of heavier elements (usually referred to as *metals*). The cloud collapses due to its own gravity, and as gravitational potential energy is released and converted into heat, the pressure, density and temperature of the material increase. This causes the cloud to begin glowing, initially at infrared and later at optical wavelengths. Ultimately, if the cloud core reaches sufficiently high temperatures – in excess of several million kelvin – nuclear reactions begin. These thermonuclear reactions provide a non-gravitational source of energy release, whose heating provides pressure to support the gas against further collapse. This changes the behaviour of the object for which, previously, gravitation was the dominant factor, and marks the transition of the object to a star. By definition, a *star* is an object in which nuclear reactions *are* (or *have been*) sufficient to balance surface radiation losses.

1.1 The Hertzsprung–Russell diagram

When we observe stars, the two most obvious characteristics are their brightness and colour. Using a star's distance to calculate its luminosity (L) from its brightness, and determining its temperature (T) from its colour, reveals that distinct relationships are found between luminosity and temperature as illustrated in the **Hertzsprung–Russell** (H–R) diagram (e.g. Figure 1.1 overleaf). Stars are not distributed randomly in this diagram. The location of a star in the H–R diagram reflects its mass, radius, age, evolutionary state and chemical composition. The most densely populated regions of the figure are where stars spend most of their lives, whereas the probability of finding a star in a short-lived phase is much lower, so fewer stars are seen in such states. Note that the H–R diagram is a snapshot – a view at a single instant – of the evolutionary states of *many* stars. It is not the evolutionary track that an *individual* star traces out over its lifetime.

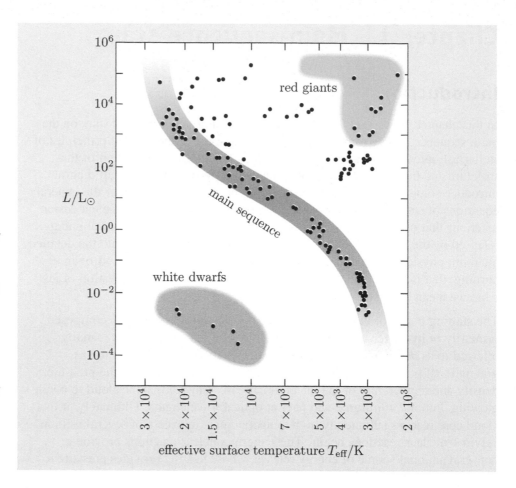

Figure 1.1   A schematic Hertzsprung–Russell diagram. This provides a snapshot of the luminosity L and effective surface temperature T_{eff} of many stars at different stages of their evolution. Note that the temperature increases from right to left. Most of the stars lie along a diagonal band, from upper-left (hot and luminous) to lower-right (cooler and fainter), called the main sequence. The second most populated region of the diagram, top right, is called the red-giant branch. The stars toward the lower left lie on the white-dwarf branch.

The H–R diagram can be presented in different ways. The original, observational diagram showed star brightness (measured as the **absolute visual magnitude**, M_V) on the vertical axis and stellar **spectral type** on the horizontal axis. Spectral type is very closely related to temperature and colour. Temperatures cannot be measured *directly* for many stars, but colours can be, so the *observational* H–R diagram is often a plot of M_V against **colour index** (such as $B - V$). The *theoretical* H–R diagram uses quantities more closely related to computations of stellar models, and plots luminosity, L, (energy radiated per unit time) against effective surface temperature, T_{eff}.

The **Stefan–Boltzmann law** states that the radiant flux F (radiant energy per square metre of surface area per second) from a **black body** of temperature T is given by $F = \sigma T^4$, where $\sigma = 5.671 \times 10^{-8}$ W m^{-2} K^{-4} and is called the Stefan–Boltzmann constant.

From now on radiant flux will be referred to simply as *flux*.

The radiant flux F passing through the surface of a spherical object of luminosity L and radius R (whose surface area is $4\pi R^2$) is simply

$$F = \frac{L}{4\pi R^2}.$$

If the object is a black body, then its luminosity L, radius R and temperature T are related by the equation

$$\frac{L}{4\pi R^2} = \sigma T^4.$$

How does this equation for a black body relate to a star? Moreover, if the Sun is a ball of gas, what do we *mean* by its *surface*? Light is continually emitted and reabsorbed by the hot gas that is the Sun. Near its outer layers, the fog-like gas eventually becomes transparent enough that some of the light escapes without being reabsorbed. However, the transition from being opaque to being transparent happens over a considerable distance. This zone, which is almost 500 km thick, is called the **photosphere**, and is the best definition we have for a surface for the Sun. The same is true in other stars, though the thickness of the photosphere is even greater in giants.

Because a star does not have an opaque, solid surface, light reaching an observer comes from a *range* of depths in its gaseous outer layers, each layer having a different temperature. What temperature should be used to characterize the surface? A useful *convention* is to refer to the temperature which a black body of the same luminosity and radius would have. This is called the **effective surface temperature** T_{eff}, and is defined as follows:

$$L = 4\pi R^2 \sigma T_{\mathrm{eff}}^4. \tag{1.1}$$

From the definition, we see that the effective surface temperature is related to the flux through the surface of the star, but how does it compare to the temperatures of the gas in its outer layers? Fortunately, stars are very good approximations to black bodies, since they absorb any light falling on them. Consequently, *the effective surface temperature coincides with the temperature at an intermediate depth in the photosphere* of the star. Deeper layers have temperatures higher than T_{eff}, while shallower layers in the photosphere have temperatures lower than T_{eff}. We will use the effective surface temperature to characterize the outer layers or surface of the star, even though a star does not have a solid surface in the conventional sense.

Note that Equation 1.1 involves three variables, L, R and T_{eff}, so if two are specified then the third can be calculated. Moreover, two of these are the axes of the theoretical H–R diagram, T_{eff} and L, so the third variable, R, can also be calculated for the H–R diagram.

- Equation 1.1 gives the relationship between the luminosity, effective surface temperature and radius of stars. How should it be changed for white dwarfs, which are fading away, or young protostars, which have yet to begin thermonuclear burning?

- Don't mess with that equation! It works just fine as it is, for any black body. It therefore applies to *all* stages of stellar evolution.

Exercise 1.1 Draw curves on Figure 1.1 to show where stars of radii $100\,\mathrm{R_\odot}$, $10\,\mathrm{R_\odot}$, $1\,\mathrm{R_\odot}$ and $0.1\,\mathrm{R_\odot}$ lie. To do this, calculate the luminosities of stars having radii of $100\,\mathrm{R_\odot}$, $10\,\mathrm{R_\odot}$, $1\,\mathrm{R_\odot}$ and $0.1\,\mathrm{R_\odot}$, for six values of effective surface temperature: 2000 K, 4000 K, 6000 K, 10 000 K, 20 000 K and 40 000 K. Use the values for the solar luminosity and radius, and the Stefan–Boltzmann constant σ given at the end of this book. You could use a calculator to do this, but it will be less tedious if you use a spreadsheet. ■

Figure references beginning with 'S' are to be found in the Solutions to exercises, at the end of the book.

Make sure you understand the spacing and shape of the curves in Figure S1.1, which shows the result of the exercise above. The curves are straight lines, separated by equal amounts, and the one corresponding to $1\,R_\odot$ passes through the datum point for the Sun, although this is not shown. A key thing to note is that both Figure 1.1 and Figure S1.1 are plotted using log–log scales, rather than simple linear scales. The spacing and shapes of the curves can be understood by taking logarithms of the equation $L = 4\pi R^2 \sigma T_{\text{eff}}^4$, giving

$$\log_{10} L = \log_{10}(4\pi\sigma) + 2\log_{10} R + 4\log_{10} T_{\text{eff}}.$$

This is the equation for a straight line where $\log_{10} L$ is on the y-axis and $\log_{10} T_{\text{eff}}$ is on the x-axis. The coefficient of the x-axis term, i.e. the slope, is 4. The intercept is $\log_{10}(4\pi\sigma) + 2\log_{10} R$, which clearly depends on the value of R. Increasing or decreasing R by a factor of 10 changes $2\log_{10} R$ (and hence $\log_{10} L$) by $+2$ or -2 respectively, thus offsetting the curves of the $0.1\,R_\odot$ and $10\,R_\odot$ stars by equal amounts, but in opposite directions, from the $1\,R_\odot$ line.

1.2   The Sun as a typical star

Although it would be wrong to get the idea that all main-sequence stars are like the Sun, it is the star that we know best, and so it is a useful reference point. We will therefore consider the structure of the Sun, to help illustrate the properties of main-sequence stars in general.

Only the surface properties of the Sun are directly observable, but these measurements may be combined with theoretical models of the Sun's interior to predict its physical characteristics. The principal physical properties of the Sun are listed in Table 1.1

Table 1.1   The physical properties of the Sun.

Measured property	Value
Mass	$M_\odot = 1.99 \times 10^{30}$ kg
Radius	$R_\odot = 6.96 \times 10^8$ m
Luminosity	$L_\odot = 3.83 \times 10^{26}$ W
Effective surface temperature	$T_{\odot,\text{eff}} = 5780$ K
Calculated property	Value
Age	$t_\odot = 4.55 \times 10^9$ years
Core density	$\rho_{\odot,c} = 1.48 \times 10^5$ kg m^{-3}
Core temperature	$T_{\odot,c} = 15.6 \times 10^6$ K
Core pressure	$P_{\odot,c} = 2.29 \times 10^{16}$ Pa

In order to deduce further physical parameters of the Sun (and of other stars) it is necessary to know a little more about how the gravitational potential energy of a ball of gas is related to its other parameters, such as its mass, radius and internal pressure. We explore the details of such relationships in Chapter 2, but for now simply note the results in the box below.

The gravitational potential energy of a star

The astrophysics of stars is intimately concerned with self-gravitating balls of plasma. Gravity, which acts to make a body collapse, can be opposed by internal pressure if the pressure is greater in the core, i.e. if a pressure gradient exists, with pressure decreasing outwards. **Hydrostatic equilibrium** exists when gravity is just balanced by the pressure gradient, so the body neither contracts nor expands.

A very useful result describing the properties of such a system is encapsulated in the **virial theorem**. This states that the volume-averaged pressure needed to support a self-gravitating body in hydrostatic equilibrium is minus one-third of the gravitational potential energy density (i.e. the gravitational energy per unit volume). In symbols

$$\langle P \rangle = -\frac{1}{3}\frac{E_{\mathrm{GR}}}{V}. \qquad (1.2)$$

The negative sign is required because the gravitational potential energy is defined as negative, while the pressure must be positive.

The gravitational potential energy of a sphere of *uniform density* may be calculated by integrating over the mass contained within it. The procedure is straightforward, but an unnecessary distraction, so we simply state the result:

$$E_{\mathrm{GR}} = -\frac{3GM^2}{5R}, \qquad (1.3)$$

where M and R are the mass and radius of the star respectively. Real stars will, of course, not have a uniform density, but the relationship above provides a useful approximation in many cases.

The mean pressure inside the Sun (and indeed within any star in hydrostatic equilibrium) is given by the virial theorem (see the box above). Assuming the Sun to have a uniform density, combining Equations 1.2 and 1.3, and using solar values for the mass and radius, gives

$$\langle P_\odot \rangle = \left(-\frac{1}{3V_\odot}\right) \times \left(-\frac{3GM_\odot^2}{5R_\odot}\right) = \frac{GM_\odot^2}{5V_\odot R_\odot}.$$

Since the Sun is a sphere, its volume is

$$V_\odot = \frac{4}{3}\pi R_\odot^3 \qquad (1.4)$$

so the virial theorem for the average pressure inside the Sun may be rewritten as

$$\langle P_\odot \rangle = \frac{3GM_\odot^2}{20\pi R_\odot^4}. \qquad (1.5)$$

● Using the values from Table 1.1, calculate the mean pressure inside the Sun.

○ The mean pressure inside the Sun $\langle P_\odot \rangle$ is

$$\langle P_\odot \rangle = \frac{3 \times 6.673 \times 10^{-11}\ \mathrm{N\ m^2\ kg^{-2}} \times (1.99 \times 10^{30}\ \mathrm{kg})^2}{20\pi \times (6.96 \times 10^8\ \mathrm{m})^4}$$

$$= 5.38 \times 10^{13}\ \mathrm{Pa}.$$

This is about 400 times smaller than the Sun's core pressure.

The mean density of the Sun is clearly just its mass divided by its volume. Since the Sun is a sphere this is simply

$$\langle \rho_\odot \rangle = \frac{M_\odot}{\frac{4}{3}\pi R_\odot^3} = \frac{3M_\odot}{4\pi R_\odot^3}.$$

(1.6)

● Using the values from Table 1.1, calculate the mean density of the Sun.

○ The mean density inside the Sun $\langle \rho_\odot \rangle$ is

$$\langle \rho_\odot \rangle = \frac{3 \times 1.99 \times 10^{30} \text{ kg}}{4\pi \times (6.96 \times 10^8 \text{ m})^3}$$
$$= 1.41 \times 10^3 \text{ kg m}^{-3}.$$

This is about 100 times smaller than its core density.

In order to calculate the mean temperature inside the Sun, we need to consider the particles of which it is composed, since the mass of the particles determines how the temperature and pressure are linked, via the ideal gas law. To do this, we need to find a way of describing the average composition in terms of the masses of the component particles. This is described in the following box.

Mean molecular mass

The composition of a star can be described in several ways, such as by stating the mass fractions X_Z of each of its constituent elements. However, often a more convenient measure is the mean molecular mass of the material. The **mean molecular mass**, μ, is the mean mass in amu (u) of the particles making up a gas. The amu scale itself is defined such that the mass of a carbon-12 atom, $m(^{12}\text{C}) = 12$ amu exactly, so

1 amu $= u = 1.661 \times 10^{-27}$ kg.

Hence, the mean molecular mass is given by the sum of the mass of the particles in amu divided by the total number of particles. In symbols, this is

$$\mu = \frac{\sum_i n_i \frac{m_i}{u}}{\sum_i n_i},$$

(1.7)

where n_i is the number of particles of a particular type, and m_i is the mass of each of those particles of that type. The word *molecular* is a little misleading, because the interiors of stars are too hot for molecules to exist, and even most atoms are completely ionized, but the expression has survived from its use in cooler environments. The mean molecular mass is sensitive to both the chemical composition of the gas *and* its degree of dissociation and ionization.

A related concept is the mean molecular mass in units of kilograms. This is usually represented by the symbol $\overline{m}$ ('m-bar') and is simply

$$\overline{m} = \mu u.$$

(1.8)

In most cases in stellar astrophysics it is sufficient to assume that the mass of a hydrogen atom or ion is $m_H \approx m_p \approx 1u$ and the mass of a helium atom or ion is $m_{He} \approx m_{He^{2+}} \approx 4u$. Since the mass of an electron is almost 2000 times smaller than the mass of a proton it is usually sufficient to assume $m_e/u \approx 0$.

For example, the mean molecular mass of a neutral gas of pure molecular hydrogen (H_2) is $\mu_{H_2} \approx 2$, and the mean molecular mass of a neutral gas of pure atomic hydrogen is $\mu_H \approx 1$. Finally, if we consider a neutral gas of completely ionized hydrogen, there are two different types of particles present: protons *and* electrons in a one-to-one ratio. The mean molecular mass is therefore

$$\mu_{H+} = \frac{N_p(m_p/u) + N_e(m_e/u)}{N_p + N_e}.$$

But since $N_p = N_e$ for a neutral gas and $m_p/u \approx 1$ and $m_e/u \approx 0$, we can write

$$\mu_{H+} \approx \frac{N_p}{N_p + N_p}$$
$$\approx 0.5.$$

Note that the mean molecular mass of ionized hydrogen is half that of neutral hydrogen, which is half that of molecular hydrogen, even though exactly the same numbers of protons and electrons are involved in each sample!

● What is the mean molecular mass of a neutral plasma containing fully ionized atoms whose atomic number is of order $\sim$ 20 to 30?

○ Atoms with atomic numbers $\sim$ 20 to 30 will typically contain as many neutrons as protons. So a fully ionized atom of a given species with atomic number Z will have a mass of $m_Z \sim 2Zu$. Each ion will also produce Z electrons in the plasma, each of mass m_e. The mean molecular mass of a plasma containing N_Z ions is therefore

$$\mu_Z \sim \frac{(N_Z(2Zu)/u) + (ZN_Z(m_e)/u)}{N_Z + ZN_Z}$$
$$\sim \frac{(2Z + 0)N_Z}{(Z+1)N_Z} \sim \frac{2Z}{Z+1}.$$

Since $Z \sim 20$ to 30 then an approximate answer is $\mu_Z \sim 2$. Hence any neutral plasma of fully ionized heavy atoms will have a mean molecular mass of about 2.

Exercise 1.2   Assume the Sun is fully ionized (and neutral) and comprises 92.7% hydrogen ions and 7.3% helium ions, plus the appropriate number of electrons. Calculate the mean molecular mass of the particles in the Sun.   ■

Now, if we assume that the gas of which the Sun is composed obeys the ideal gas law, then we may write its average pressure as

$$\langle P_\odot \rangle = \frac{\langle \rho_\odot \rangle \, kT_{\mathrm{I}}}{\overline{m}}, \tag{1.9}$$

where T_{I} is a typical temperature inside the Sun and $\overline{m}$ is the mean mass of the particles. Combining this with Equations 1.6 and 1.5, we have

$$T_{\mathrm{I}} = \frac{\overline{m}G\mathrm{M}_\odot}{5k\mathrm{R}_\odot}. \tag{1.10}$$

● Assuming that the average mass of the particles (nuclei and electrons) in the Sun is $\overline{m} \approx 0.6u$, use the values from Table 1.1 to calculate the typical temperature T_{I} inside the Sun.

○ The typical temperature inside the Sun T_{I} is

$$T_{\mathrm{I}} = \frac{0.6 \times 1.661 \times 10^{-27}\ \mathrm{kg} \times 6.673 \times 10^{-11}\ \mathrm{N\ m^2\ kg^{-2}} \times 1.99 \times 10^{30}\ \mathrm{kg}}{5 \times 1.381 \times 10^{-23}\ \mathrm{J\ K^{-1}} \times 6.96 \times 10^{8}\ \mathrm{m}}$$
$$= 2.75 \times 10^{6}\ \mathrm{K}.$$

This is about 6 times smaller than the Sun's core temperature.

1.3   The equations of stellar structure

There is a set of equations which can be used to describe the conditions inside the Sun, and indeed within any main-sequence star. These are known as the **equations of stellar structure**, and comprise four first-order differential equations which describe the way in which a star's mass distribution, pressure, luminosity and temperature each varies as a function of distance from the star's core. They are supplemented by a sct of four linking equations, which relate a star's pressure, luminosity, opacity and energy generation rate to its temperature and density. You have already met the first of these linking equations above: Equation 1.1 links a star's luminosity to its effective surface temperature using the Stefan–Boltzmann law.

A second of the linking equations is provided by a gas law. Clearly, the temperature, density and pressure within a star are not constant, and in fact each of these quantities will increase towards the core. In main-sequence stars, we can assume that the pressure, density and temperature inside the star are linked by the ideal gas law, which can be written as

$$P(r) = \frac{\rho(r)kT(r)}{\overline{m}}, \tag{1.11}$$

where $\overline{m}$ is the mean mass per particle inside the star. Recall that the notation $\rho(r)$ means that the density ρ is a function of the radial coordinate r in the star. It is also worth noting that $\overline{m}$ may also vary as a function of radius, and such abundance gradients are important in the post-main sequence stages of stellar evolution. For now, however, we assume this quantity may be treated as a constant.

We now proceed with the four differential equations of stellar structure, at this stage simply presenting them with a little justification. Clearly the Sun does not appear to change in size, and moreover there is no evidence that it has changed much in size over its recent lifetime. Hence, it is safe to assume that the Sun and other stars on the main sequence, are in a state of hydrostatic equilibrium. The pressure gradient at any point in the Sun must therefore be balanced by the gravitational force per unit volume, and we can write:

$$\frac{dP(r)}{dr} = -\frac{G\,m(r)\,\rho(r)}{r^2}, \tag{1.12}$$

where $m(r)$ is the mass interior to a sphere of radius r (sometimes referred to as the **enclosed mass** at radius r) and $\rho(r)$ is the density at radius r. The minus sign reminds us that the pressure gradient increases *inwards* (i.e. the pressure is greater at smaller radii) whilst the radial coordinate (r) increases *outwards*.

The mass and density are related by the equation of **mass continuity**. This is the second of the four differential equations, and in a spherical geometry may be written as:

$$\frac{dm(r)}{dr} = 4\pi r^2 \rho(r). \tag{1.13}$$

For stars in which the majority of the energy transport is provided by radiation (rather than convection), another equation linking various properties is that of **radiative diffusion**. The temperature gradient $dT(r)/dr$ at some radius r within a star depends on the temperature $T(r)$, luminosity $L(r)$, density $\rho(r)$ and opacity $\kappa(r)$ (see the box below about opacity) at that radius according to:

$$\frac{dT(r)}{dr} = -\frac{3\kappa(r)\,\rho(r)\,L(r)}{(4\pi r^2)(16\sigma)\,T^3(r)} \tag{1.14}$$

where σ is the Stefan–Boltzmann constant. The minus sign again indicates that the temperature increases *inwards* (i.e. the temperature is greater at smaller r). In the case of the Sun, and other stars of similar mass, radiative diffusion will dominate in the region known (not surprisingly) as the radiative zone. In the Sun's case this lies between about 0.2 and 0.7 solar radii from the centre.

Finally, we note that the luminosity increases outwards from the core of a star, as new sources of energy are encountered. The energy generation equation describes the increase in luminosity L as a function of radius r:

$$\frac{dL(r)}{dr} = 4\pi r^2\,\varepsilon(r) \tag{1.15}$$

where $\varepsilon(r)$ is the energy generation rate per unit volume.

Equations 1.12 to 1.15 comprise the fundamental differential equations of stellar structure. They are based on simple assumptions of hydrostatic equilibrium and spherical symmetry, and assume that energy is transported within the star only by radiative diffusion. Modifications have to be made to these equations in cases where convective energy transport is important (such as in the outer layers of the Sun and other low-mass main-sequence stars), but in general they provide a means of calculating the structures of typical stars.

In order to solve these differential equations, other information is needed which relates the luminosity $L(r)$, pressure $P(r)$, opacity $\kappa(r)$ and energy generation rate $\varepsilon(r)$ to the stellar density and temperature. The Stefan–Boltzmann law via Equation 1.1 and the ideal gas law via Equation 1.11 provide this information in the first and second cases, whilst the Kramers opacity approximation (Equation 1.16 below) can provide the link in the third case. You will meet the final linking equation, namely an equivalent expression for the energy generation rate, in Chapter 3. In order to then solve this complete set of equations, a set of boundary conditions must also be selected. These quantify the behaviour of the variable quantities at the extremes where $r = 0$ and $r = R$.

Opacity

The opacity, κ (the Greek lower case letter kappa), of some material is its ability to block radiation. It is expressed as an absorption cross-section per unit mass, so has the unit $m^2\,kg^{-1}$. Perfectly transparent matter would have an opacity of zero (see Figure 1.2).

The opacity of material is caused by several physical processes, the four most common being: **bound–bound** atomic transitions, **bound–free** transitions (i.e. ionization), **free–free** (thermal **bremsstrahlung**) interactions, and **electron scattering** (especially in fully ionized [hot and/or low-density] material where there are a lot of free electrons). To calculate the opacity of some material, the contribution of each process must be computed, and all contributions added.

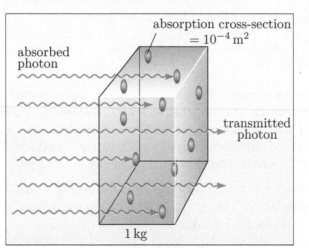

Figure 1.2 Schematic of opacity: if 1 kg of material contains 11 particles, each presenting an absorption cross-section of $10^{-4}\ m^2$, the opacity of the material is $11 \times 10^{-4}\ m^2\,kg^{-1}$. In practice, the number of particles will be $\gg 11$, and the absorption cross-section of each will be $\ll 10^{-4}\ m^2$.

Fortunately, these calculations have been done for us, and two useful generalizations can be made:

1. For stellar-composition material at temperatures $T \geq 30\,000$ K, the opacity κ is dominated by free–free and bound–free absorption, for which an approximate form is $\kappa \propto \rho T^{-3.5}$ (see Figure 1.3). Any opacity of this form,

$$\kappa(r) = \kappa_0 \rho(r) T^{-3.5}(r), \tag{1.16}$$

    is called a **Kramers opacity**, after the Dutch physicist Hendrik (Hans) Kramers who first found this solution in 1923. The Kramers opacity is a very useful approximation, but you should note that it is just that, and not a fundamental physical law.

2. In low-density environments and at very high temperature, scattering by free electrons dominates. In electron scattering, the absorption per free electron is independent of temperature or density, so this opacity source has an almost constant absorption per electron, and therefore per unit mass of ionized material. It is responsible for the flat tail at high temperature in Figure 1.3, which sets a lower limit on the opacity at high temperature and low density.

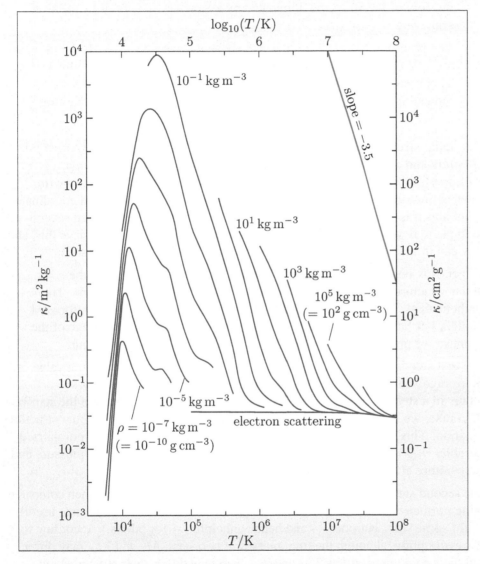

Figure 1.3 Opacity for solar composition material, as a function of temperature (on the horizontal axis) and density, in a log–log plane. Each curve is labeled by the value of the density ρ in kg m^{-3}. In the regime $T \geq 30\,000$ K, i.e. $T \geq 10^{4.5}$ K, the curves (in the log–log plane) are roughly linear with slope -3.5, and more or less uniformly spaced for equal increments of log density, confirming that $\log_{10} \kappa = \mathrm{const} + \log_{10} \rho - 3.5 \log_{10} T$, or equivalently $\kappa \propto \rho T^{-3.5}$. Electron scattering has an almost constant cross-section per unit mass, and is the dominant opacity at high temperatures and low densities, where it sets a lower bound on the opacity values.

● If the absorbing particles in Figure 1.2 exhibit a Kramers opacity, and their temperature is doubled, what happens to the opacity and their absorption cross-sections? If the particles are electrons, so that the opacity is due entirely to electron scattering, what happens to the opacity if the temperature is doubled?

○ The opacity of particles exhibiting a Kramers opacity decreases if the temperature is increased. The overall opacity would decrease by a factor $2^{3.5} \approx 11$, to $\approx 10^{-4}$ m^2 kg^{-1}. Since there are still 11 particles, we infer that their individual absorption cross-sections must also decrease by a factor of ≈ 11, to $\approx 9 \times 10^{-5}$ m^2. Electron scattering, on the other hand, depends only on the *number* of free electrons, *not* their *temperature*, so in the second case the opacity would be unchanged.

● Table 1.1 gives the core temperature and core density of the Sun. Use these values to mark a cross on Figure 1.3 indicating the conditions in the solar core. Based on where it lies, do you expect the opacity of the material in the core of the Sun to be dominated by a Kramers opacity or by electron scattering?

○ The point for the solar core lies just above the curve for $\rho = 10^5$ kg m^{-3}, at $\log_{10}(T/\text{K}) = 7.19$, where the opacity is seen to be $\approx 2 \times 10^{-1}$ m^2 kg^{-1}. This is still on the sloping part of the opacity–temperature curves, indicating that Kramers opacity still dominates over electron scattering.

The stellar structure equations describe the radial gradients $\mathrm{d}P(r)/\mathrm{d}r$, $\mathrm{d}m(r)/\mathrm{d}r$, $\mathrm{d}T(r)/\mathrm{d}r$ and $\mathrm{d}L(r)/\mathrm{d}r$ in terms of a number of variables including $m(r)$, $P(r)$, $\rho(r)$, $\kappa(r)$, $L(r)$, $T(r)$, $\varepsilon(r)$ and r. Ideally one would like to be able to write down expressions for all of these quantities as a function of the radial coordinate r, but alas it is rather difficult to do so. Over the years there have been several attempts to develop simplistic models that use approximations to achieve this, and they can in many cases provide valuable insights into the physics of stars.

In fact, it is possible to infer some important relationships between the stellar structure variables by making just a few simple approximations. The mathematical treatment that we will follow to illustrate this is straightforward enough, but rather tedious and would serve as a distraction at this stage of the book, so we provide just a few insights rather than mathematical rigour.

The first step is to divide each stellar structure variable by its maximum value or its average value in the star. For example, whenever the enclosed mass $m(r)$ occurs in a stellar structure equation, we divide it by the total mass of the star M. Of course, we must perform the same division on both sides of the stellar structure equation. This step recasts the stellar structure equations in terms of normalized variables $m(r)/M$, $P(r)/P_\mathrm{c}$, $T(r)/T_\mathrm{c}$ etc., where P_c and T_c are the pressure and temperature at the core of the star.

The second step is to assume that all stars have the same structure when compared using normalized variables. (This assumption is reasonable — and hence useful — for some stars, but poor — and hence unhelpful — for others.) According to this assumption, although the total masses of two stars, M_1 and M_2, may differ, and their core temperatures $T_{\mathrm{c},1}$ and $T_{\mathrm{c},2}$ also may differ, their structures are similar enough that the variation of normalized temperature $T(r)/T_\mathrm{c}$ with normalized enclosed mass $m(r)/M$ is the same for both objects.

The outcome of taking the two steps described above is that we can write a *companion equation* for each of the stellar structure equations given above. The companion equation differs from the original in four respects:

1. each occurrence of a derivative is replaced by a simple ratio of the corresponding variables,

2. each occurrence of a stellar structure variable is replaced by either its maximum or average value,

3. the equals sign = is replaced by the symbol $\propto$ meaning 'is proportional to', and

4. any physical constants can be dropped.

A worked example is given below to make this clearer.

Worked Example 1.1

As an illustration of the use of the stellar structure equations, we use them to derive an approximate mass–luminosity relationship for high-mass stars on the main sequence. In such stars, the main source of opacity is electron scattering, so for them the opacity is roughly constant (i.e. independent of temperature or density).

Solution

In the following, we use M and R to represent the star's total mass and radius, L for its surface luminosity, P_c for its core pressure, T_c for its core temperature, $\bar{\rho}$ for its mean density and $\bar{\kappa}$ for its mean opacity.

We begin by writing the companion equation to Equation 1.13 for mass continuity. In this case we can replace $dm(r)/dr$, by the simple ratio, M/R. We can also replace r by the radius R and $\rho(r)$ by the mean density $\bar{\rho}$. So, the companion equation to the mass continuity equation is simply $M/R \propto R^2\bar{\rho}$ which may be re-written as (i) $\bar{\rho} \propto M/R^3$. This of course makes sense, since the average density is just $\bar{\rho} = M/\frac{4}{3}\pi R^3$.

The equation of hydrostatic equilibrium (Equation 1.12) gives rise to the companion equation $P_c/R \propto M\bar{\rho}/R^2$ which using (i) from above may be re-written as (ii) $P_c \propto M^2/R^4$.

For an unchanging chemical composition, the ideal gas law leads to the expression $P_c \propto \bar{\rho}T_c$. Substituting for $\bar{\rho}$ and P_c using (i) and (ii) from above, this becomes (iii) $T_c \propto M/R$.

Finally, the temperature gradient equation leads to $T_c/R \propto \bar{\kappa}\bar{\rho}L/R^2T_c^3$, which can immediately be re-arranged as $T_c^4 \propto \bar{\kappa}\bar{\rho}L/R$. Now substituting for $\bar{\rho}$ using (i) this becomes $T_c^4 \propto \bar{\kappa}ML/R^4$ and then substituting for T_c using (iii) we have (iv) $L \propto M^3/\bar{\kappa}$.

Now, as noted at the beginning of this example, in a high-mass main-sequence star, the opacity is constant, so the mass–luminosity relationship for such stars is approximately $L \propto M^3$.

Exercise 1.3 Following a similar process to that in the above worked example, what would be an approximate relationship between the mass, luminosity and radius of a low-mass main-sequence star, in which the opacity may be represented by a Kramers opacity? ∎

The two mass–luminosity relationships derived in the previous worked example and exercise may be summarized as:

$$L \propto M^3 \quad \text{for high-mass main-sequence stars} \tag{1.17}$$
$$L \propto M^{5.5} R^{-0.5} \quad \text{for low-mass main-sequence stars.} \tag{1.18}$$

The latter equation is known as the Eddington mass–luminosity–radius relationship. Note, these equations are only approximations to the real situation, for the reasons mentioned, and we have ignored factors such as differing chemical composition, for instance. Nonetheless, they are good approximations for real stars and demonstrate the power of this simple analysis in tackling complex situations.

1.4 The proton–proton (p–p) chain

With hydrogen being the most abundant element in the Universe it is understandable that hydrogen is both the starting point for stellar nuclear burning and, as it turns out, the longest-lasting nuclear fuel. In this section we examine the first of two hydrogen-burning processes, beginning with that which dominates in the Sun and other low-mass main-sequence stars.

As a contracting pre-main-sequence star heats up, it reaches a temperature of about 10^6 K at which protons can undergo fusion with any pre-existing light nuclei, such as deuterium (D), lithium (Li), beryllium (Be) and boron (B). However, these reactions are extremely rapid and release only a limited amount of energy because the light nuclei are present in such small quantities. In order to begin life properly as star, a different reaction must be initiated, one in which protons are combined with each other to make heavier nuclei. If one could simply combine two protons to make a nucleus of ^2_2He, then hydrogen burning would be very rapid. However, a nucleus of helium containing no neutrons is not stable, so this is not how hydrogen fusion proceeds.

Instead, at a temperature of about 10^7 K, the first step in the **proton–proton chain** occurs when two protons react to form a nucleus of deuterium, which is called a **deuteron**:

$$\text{p} + \text{p} \longrightarrow \text{d} + \text{e}^+ + \nu_e,$$

where the deuteron (d) comprises a proton (p) and a neutron (n). This first step is therefore a weak interaction in which a proton has been converted into a neutron, with the release of a positron (e^+) and an electron neutrino (ν_e). It is important to note that this first step is incredibly slow. On average, an individual proton in the core of the Sun will have to wait for about 5 billion years before it undergoes such a reaction. (Although, since the core of the Sun contains a huge number of protons, some of them will react as soon as the conditions are right for them to do so.) This reaction therefore effectively governs the length of time for which a star will undergo hydrogen fusion – its main sequence lifetime.

Once a deuterium nucleus has formed it will rapidly capture another proton to form a nucleus of helium-3, in the reaction:

$$\text{d} + \text{p} \longrightarrow {}^3_2\text{He} + \gamma.$$

However, from here, the helium-3 nucleus can undergo one of two reactions. In the Sun, 85% of the time, the helium-3 nucleus will react with another helium-3 nucleus to form a helium-4 nucleus directly, in the reaction:

$$^{3}_{2}\text{He} + {}^{3}_{2}\text{He} \longrightarrow {}^{4}_{2}\text{He} + \text{p} + \text{p}.$$

This set of reactions constitutes the first branch of the p–p chain, (ppI).

In the rest of the cases, the helium-3 nucleus will react with an existing helium-4 nucleus to form a nucleus of beryllium-7 as follows:

$$^{3}_{2}\text{He} + {}^{4}_{2}\text{He} \longrightarrow {}^{7}_{4}\text{Be} + \gamma.$$

Virtually every beryllium-7 nucleus will subsequently capture an electron to form lithium-7, which then reacts with a further proton to produce two helium-4 nuclei:

$$\text{e}^{-} + {}^{7}_{4}\text{Be} \longrightarrow {}^{7}_{3}\text{Li} + \nu_{e}$$
$$\text{p} + {}^{7}_{3}\text{Li} \longrightarrow {}^{4}_{2}\text{He} + {}^{4}_{2}\text{He}.$$

This set of reactions constitutes the second branch of the p–p chain, (ppII).

In a very tiny proportion of cases, the beryllium-7 nucleus will instead react directly with a proton forming a nucleus of boron-8 which subsequently undergoes beta-plus decay to beryllium-8, which in turn splits into two helium-4 nuclei:

$$\text{p} + {}^{7}_{4}\text{Be} \longrightarrow {}^{8}_{5}\text{B}$$
$$^{8}_{5}\text{B} \longrightarrow {}^{8}_{4}\text{Be} + \text{e}^{+} + \nu_{e}$$
$$^{8}_{4}\text{Be} \longrightarrow {}^{4}_{2}\text{He} + {}^{4}_{2}\text{He}.$$

This set of reactions constitutes the third branch of the p–p chain, (ppIII).

All three branches are summarized in Figure 1.4 (overleaf). In stars with higher core temperatures, the proportions of nuclei undergoing fusion via the ppII and ppIII branches are higher than in the Sun.

● The proton–proton chain comprises nine reactions in three branches, but just one reaction determines the overall rate at which hydrogen burning proceeds. Which reaction is the bottleneck and why?

○ The slowest reaction in the series is the first one, $\text{p} + \text{p} \longrightarrow \text{d} + \text{e}^{+} + \nu_{e}$. This is slowest because during the collision of the protons, one of them must undergo a β^{+}-decay, $\text{p} \longrightarrow \text{n} + \text{e}^{+} + \nu_{e}$. The β^{+}-decay is mediated by the weak nuclear force, so has a very low probability of occurring.

1.5 The mass defect

The proton–proton chain converts four hydrogen nuclei (protons) into a $^{4}_{2}\text{He}$ nucleus, two positrons that quickly collide with electrons and are annihilated, and two neutrinos. Hence, branch I of the p–p chain may be summarized as:

$$2\text{e}^{-} + 4\text{p} \longrightarrow {}^{4}_{2}\text{He} + 2\nu_{e} + 2\gamma_{\text{pd}} + 4\gamma_{e}.$$

(The *unconventional* subscripts used here on the γ-rays are to distinguish the γ-rays from the p + d reaction from those from the electron–positron annihilation.)

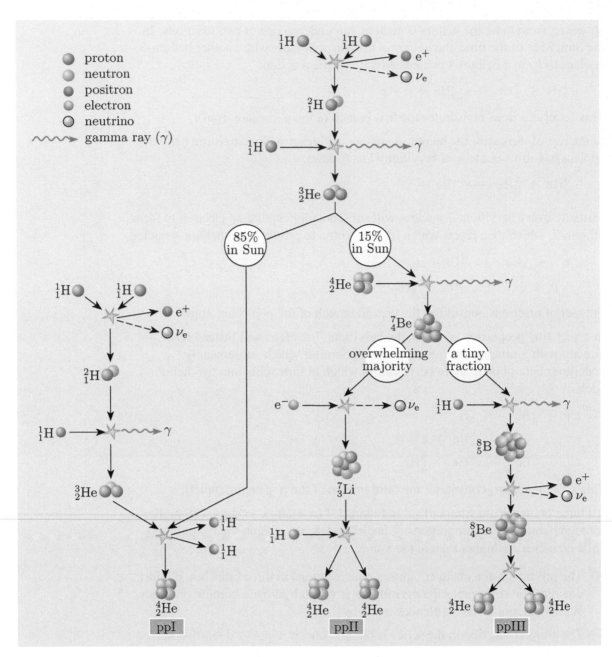

Figure 1.4   The p–p chain consists of three branches (ppI, ppII and ppIII). Each of them results in the creation of nuclei of helium-4 from four protons. The relative proportions of each branch occurring in the Sun are shown.

The energy released in these reactions can be assessed from the differences in the masses of each particle, and the masses of the nuclei can be obtained from the atomic masses of their isotopes. We use the atomic mass unit (amu) scale where $m(^{12}C) = 12$ amu exactly, so 1 amu $= 1.660\,540 \times 10^{-27}$ kg. Isotope tables give the atomic masses of hydrogen and helium-4 as:

$$m(^{1}_{1}H) = 1.007\,825 \text{ amu} = 1.673\,534 \times 10^{-27} \text{ kg}$$
$$m(^{4}_{2}He) = 4.002\,60 \text{ amu} = 6.646\,478 \times 10^{-27} \text{ kg}.$$

These include one electron for hydrogen and two electrons for helium, where

$m(e^-) = 9.109 \times 10^{-31}$ kg, so subtracting the electronic component gives nuclear masses $m_n(^1_1\text{H}) = 1.672\,623 \times 10^{-27}$ kg for the ^1_1H *nucleus* and $m_n(^4_2\text{He}) = 6.644\,656 \times 10^{-27}$ kg for the ^4_2He *nucleus*.

The reaction sequence has the following mass difference:

$$\Delta m = \text{initial mass} - \text{final mass}$$
$$= 2m(e^-) + m(4p) - m_n(^4_2\text{He}) - m(2\nu_e) - m(2\gamma_{pd}) - m(4\gamma_e)$$
$$= (2 \times 9.109 \times 10^{-31} \text{ kg}) + (4 \times 1.672623 \times 10^{-27} \text{ kg})$$
$$- 6.644656 \times 10^{-27} \text{ kg} - 0 - 0 - 0$$
$$= 4.7658 \times 10^{-29} \text{ kg}.$$

This mass difference is called the **mass defect**. Note for future reference that the mass defect, discounting the contribution of the electron–positron annihilations, corresponds to the fraction ≈ 0.0066 of the original mass of 4 protons.

Einstein's famous equation $E = mc^2$ expresses the mass–energy equivalence and gives the energy released from this change in mass. The energy associated with the mass defect is given the symbol ΔQ, and in our example is $\Delta Q = (\Delta m)c^2 = 4.2833 \times 10^{-12}\,\text{J} = 26.74$ MeV. Some of this is in the form of the kinetic energy of reaction products, and some is in the form of the two γ_{pd}-rays and the four γ_e-rays. Most of this energy is quickly absorbed by the surrounding particles and thus appears as the increased kinetic energy of the gas. However, the two neutrinos do not interact with the local gas, and escape from the star unimpeded. They carry off a small amount of energy, on average 0.26 MeV each for the neutrinos in the first branch of the p–p chain, reducing the effective energy contribution to the star to $26.74\,\text{MeV} - (2 \times 0.26\,\text{MeV}) = 26.22$ MeV.

● In most of the exercises in this book only four significant figures are given for the physical constants. However, in the calculation above we have used seven for the atomic and nuclear masses. Why?

○ Whenever you have to subtract nearly equal quantities, the number of significant figures decreases. Consider

$$4m(p) - m_n(^4_2\text{He}) = 6.690\,492 \times 10^{-27}\,\text{kg} - 6.644\,656 \times 10^{-27}\,\text{kg}$$
$$= 0.045\,836 \times 10^{-27}\,\text{kg}.$$

Although the nuclear masses are quoted to 7 significant figures, the numbers are so similar that only 5 significant figures remain after the subtraction. If we had begun with only 4 figures, we would have been left with only 2, and numerical accuracy would have been lost.

Exercise 1.4   Following a similar procedure to that outlined above for the first branch of the proton–proton chain, what is the energy released by the reactions comprising the second branch of the proton–proton chain? (Note: Of the neutrinos released in the reaction where a beryllium-7 nucleus captures an electron, 90% of them carry away an energy of 0.86 MeV, whilst the remaining 10% carry away an energy of 0.38 MeV, depending on whether or not the beryllium-7 is created in an excited state.)   ■

You have seen that each instance of the first branch of the p–p chain contributes 26.22 MeV of energy to the star and each instance of the second branch of the p–p chain contributes 25.67 MeV of energy. Since the first branch occurs 85% of the time and involves two proton–proton reactions, and the second branch occurs 15% of the time but involves only one proton–proton reaction, the average energy released for each reaction between two protons is $(0.85 \times 26.22\,\text{MeV})/2 + (0.15 \times 25.67\,\text{MeV}) = 15.0\,\text{MeV}$.

1.6   The carbon–nitrogen–oxygen (CNO) cycle

The proton–proton chain is the main fusion reaction in the Sun, but it is *not* the main one in many *more-massive* main-sequence stars. Here there are two closed cycles of reactions which not only convert hydrogen into helium, but also convert carbon-12 nuclei (formed in an earlier generation of stars and incorporated when the star formed) into carbon-13 and various isotopes of nitrogen and oxygen.

The **carbon–nitrogen (CN) cycle** uses $^{12}_{6}\text{C}$ as a catalyst. That is, *if* the CN cycle completes, any $^{12}_{6}\text{C}$ involved in the reaction sequence is returned at the end without being consumed. An illustrative view, emphasizing the cyclical nature, and also showing the associated **oxygen–nitrogen (ON) cycle** which requires higher temperatures, is shown in Figure 1.5. Collectively these two cycles are known as the carbon–nitrogen–oxygen cycle or **CNO cycle**.

The CN-cycle can be considered as starting with a carbon-12 nucleus. This captures a proton to form nitrogen-13, which then undergoes beta-plus decay to carbon-13. This captures two further protons to form nitrogen-14 and then oxygen-15. A beta-plus decay to nitrogen-15 is then followed by the capture of a proton, emission of an alpha-particle (i.e. helium-4 nucleus) and the recovery of the original carbon-12 nucleus.

Figure 1.5   The CNO cycle is made up of two parts: the CN cycle which operates at lower temperatures, and the ON cycle which becomes more important at higher temperatures. The cycles contain essentially only three types of reactions: The notation p,γ indicates the capture of a proton with the emission of a γ-ray, which increase the atomic number and atomic mass number by one, the notation $\beta^+\nu_e$ indicates a beta-plus decay accompanied by neutrino emission, and the notation p,α indicates a reaction that closes each cycle with the capture of a fourth proton, followed by the emission of a helium nucleus (α-particle) and the recovery of the starting nucleus.

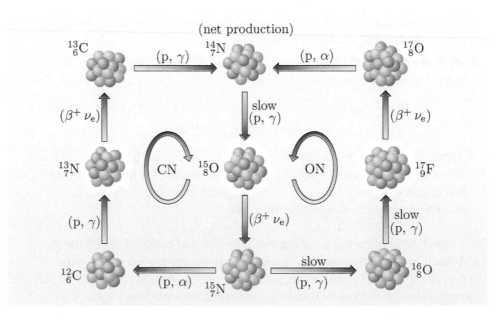

Similarly, the ON-cycle can be considered as starting with a nitrogen-14 nucleus. This captures a proton to form oxygen-15, which then undergoes beta-plus decay to nitrogen-15. This captures two further protons to form oxygen-16 and then fluorine-17. A beta-plus decay to oxygen-17 is then followed by the capture of a proton, emission of an alpha-particle (i.e. helium-4 nucleus) and the recovery of the original nitrogen-14 nucleus.

We shall return to the physics of the p–p chain and the CNO-cycle in Chapter 3.

Summary of Chapter 1

1. The Hertzsprung–Russell diagram is of fundamental importance for describing the properties of stars and for tracking their evolution. It has two forms: the observational H–R diagram plots absolute visual magnitude M_V versus colour index (often $B - V$) or spectral type; the theoretical H–R diagram plots luminosity L versus effective surface temperature T_{eff} on a log–log scale.

2. $L = 4\pi R^2 \sigma T_{\text{eff}}^4$ relates stellar radius, luminosity and effective surface temperature. Loci of constant radius in the H–R diagram are diagonal lines from upper left to lower right.

3. The virial theorem, which is derived from the condition for hydrostatic equilibrium, concludes that the average pressure needed to support a self-gravitating system is minus one-third of the gravitational potential energy density $\langle P \rangle = -E_{\text{GR}}/3V$.

4. The gravitational potential energy of a spherical cloud of gas of uniform density is $E_{\text{GR}} = -3GM^2/5R$. This is a useful approximation for most stars.

5. The mean molecular mass of a sample of gas is given by the sum of the mass of the particles in amu divided by the total number of particles. The mean molecular masses for three forms of hydrogen are: $\mu_{\text{H}_2} \approx 2$; $\mu_{\text{H}} \approx 1$ and $\mu_{\text{H+}} \approx 0.5$. The mean molecular mass of the Sun is $\mu_\odot \approx 0.6$.

6. We can express some average properties of a star using the following equations:

   - the mean density: $\langle \rho \rangle = 3M/4\pi R^3$;
   - the mean (volume-averaged) pressure:
     $\langle P \rangle = -E_{\text{GR}}/3V \approx 3GM^2/20\pi R^4$
   - and the typical internal temperature: $T_{\text{I}} \approx GM\overline{m}/5kR$
   - where, by the ideal gas law: $\langle P \rangle = \langle \rho \rangle kT_{\text{I}}/\overline{m}$.

7. There are four differential equations which characterize stellar structure in terms of the star's pressure gradient, mass distribution, temperature gradient and luminosity gradient, each as a function of radius:

$$\frac{\mathrm{d}P(r)}{\mathrm{d}r} = -\frac{G\,m(r)\,\rho(r)}{r^2} \qquad \text{(Eqn 1.12)}$$

$$\frac{\mathrm{d}m(r)}{\mathrm{d}r} = 4\pi r^2\,\rho(r) \qquad \text{(Eqn 1.13)}$$

$$\frac{\mathrm{d}T(r)}{\mathrm{d}r} = -\frac{3\kappa(r)\,\rho(r)\,L(r)}{(4\pi r^2)(16\sigma)\,T^3(r)} \qquad \text{(Eqn 1.14)}$$

$$\frac{dL(r)}{dr} = 4\pi r^2 \,\varepsilon(r). \qquad\qquad \text{(Eqn 1.15)}$$

They can be solved by combining them with four linking equations which relate a star's pressure, luminosity, opacity and energy generation rate to its temperature and density.

8. Useful approximations to the structure of stars may be obtained by rewriting the stellar structure equations and the linking equations as simple proportionalities between the maximum or average values of the various stellar parameters.

9. For stellar-composition material at temperatures $T \geq 30\,000$ K, the opacity κ is dominated by free–free and bound–free absorption, for which $\kappa(r) \propto \rho(r)T^{-3.5}(r)$. Any opacity of this form is called a Kramers opacity. In low-density environments and at very high temperature, scattering by free electrons dominates. In this case the absorption per free electron is independent of temperature or density, so this opacity source has an almost constant absorption per electron.

10. Based on a simple analysis, the stellar structure equations and the two different opacity relations indicate the following mass–luminosity relationships:

$$L \propto M^3 \quad \text{for high-mass main-sequence stars}$$
$$L \propto M^{5.5} R^{-0.5} \quad \text{for low-mass main-sequence stars.}$$

11. Thermonuclear burning begins at $T_c \approx 10^6$ K to 10^7 K with the most abundant and least complex nucleus, the proton (hydrogen nucleus). The main branch (ppI) of the proton–proton chain can be written:

$$p + p \longrightarrow d + e^+ + \nu_e$$
$$p + d \longrightarrow {}^3_2\text{He} + \gamma$$
$$ {}^3_2\text{He} + {}^3_2\text{He} \longrightarrow {}^4_2\text{He} + 2p.$$

A crucial step in the $p + p \longrightarrow d + e^+ + \nu_e$ reaction is a β^+-decay, $p \longrightarrow n + e^+ + \nu_e$. This is the bottleneck in the reaction chain.

12. The mass defect – the difference in the masses of the reactants and products of the nuclear reactions – quantifies the energy liberated into the gas, via $E = mc^2$. The main branch of the p–p chain liberates ≈ 26.7 MeV, of which ≈ 0.5 MeV is carried away from the star by the neutrinos from the two $p \longrightarrow n + e^+ + \nu_e$ reactions.

13. There are three branches to the p–p chain, each delivering slightly different energy per event and occurring with different frequency. In the Sun, ppI occurs 85% of the time, ppII accounts for almost all of the rest, and ppIII occurs in a very tiny proportion of cases. The average energy released per proton–proton fusion event is ≈ 15 MeV.

14. The second main hydrogen-burning process is the CNO cycle, in which carbon, nitrogen and oxygen nuclei are used as catalysts while synthesizing helium from hydrogen. The CN cycle operates first, and the ON cycle comes into play at higher temperatures.

Chapter 2   Gravitational contraction

Introduction

Stars are self-gravitating balls of gas. In order to understand how stars evolve it is therefore vital to understand the physics of how such balls of gas behave. In fact, gravitational contraction, not nuclear fusion as you may have thought, is the ultimate driver of stellar evolution. In this chapter we examine the physics of gravitational contraction then use some of the results of this to derive the lower and upper mass limits for a star on the main sequence. We conclude the chapter by considering the different roles of gravitational contraction and nuclear fusion in the evolution of stars.

2.1   A self-gravitating gas cloud

The gravitational collapse of a gas cloud to form a star begins with a system that is out of equilibrium – otherwise it would not collapse – and ends with an almost static system where an equilibrium exists between the tendency to collapse due to self-gravity, and the tendency to expand due to huge internal pressure. We will follow the change from a state of disequilibrium to one of equilibrium, but in order to do so, we need to understand the competing forces and know what determines whether or not equilibrium has been achieved. The next few subsections deal with those topics first.

2.1.1   Self-gravity versus internal pressure

The gravitational force which makes a gas cloud contract is the mutual gravity of all of its particles. It would be difficult to add up the forces exerted by all of the constituent particles for a randomly shaped gas cloud, but by assuming spherical symmetry, the problem is greatly simplified and becomes manageable. Although gas clouds are not spherical at the start of the collapse, they certainly are by the end, and it is a reasonable assumption to make throughout the process.

Consider a spherical region of gas with total mass M and radius R, in which the only forces acting are those of gravity (due to the masses of the particles of which the cloud is composed) and pressure (due to the thermal motion of the particles of which the cloud is composed). The density and pressure can be assumed to vary radially with distance from the centre of the cloud and are denoted by $\rho(r)$ and $P(r)$ respectively.

It is instructive to calculate the total amount of matter enclosed by a sphere of radius r. In order to do this, let us assume the sphere can be described as a series of thin, concentric shells of radius x each of which has minuscule thickness δx (Figure 2.1), and each of which has a different density $\rho(x)$. The thickness of each shell is chosen to be small enough, that the variation in density across each shell, or equivalently in adjacent shells, is negligible. The mass δm of each thin shell is approximated by the thin-shell volume multiplied by the density at that radius, i.e. $\delta m \approx$ (surface area × thickness) × density $\approx 4\pi x^2 \times \delta x \times \rho(x)$.

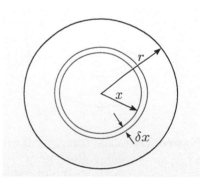

Figure 2.1   A sphere of radius r may be treated as the combination of many very thin shells. One such thin shell, having radius x, thickness δx and density $\rho(x)$ is shown. The mass of that shell of radius x is $\delta m \approx 4\pi x^2 \times \delta x \times \rho(x)$.

It is conventional to write the minuscule increment, in this case δx, at the end of the equation, giving $\delta m \approx 4\pi \rho(x) x^2\, \delta x$.

If we now suppose that the sphere is made up of n such shells and we label each shell with the subscript i, then the total mass in the sphere of radius r is simply the sum of the masses in each of the n thin shells, so

$$m(r) = \sum_{i=1}^{n} \delta m_i \approx \sum_{i=1}^{n} 4\pi \rho(x_i)\, x_i^2\, \delta x_i,$$

where δx_i is the thickness of the shell at radius x_i, with δm_i being the amount of mass in that shell and $\rho(x_i)$ its density.

In the limit when the shell thickness becomes infinitesimal ($\delta x_i \to 0$), the approximation becomes exact. In this case, we replace the symbol δ (signifying a minuscule but finite increment) with 'd' (signifying an infinitesimal increment), drop the index i altogether, and replace the summation sign $\sum$ for discrete steps with the integration sign $\int$ for infinitesimal intervals. We write

$$m(r) = \int_{x=0}^{x=r} \mathrm{d}m = \int_{x=0}^{x=r} 4\pi\, \rho(x)\, x^2\, \mathrm{d}x \tag{2.1}$$

where the sum is now written using the integral symbol. The limits of the sum are also modified, so that instead of describing the sum as being from the shell labeled $i = 1$ to the shell labeled $i = n$, we now write the radii of the shells over which we integrate, which in this case is from the shell at radius $x = 0$ to the shell at radius $x = r$.

If we are to make further progress and evaluate this enclosed mass, we have to know *how* the density varies with radius, i.e. we need an expression for $\rho(x)$. For instance, if the density is constant, i.e. $\rho(x) = \rho_0$, Equation 2.1 may be evaluated simply to give $m = \frac{4}{3}\pi \rho_0 r^3$ which is the expected relationship between the mass and radius of a sphere of uniform density.

Now, according to Newton's theorem, the matter in a sphere (with mass given by Equation 2.1) acts gravitationally as a single point mass m situated at the centre of the sphere. Hence, the mass $m(r)$ enclosed by a sphere of radius r gives rise to an inward gravitational acceleration at the surface of the sphere of magnitude

$$g(r) = \frac{G\, m(r)}{r^2}. \tag{2.2}$$

The force opposing this gravitational acceleration is provided by the gas pressure. In order to determine this force, consider a small packet of the gas situated between radii r and $r + \Delta r$. The packet of gas has a cross-sectional area of ΔA and therefore has a volume of $\Delta r\, \Delta A$ (see Figure 2.2).

The mass of the volume element is just its density multiplied by its volume, i.e. $\Delta M = \rho(r)\, \Delta r\, \Delta A$, so the inward gravitational force acting on it has a magnitude $F_{\mathrm{grav}}(r) = g(r)\, \Delta M = g(r)\, \rho(r)\, \Delta r\, \Delta A$.

Now, let us assume that the pressure below the small volume element is $P(r)$ and the pressure above the volume element is $P(r) + \Delta P$. So the forces due to the gas pressure are $(P(r) + \Delta P)\, \Delta A$ acting inwards from above and $P(r)\, \Delta A$ acting outwards from below. We can write the increase in pressure in terms of the pressure gradient as

$$\Delta P = \frac{\mathrm{d}P(r)}{\mathrm{d}r} \times \Delta r.$$

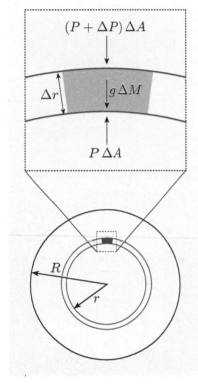

Figure 2.2   The forces acting on a small volume element of the gas in a spherical cloud. The gravitational force acting on the element is $g(r)\, \Delta M$, and the pressure force is the difference between the pressures above and below the volume element.

The net force due to gas pressure acting inwards on the small volume element has a magnitude

$$F_{\text{gas}}(r) = (P(r) + \Delta P)\,\Delta A - P(r)\,\Delta A = \Delta P\,\Delta A = \frac{\mathrm{d}P(r)}{\mathrm{d}r}\,\Delta r\,\Delta A.$$

Therefore, the total force acting on the small volume element is the sum of the forces due to gravity and gas pressure. The magnitude of this inward force is

$$F(r) = g(r)\,\rho(r)\,\Delta r\,\Delta A + \frac{\mathrm{d}P(r)}{\mathrm{d}r}\,\Delta r\,\Delta A.$$

But since $\Delta M = \rho(r)\,\Delta r\,\Delta A$, we can rewrite this as

$$F(r) = g(r)\,\Delta M + \frac{\mathrm{d}P(r)}{\mathrm{d}r}\frac{\Delta M}{\rho(r)}.$$

The magnitude of the acceleration of the small volume element is simply $\mathrm{d}^2 r/\mathrm{d}t^2 = F(r)/\Delta M$. So the inward acceleration of any mass element at distance r from the centre, due to the combined effects of gravity and pressure is

$$-\frac{\mathrm{d}^2 r}{\mathrm{d}t^2} = g(r) + \frac{1}{\rho(r)}\frac{\mathrm{d}P(r)}{\mathrm{d}r} \tag{2.3}$$

where $g(r) = Gm(r)/r^2$. In order for the pressure force to act in the opposite direction to the gravitational force, the pressure must increase inwards and $\mathrm{d}P(r)/\mathrm{d}r$ must be negative.

This equation captures a crucial relationship for the formation and evolution of a star. It shows that the net contraction or expansion of a star depends on competition between gravity, which acts to make the star collapse, and internal pressure, which tends to support it against gravity. (Often we refer to the internal pressure as the *pressure support*.)

● How should Equation 2.3 be modified for white dwarfs, for which nuclear reactions have ceased?

○ No change is required; the equation is general. Many of the equations for the structure and evolution of stars were worked out long before nuclear reactions and the proton–proton chain were known about. Nuclear reactions provide the energy that *prolongs* a star's life, but gravity *drives* their evolution.

Over the star's lifetime, the relative strengths of the gravitational force and the internal pressure force change greatly: during star formation, gravity dominates; during its stable phase the forces are balanced; and at the end of its life the forces may differ throughout the star, causing the core to collapse while the outermost layers expand at thousands of kilometres per second. The sequence of events and their duration depends on *how* the pressure support is provided.

Competition between the opposing forces provided by gravity and pressure support are what determines the evolution and the ultimate fate of a star.

There are two interesting limiting cases of Equation 2.3.

- The first is that of unimpeded collapse, where there is no pressure support. This is called **free fall**. It is found by setting the pressure gradient to zero: $\mathrm{d}P(r)/\mathrm{d}r = 0$.

- The second case is where the pressure gradient just balances the gravitational acceleration, so there is no net acceleration. This is called **hydrostatic equilibrium**, and is found by setting the net acceleration to zero: $\mathrm{d}^2r/\mathrm{d}t^2 = 0$.

We briefly examine these two cases in the next two subsections.

2.1.2   Free fall

If there is no pressure support, Equation 2.3 becomes simply

$$-\frac{\mathrm{d}^2r}{\mathrm{d}t^2} = g(r) = \frac{G\,m(r)}{r^2} \qquad (2.4)$$

i.e. each volume element of the gas cloud accelerates towards the centre and its kinetic energy increases as its gravitational potential energy decreases.

Worked Example 2.1

What is the time for a spherical cloud of gas to collapse under free fall if there is no pressure support?

Solution

In order to calculate the time for collapse, we must first calculate the speed with which the cloud will collapse as a function of radius, and then integrate over all radii to find the total time. In order to do this we consider how the kinetic energy of a mass element increases at the expense of its gravitational potential energy, as a result of collapse.

We can assume that the cloud is initially at rest at a radius r_0, and that the mass enclosed within this radius remains constant with a value m_0. Then the increase in kinetic energy of a mass element δm is just the decrease of its gravitational potential energy, $\Delta E_\mathrm{K} = -\Delta E_\mathrm{GR}$, so we may write

$$E_\mathrm{K}(r_\mathrm{final}) - E_\mathrm{K}(r_0) = -\left(\frac{-Gm_0\,\delta m}{r_\mathrm{final}} - \frac{-Gm_0\,\delta m}{r_0}\right)$$

$$\frac{1}{2}\,\delta m\,v_\mathrm{final}^2 - \frac{1}{2}\,\delta m\,v_0^2 = \frac{Gm_0\,\delta m}{r_\mathrm{final}} - \frac{Gm_0\,\delta m}{r_0}.$$

Since the cloud is initially at rest, $v_0 = 0$, and the mass δm cancels out leaving

$$v_\mathrm{final}^2 = \frac{2Gm_0}{r_\mathrm{final}} - \frac{2Gm_0}{r_0}$$

or, in general,

$$v(r) = \left(\frac{2Gm_0}{r} - \frac{2Gm_0}{r_0}\right)^{1/2}.$$

The **free-fall time** for the cloud to collapse (from radius $r = r_0$) into the centre (at radius $r = 0$) is then found by integrating as follows

$$t_\mathrm{ff} = \int_{r=r_0}^{r=0} \frac{\mathrm{d}t}{\mathrm{d}r}\,\mathrm{d}r$$

but since $v(r) = -\mathrm{d}r/\mathrm{d}t$ (the minus sign indicating that v increases as r decreases), we can replace $\mathrm{d}t/\mathrm{d}r$ in the above by $-1/v(r)$, so the free-fall time is

$$t_{\mathrm{ff}} = -\int_{r_0}^0 \left[\frac{2Gm_0}{r} - \frac{2Gm_0}{r_0}\right]^{-1/2} \mathrm{d}r$$

$$= -\left(\frac{2Gm_0}{r_0}\right)^{-1/2} \int_{r_0}^0 \left[\frac{r_0}{r} - 1\right]^{-1/2} \mathrm{d}r$$

$$= -\left(\frac{r_0}{2Gm_0}\right)^{1/2} \int_{r_0}^0 \left[\frac{r}{r_0 - r}\right]^{1/2} \mathrm{d}r.$$

If we then make the substitution $r = xr_0$, then $\mathrm{d}r/\mathrm{d}x = r_0$. The limits for integration, $r = r_0$ and $r = 0$, then become $x = 1$ and $x = 0$ respectively, and we can rewrite the integration as

$$t_{\mathrm{ff}} = -\left(\frac{r_0^3}{2Gm_0}\right)^{1/2} \int_1^0 \left[\frac{x}{1-x}\right]^{1/2} \mathrm{d}x.$$

Finally, multiplying through by minus one changes the order of the integration limits, and we have

$$t_{\mathrm{ff}} = \left(\frac{r_0^3}{2Gm_0}\right)^{1/2} \int_0^1 \left[\frac{x}{1-x}\right]^{1/2} \mathrm{d}x.$$

Now, the integral above may be solved by making the further substitution $x = \sin^2\theta$, so that $\mathrm{d}x/\mathrm{d}\theta = 2\sin\theta\cos\theta$. The limits for integration, $x = 0$ and $x = 1$, then become $\theta = 0$ and $\theta = \pi/2$ respectively, and we can rewrite the integration as

$$t_{\mathrm{ff}} = \left(\frac{r_0^3}{2Gm_0}\right)^{1/2} \int_0^{\pi/2} \left[\frac{\sin^2\theta}{1-\sin^2\theta}\right]^{1/2} 2\sin\theta\cos\theta\,\mathrm{d}\theta$$

$$= \left(\frac{r_0^3}{2Gm_0}\right)^{1/2} \int_0^{\pi/2} 2\sin^2\theta\,\mathrm{d}\theta.$$

We now make use of the trigonometric identity $2\sin^2\theta = 1 - \cos 2\theta$ which gives a function that is easily integrated.

$$t_{\mathrm{ff}} = \left(\frac{r_0^3}{2Gm_0}\right)^{1/2} \int_0^{\pi/2} (1 - \cos 2\theta)\,\mathrm{d}\theta$$

$$= \left(\frac{r_0^3}{2Gm_0}\right)^{1/2} \left[\theta - \frac{\sin 2\theta}{2}\right]_0^{\pi/2}$$

$$= \left(\frac{r_0^3}{2Gm_0}\right)^{1/2} \left[\frac{\pi}{2} - \frac{\sin \pi}{2} - 0 + \frac{\sin 0}{2}\right]$$

$$= \left(\frac{r_0^3}{2Gm_0}\right)^{1/2} \left[\frac{\pi}{2}\right] = \left(\frac{\pi^2 r_0^3}{8Gm_0}\right)^{1/2}.$$

We can then recognize that the initial density of the spherical cloud is $\rho = 3m_0/4\pi r_0^3$, so the final expression for the free-fall time for a spherical cloud of density ρ to collapse under the influence of gravity if unopposed by gas pressure is:

$$\tau_{\mathrm{ff}} = \left(\frac{3\pi}{32G\rho}\right)^{1/2}. \tag{2.5}$$

However, it is never the case that gravitational collapse is completely unopposed by gas pressure. In reality, the loss of gravitational energy usually increases the thermal energy of the gas, which then increases the gas pressure and therefore opposes the collapse. The free-fall situation is a good approximation to reality though in cases where a gas cloud is transparent to its own radiation (i.e. when it is optically thin). In this case, energy can be radiated away and collapse does not heat up the cloud.

Exercise 2.1   If the Sun had no pressure support, what would be its free-fall time? (Assume the mean density of the Sun is 1.41×10^3 kg m^{-3} as derived in Chapter 1.)   ∎

2.1.3   Hydrostatic equilibrium

If the pressure gradient in a gas cloud just balances the gravitational acceleration, there is no net acceleration. In this case Equation 2.3 becomes

$$g(r) + \frac{1}{\rho(r)}\frac{\mathrm{d}P(r)}{\mathrm{d}r} = 0$$

or substituting for $g(r)$ from Equation 2.2, we obtain

$$\frac{\mathrm{d}P(r)}{\mathrm{d}r} = -\frac{G\,m(r)\,\rho(r)}{r^2}. \tag{Eqn 1.12}$$

(Remember that $\mathrm{d}P(r)/\mathrm{d}r$ is negative.) The equation of hydrostatic equilibrium is one of the fundamental equations of stellar structure which we introduced in Chapter 1.

Worked Example 2.2

The virial theorem, which you also met in Chapter 1, may be derived from the equation of hydrostatic equilibrium as follows.

Solution

First divide both sides of Equation 1.12 by $4\pi r^2\,\rho(r)$, to give

$$\frac{\mathrm{d}P(r)}{4\pi r^2 \rho(r)\,\mathrm{d}r} = -\frac{G\,m(r)}{4\pi r^4}$$

then note that $4\pi r^2\,\rho(r)\,\mathrm{d}r = \mathrm{d}m(r)$, so

$$\frac{\mathrm{d}P(r)}{\mathrm{d}m(r)} = -\frac{G\,m(r)}{4\pi r^4}$$

so,

$$4\pi r^3 \, \mathrm{d}P(r) = -\frac{G\,m(r)}{r} \, \mathrm{d}m(r)$$

and since the volume of a sphere of radius r is $V(r) = \frac{4}{3}\pi r^3$, we have

$$3V(r) \, \mathrm{d}P(r) = -\frac{G\,m(r)}{r} \, \mathrm{d}m(r).$$

Now integrating this equation over the entire volume of the cloud gives

$$3 \int_{r=0}^{r=R} V(r) \, \mathrm{d}P(r) = -\int_{0}^{M} \frac{G\,m(r)}{r} \, \mathrm{d}m(r)$$

where $r = R$ and $r = 0$ indicate the surface and centre of the cloud respectively. The right-hand side of the above is simply the gravitational potential energy of the cloud, E_{GR}, and the left-hand side may be integrated by parts to give

$$3 \left[P(r)V(r) \right]_{r=0}^{r=R} - 3 \int_{0}^{V} P(r) \, \mathrm{d}V(r) = E_{\mathrm{GR}}.$$

The first term on the left-hand side is zero in both limits, since the enclosed volume is zero at the centre and the pressure may be assumed to be zero at the surface. The second term on the left-hand side may be expressed as $-3 \langle P \rangle V$, where $\langle P \rangle$ is the average pressure in the cloud. This therefore yields the virial theorem, namely that the average pressure of a gas cloud in hydrostatic equilibrium is equal to minus one-third of its gravitational energy per unit volume:

$$\langle P \rangle = -\frac{E_{\mathrm{GR}}}{3V}. \tag{Eqn 1.2}$$

2.2 Stability and the virial theorem

Having studied the competing forces of gravitation and pressure support in a gas cloud, we now look more closely at the conditions for the stability of the gas under two distinct conditions: a classical ideal gas, and a gas in which the particles providing the pressure support move with relativistic velocities, i.e. $v \approx c$. As you will see in later sections of this chapter, the former is important in determining the lower mass limit for stars on the main sequence, and the latter is important for determining the upper mass limit for stars on the main sequence.

The pressure of an ideal gas can be written in terms of the momentum p and velocity v of its constituent particles as,

$$P = \frac{1}{3}\frac{N}{V}pv, \tag{2.6}$$

where we have assumed that all the constituent particles possess the same momentum and velocity. In reality, particles in the gas will have a range of these quantities, and it is more appropriate to consider an average value for each. We return to this point in the following section. The ratio N/V is just the number of particles per unit volume, i.e. the number density n.

For *non-relativistic* particles, the momentum of each particle is $p = mv$ and hence $pv = mv^2 = 2 \times \frac{1}{2}mv^2$. Then Equation 2.6 can be expressed as

$$P = \frac{1}{3}\frac{N}{V} \times 2 \times \frac{1}{2}mv^2 = \frac{2}{3} \times \frac{1}{V} \times N\frac{1}{2}mv^2$$

where $\frac{1}{2}mv^2$ is the kinetic energy *per particle*, and $N\frac{1}{2}mv^2$ is the kinetic energy E_K of the ensemble of N particles, i.e.

$$\text{for non-relativistic particles}\quad P_{\mathrm{NR}} = \frac{2}{3}\frac{E_K}{V}. \tag{2.7}$$

For *ultra-relativistic* particles, the velocity approaches the speed of light, so $pv \approx pc$. Then Equation 2.6 becomes

$$P = \frac{1}{3}\frac{N}{V}pc.$$

For ultra-relativistic particles, the kinetic energy *per particle* is pc, so Npc is the kinetic energy E_K of the ensemble of N particles, i.e.

$$\text{for ultra-relativistic particles}\quad P_{\mathrm{UR}} = \frac{1}{3}\frac{E_K}{V}. \tag{2.8}$$

The significance of these equations for the gas pressure becomes clear once they are combined with the virial theorem. Considering the non-relativistic case first, combining Equation 2.7 with Equation 1.2, implies that $-E_{\mathrm{GR}}/3V = 2E_K/3V$, or

$$2E_K + E_{\mathrm{GR}} = 0. \tag{2.9}$$

Since the total energy of the system is just $E_{\mathrm{TOT}} = E_K + E_{\mathrm{GR}}$, it follows that, in the non-relativistic case

$$E_{\mathrm{TOT}}(\mathrm{NR}) = -E_K \qquad \text{and} \qquad E_{\mathrm{TOT}}(\mathrm{NR}) = \frac{1}{2}E_{\mathrm{GR}}. \tag{2.10}$$

Several consequences of these relations can be stated. The binding energy $(-E_{\mathrm{TOT}})$ of a system in hydrostatic equilibrium is the kinetic energy of translational motion, so a highly bound system is very hot, and a loosely bound system is cool. If a star collapses slowly enough to remain close to hydrostatic equilibrium, then as the gravitational energy E_{GR} and total energy E_{TOT} decrease, the kinetic energy E_K must increase. That is, a self-gravitating system which collapses slowly enough to remain close to hydrostatic equilibrium heats up. As you will see in Chapter 8, this provides a good approximation to proto-stars as they approach the main sequence.

Turning to the ultra-relativistic case and combining Equation 2.8 with the virial theorem (Equation 1.2) implies that $-E_{\mathrm{GR}}/3V = E_K/3V$, or

$$E_K + E_{\mathrm{GR}} = 0. \tag{2.11}$$

Since the total energy is just $E_{\mathrm{TOT}} = E_K + E_{\mathrm{GR}}$, it follows that, in the ultra-relativistic case

$$E_{\text{TOT}}(\text{UR}) = 0. \tag{2.12}$$

That is, in the ultra-relativistic case, the system is at the boundary of being bound (i.e. $E_{\text{TOT}} < 0$) and unbound (i.e. $E_{\text{TOT}} > 0$). If a bound system approaches the ultra-relativistic limit, it moves closer towards becoming unbound.

2.3 The lower mass limit for stars

In Chapter 1 we discussed hydrogen burning in stars, but did not consider which self-gravitating objects attain stardom and which do not. If the temperature in the core of a protostar continued to increase as it contracted, ultimately the ignition temperature for hydrogen burning would be reached in every case. However, the temperature cannot always increase, due to the onset of **electron degeneracy** (see the box below). Unlike the case of an ideal gas, in a degenerate electron gas, an increase in pressure does *not* lead to an increase in temperature. If a contracting protostar becomes degenerate before it becomes hot enough for hydrogen fusion to begin, it will instead become a **brown dwarf**. The situation under which this can occur is explored in this section.

Electron degeneracy

Electron degeneracy is a condition in which the quantum nature of electrons cannot be ignored. Although it is often the case that electrons can be treated as isolated, classical particles, like all matter they also have wave-like properties. This is one example of the wave–particle duality that pervades a quantum view of the Universe, and which is beautifully demonstrated in an experiment showing that a beam of electrons can be diffracted.

If electrons are packed so densely that their separations are comparable to their de Broglie wavelengths, then the full range of their quantum properties has to be considered, including the Pauli exclusion principle which prohibits the overlapping of electrons having the same energy. This leads to some unexpected but well understood consequences, such as the decoupling of pressure and temperature in a degenerate electron gas. Other consequences of electron degeneracy will be encountered later in this book, as will consequences of neutron degeneracy.

In a cold, dense gas, the behaviour of the electrons is controlled by the laws of quantum mechanics once the typical separation between the electrons becomes comparable to their de Broglie wavelength. The de Broglie wavelength is given by $\lambda_{\text{dB}} = h/p$ where h is Planck's constant and p is the electron's momentum.

Now, in the core of a star, an electron's kinetic energy can be written as $E_{\text{K}} = \frac{3}{2}kT_{\text{c}}$, where T_{c} is the core temperature of the star. Another expression for the kinetic energy of an electron is simply $E_{\text{K}} = \frac{1}{2}m_{\text{e}}v^2$. So equating these two we obtain $3kT_{\text{c}} = m_{\text{e}}v^2$ or $v = (3kT_{\text{c}}/m_{\text{e}})^{1/2}$. (Strictly, electrons will exist with a range of speeds, but $(3kT_{\text{c}}/m_{\text{e}})^{1/2}$ is close to the average.) The momentum of an electron is given by $p = m_{\text{e}}v$, so $p = (3m_{\text{e}}kT_{\text{c}})^{1/2}$. Hence the de Broglie wavelength of an electron in the core of a star is

$$\lambda_{dB} = \frac{h}{(3m_e k T_c)^{1/2}}. \tag{2.13}$$

The core of a star consists of an ionized plasma containing equal numbers of protons and electrons. The mean separation l between the protons in the core is given by $\rho_c = m_p/l^3$ where ρ_c is the density of the plasma in the core and m_p is the mass of a proton. The mean separation between the electrons will also be l (because of overall charge neutrality), hence the plasma will become degenerate when $\lambda_{dB} \approx l$, or when

$$\left(\frac{m_p}{\rho_c}\right)^{1/3} \approx \frac{h}{(3m_e k T_c)^{1/2}}.$$

So the limiting density for the core, below which electron degeneracy can be avoided is

$$\rho_c < \frac{m_p (3m_e k T_c)^{3/2}}{h^3}. \tag{2.14}$$

In order to find out what mass of star corresponds to this density limit, we must find an expression linking the star's mass to its internal temperature. We can do that by making use of the result obtained in the last section found by combining the virial theorem with the ideal gas law.

As you saw in Section 2.2, the condition for hydrostatic equilibrium inside a star is $2E_K + E_{GR} = 0$. For an ideal gas containing N protons and N electrons, the kinetic energy of the gas is $E_K = \frac{3}{2}NkT_I + \frac{3}{2}NkT_I = 3NkT_I$ where T_I is a typical internal temperature of the star. The gravitational potential energy of the gas is $E_{GR} = -3GM^2/5R$ (for a spherical region of uniform density). Hence we can write $2 \times 3NkT_I = 3GM^2/5R$ and so the typical thermal energy inside the star is

$$kT_I = \frac{GM^2}{10NR}.$$

Now, we note that the total mass of the star can be written as $M = Nm_p + Nm_e$ where m_p and m_e are the masses of a proton and electron respectively. Since $m_p >> m_e$, we have $M \approx Nm_p$ and the number of protons (or electrons) can be written as $N = M/m_p$, so the typical thermal energy becomes

$$kT_I = \frac{GMm_p}{10R}$$

and since the average density for a sphere of mass M and radius R is $\langle\rho\rangle = M/\frac{4}{3}\pi R^3$, we can replace the radius R by $(3M/4\pi \langle\rho\rangle)^{1/3}$. Therefore the expression for the typical thermal energy becomes

$$kT_I = \frac{1}{10}\left(\frac{4\pi}{3}\right)^{1/3} Gm_p M^{2/3} \langle\rho\rangle^{1/3}. \tag{2.15}$$

So, for an ideal gas, the temperature inside a star is proportional to the two-thirds power of the mass multiplied by the one-third power of the average density.

We may therefore combine Equation 2.15 with Equation 2.14 to calculate the limiting mass below which the core of a star will be degenerate.

Worked Example 2.3

What is the mass limit between objects which become stars and those which become brown dwarfs?

(a) Combine the condition to avoid degeneracy from Equation 2.14 with the expression for the typical internal thermal energy from Equation 2.15 to find an expression for the limiting core thermal energy kT_c to avoid degeneracy, in terms of the mass of the star. As an order of magnitude approximation, you should assume that the core of the star is about 10 times hotter than its typical internal temperature T_I and that the core density is about 100 times the average density, as is the case for the Sun.

(b) Compute the minimum mass for a star to ignite hydrogen burning in its core and avoid becoming a degenerate brown dwarf. Adopt $T_{ign} = 1.5 \times 10^6$ K as the minimum ignition temperature for hydrogen burning.

Solution

(a) Assuming $T_c = 10T_I$ and $\rho_c = 100 \langle \rho \rangle$, Equation 2.15 becomes

$$\frac{1}{10}kT_c = \frac{1}{10}\left(\frac{4\pi}{3}\right)^{1/3} Gm_p M^{2/3} \left(\rho_c/100\right)^{1/3}.$$

Substituting into this the limiting core density from Equation 2.14, canceling the factors of $1/10$ and rearranging slightly gives

$$kT_c < \left(\frac{4\pi}{300}\right)^{1/3} GM^{2/3}m_p^{4/3} \left(3m_e kT_c\right)^{1/2}/h.$$

Collecting the terms in kT_c and rearranging this further gives the condition to avoid degeneracy as

$$kT_c < \left(\frac{4\pi}{300}\right)^{2/3} 3G^2 \frac{m_p^{8/3}m_e}{h^2} M^{4/3}.$$

(b) If $T_c = T_{ign}$, we require

$$M^{4/3} > \frac{kT_{ign}}{\left(\frac{4\pi}{300}\right)^{2/3} 3G^2 \frac{m_p^{8/3}m_e}{h^2}}.$$

Putting the numbers into the right-hand fraction in the denominator, we have

$$\frac{m_p^{8/3}m_e}{h^2} = \frac{\left(1.673 \times 10^{-27}\,\text{kg}\right)^{8/3} \times 9.109 \times 10^{-31}\,\text{kg}}{\left(6.626 \times 10^{-34}\,\text{J s}\right)^2}$$

$$= 8.184 \times 10^{-36}\,\text{kg}^{11/3}\,\text{J}^{-2}\,\text{s}^{-2}.$$

Therefore

$$M^{4/3} > \frac{1.381 \times 10^{-23}\,\text{J K}^{-1} \times 1.5 \times 10^6\,\text{K}}{\left(\frac{4\pi}{300}\right)^{2/3} \times 3 \times \left(6.673 \times 10^{-11}\,\text{N m}^2\,\text{kg}^{-2}\right)^2 \times 8.184 \times 10^{-36}\,\text{kg}^{11/3}\,\text{J}^{-2}\,\text{s}^{-2}}$$

$$> 1.571 \times 10^{39}\,\text{J}^3\,\text{N}^{-2}\,\text{m}^{-4}\,\text{kg}^{1/3}\,\text{s}^2.$$

Taking the (3/4)-power of both sides gives

$$M > (1.571 \times 10^{39})^{3/4}[(\mathrm{N\,m})^3\,\mathrm{N}^{-2}\,\mathrm{m}^{-4}\,\mathrm{kg}^{1/3}\,\mathrm{s}^2]^{3/4}$$
$$> (2.5 \times 10^{29})(\mathrm{N\,m}^{-1}\,\mathrm{kg}^{1/3}\,\mathrm{s}^2)^{3/4}$$
$$> (2.5 \times 10^{29})(\mathrm{kg\,m\,s}^{-2}\,\mathrm{m}^{-1}\,\mathrm{kg}^{1/3}\,\mathrm{s}^2)^{3/4}$$
$$> (2.5 \times 10^{29})(\mathrm{kg}^{4/3})^{3/4}$$
$$> 2.5 \times 10^{29}\,\mathrm{kg}.$$

Since $1\,\mathrm{M}_\odot = 1.99 \times 10^{30}$ kg, we can also express the limiting mass as $M > 2.5 \times 10^{29}$ kg$/1.99 \times 10^{30}$ kg $\mathrm{M}_\odot^{-1} \sim 0.1\,\mathrm{M}_\odot$, to the nearest order of magnitude.

The example above shows that the temperature at which degeneracy could set in depends on the mass of the star, according to $T \propto M^{4/3}$. Successful stars reach the hydrogen ignition temperature before degeneracy sets in, but objects of sufficiently low mass become degenerate whilst cooler. For the latter objects, further contraction does not increase the temperature, so they fail to initiate thermonuclear reactions at a sufficiently high rate. Detailed computations show that the limit between successful stars and unsuccessful brown dwarfs occurs at $M = 0.08\,\mathrm{M}_\odot$.

2.4   The upper mass limit for stars

The upper mass limit for stars is probably set by the situation when the radiation pressure in the star's core exceeds the gas pressure there. This makes the core unstable and liable to disruption. As you saw in Section 2.2, in the ultra-relativistic limit, the star will be at the boundary between being bound (i.e. $E_{\mathrm{TOT}} < 0$) and unbound (i.e. $E_{\mathrm{TOT}} > 0$). As photons are ultra-relativistic, stars in which a large fraction of the pressure support is provided by radiation pressure can approach this unbound limit.

The radiation pressure due to photons in the core of a star is given by

$$P_{\mathrm{rad}} = \frac{4\sigma T_{\mathrm{c}}^4}{3c} \tag{2.16}$$

where σ is the Stefan–Boltzmann constant, T_{c} is the core temperature and c is the speed of light.

The pressure due to an ideal gas can be expressed as

$$P_{\mathrm{gas}} = \frac{\rho_{\mathrm{c}} k T_{\mathrm{c}}}{\overline{m}} \tag{Eqn 1.11}$$

where ρ_{c} is the core density, T_{c} is the core temperature and $\overline{m}$ is the mean molecular mass in kilograms.

The *total pressure* in the core of a star will therefore be given by the sum of these two pressures, $P_{\mathrm{c}} = P_{\mathrm{gas}} + P_{\mathrm{rad}}$. It is useful to express the relative contributions of the two pressures in terms of a fraction f (< 1) of the total pressure where $P_{\mathrm{gas}} = f P_{\mathrm{c}}$ and $P_{\mathrm{rad}} = (1 - f) P_{\mathrm{c}}$.

Now, we can rearrange Equation 1.11 to write

$$T_c = \frac{f P_c \overline{m}}{\rho_c k}$$

and then substitute this into Equation 2.16 to get

$$(1-f)P_c = \frac{4\sigma}{3c} \left(\frac{f P_c \overline{m}}{\rho_c k} \right)^4 .$$

This may be rearranged to give an expression for the core pressure as

$$P_c = \left(\frac{3c}{4\sigma} \frac{1-f}{f^4} \right)^{1/3} \left(\frac{k\rho_c}{\overline{m}} \right)^{4/3} . \tag{2.17}$$

Models for stellar interiors have been developed which express the relationship between the core pressure and core density. One particular model, originated by Donald Clayton in the late 1960s, states that they are related by

$$P_c \approx \left(\frac{\pi}{36} \right)^{1/3} G M^{2/3} \rho_c^{4/3} . \tag{2.18}$$

Hence combining Equations 2.18 and 2.17 we have

$$\left(\frac{\pi}{36} \right)^{1/3} G M^{2/3} \rho_c^{4/3} \approx \left(\frac{3c}{4\sigma} \frac{1-f}{f^4} \right)^{1/3} \left(\frac{k\rho_c}{\overline{m}} \right)^{4/3} .$$

The core density conveniently cancels out from both sides, leaving

$$M \approx \left(\frac{36}{\pi} \frac{3c}{4\sigma} \frac{(1-f)}{f^4} \right)^{1/2} \left(\frac{k}{\overline{m}} \right)^2 \left(\frac{1}{G} \right)^{3/2} . \tag{2.19}$$

Now, we can assume that a star will likely become unstable when the radiation pressure in the core exceeds the gas pressure. In other words, the limiting situation will be when $P_{\mathrm{rad}} > P_{\mathrm{gas}}$, or when the gas pressure provides less than half the total pressure in the core, i.e. when $f < 0.5$.

Exercise 2.2   Substitute the numbers into Equation 2.19 to obtain an estimate for the upper mass limit for a star. Assume that the mean molecular mass is similar to that of the Sun, i.e. $\overline{m} \approx 0.6u$.   ∎

The result of the previous exercise, that the maximum mass for a star is of order $100 \, \mathrm{M}_\odot$, is borne out by observations of massive, eclipsing binary stars in which the masses of the component stars may be measured from their binary motion.

2.5   The Kelvin–Helmholtz contraction timescale

In this penultimate section of the Chapter we draw together the ideas developed earlier in order to quantify the role of gravitational contraction in the evolution of stars.

You know from the virial theorem that for a gas of non-relativistic particles in hydrostatic equilibrium, $2E_K + E_{\mathrm{GR}} = 0$. This can be rewritten as $E_K = -\frac{1}{2}E_{\mathrm{GR}}$. So, as gravitational potential energy is liberated during contraction of a gas cloud, one half of the liberated gravitational potential energy is stored as kinetic energy (heat), whilst the other half is radiated away. We can therefore ask the following: if the Sun had been radiating at its current luminosity fueled solely by gravitational collapse, how long would it have taken to collapse to its current size?

The gravitational potential energy liberated, treating it as a *uniform* density sphere (see Equation 1.3), is

$$E_{\text{lib}} = -\Delta E_{\text{GR}} = \frac{3}{5}\frac{G\text{M}_\odot^2}{\text{R}_\odot} - \frac{3}{5}\frac{G\text{M}_\odot^2}{R_{\text{initial}}} = \frac{3}{5}G\text{M}_\odot^2\left(\frac{1}{\text{R}_\odot} - \frac{1}{R_{\text{initial}}}\right).$$

Since $R_{\text{initial}} \gg \text{R}_\odot$, so $1/R_{\text{initial}} \ll 1/\text{R}_\odot$, and therefore $E_{\text{lib}} \approx 3G\text{M}_\odot^2/5\text{R}_\odot$. The time τ taken to radiate half of this, at the Sun's current luminosity of $\text{L}_\odot$, is

$$\tau = \frac{0.5E_{\text{lib}}}{\text{L}_\odot} = \frac{3}{10}\frac{G\text{M}_\odot^2}{\text{R}_\odot\text{L}_\odot}.$$

Recall, however, that the constant in this expression, 3/10, comes from the assumption of a *uniform* density sphere, which causes the gravitational potential energy of the sphere, the liberated energy, and the timescale all to be underestimated. Acknowledging this uncertainty, the portion *excluding* the uncertain constant is called the **Kelvin–Helmholtz contraction time** *of the Sun*, given by

$$\tau_{\text{KH},\odot} = \frac{G\text{M}_\odot^2}{\text{R}_\odot\text{L}_\odot}. \tag{2.20}$$

Notice the distinction between the Kelvin–Helmholtz contraction time and the free-fall timescale. The Kelvin–Helmholtz timescale includes consideration of the rate at which the liberated gravitational potential energy is radiated, whereas the free-fall timescale is for an unopposed collapse.

Exercise 2.3   Evaluate the Kelvin–Helmholtz time for the Sun, in seconds and years.   ■

● You have now shown that the Kelvin–Helmholtz contraction time for the Sun is $\approx 3 \times 10^7$ yr. What does this indicate about the possible lifetime of the Sun?

○ Even without considering nuclear energy sources, we can derive a possible lifetime for the Sun of ~30 million years. In fact, astronomers believed for a time that the Sun was powered by slow gravitational collapse, and that it would last for about this length of time. However, that belief pre-dated the discovery, from radioactive-decay dating, that some rocks on the Earth were orders of magnitude older. This forced astronomers to acknowledge that some other energy source was delaying the gravitational collapse of the Sun. That energy source was later revealed with the discovery of thermonuclear fusion.

The Kelvin–Helmholtz time can be generalized to stars of other mass by maintaining the same parameter dependence, and normalizing to our result for the

Sun, i.e. multiplying by 1 and then rearranging

$$\tau_{\mathrm{KH}} = \frac{GM^2}{RL} = \frac{GM^2}{RL} \times \frac{\left(\frac{G\mathrm{M}_\odot^2}{\mathrm{R}_\odot \mathrm{L}_\odot}\right)}{\left(\frac{G\mathrm{M}_\odot^2}{\mathrm{R}_\odot \mathrm{L}_\odot}\right)} = \left(\frac{G\mathrm{M}_\odot^2}{\mathrm{R}_\odot \mathrm{L}_\odot}\right) \times \frac{\left(\frac{GM^2}{RL}\right)}{\left(\frac{G\mathrm{M}_\odot^2}{\mathrm{R}_\odot \mathrm{L}_\odot}\right)}$$

$$= 3 \times 10^7 \,\mathrm{yr} \times \left(\frac{M}{\mathrm{M}_\odot}\right)^2 \frac{1}{R/\mathrm{R}_\odot} \frac{1}{L/\mathrm{L}_\odot}.$$

This can be reduced further to just an explicit dependence on mass by making some approximations.

First, from the H–R diagram, it can be seen that the plot of $\log_{10} L$ versus $\log_{10} T_{\mathrm{eff}}$ has a main sequence slope for stars near $1\,\mathrm{M}_\odot$ of ≈ 8, from which we infer that, for solar-type stars on the main sequence, $L \propto T_{\mathrm{eff}}^8$. Next, you know that $L = 4\pi R^2 \sigma T_{\mathrm{eff}}^4 \propto R^2 T_{\mathrm{eff}}^4$, so we can substitute for T_{eff} using the main-sequence slope relationship, $T_{\mathrm{eff}} \propto L^{1/8}$, and write $L \propto R^2 L^{4/8}$, i.e. $R^2 \propto L^{1-4/8} = L^{1/2}$, or $R \propto L^{1/4}$ (on the main sequence, near to the mass of the Sun).

Our expression for the Kelvin–Helmholtz time then becomes

$$\tau_{\mathrm{KH}} = 3 \times 10^7 \,\mathrm{yr} \times \left(\frac{M}{\mathrm{M}_\odot}\right)^2 \left(\frac{1}{L/\mathrm{L}_\odot}\right)^{1/4} \frac{1}{L/\mathrm{L}_\odot}$$

$$= 3 \times 10^7 \,\mathrm{yr} \times \left(\frac{M}{\mathrm{M}_\odot}\right)^2 \left(\frac{L}{\mathrm{L}_\odot}\right)^{-5/4}.$$

We now use a relationship between mass and luminosity to write this solely in terms of the mass. For solar-mass stars, the relation $L \propto M^{3.5}$ is usually adopted. Substituting for the luminosity then approximates the Kelvin–Helmholtz time for solar-type stars as

$$\tau_{\mathrm{KH}} = 3 \times 10^7 \,\mathrm{yr} \times \left(\frac{M}{\mathrm{M}_\odot}\right)^2 \left(\left(\frac{M}{\mathrm{M}_\odot}\right)^{7/2}\right)^{-5/4}$$

$$= 3 \times 10^7 \,\mathrm{yr} \times \left(\frac{M}{\mathrm{M}_\odot}\right)^{2-\left(\frac{7}{2}\times\frac{5}{4}\right)}$$

$$\approx 3 \times 10^7 \,\mathrm{yr} \times \left(\frac{M}{\mathrm{M}_\odot}\right)^{-2.4}.$$

(Remember that this is only an approximation, as both the main-sequence slope $\log_{10} L$ versus $\log_{10} T_{\mathrm{eff}}$ and the mass–luminosity relation vary slowly with mass.) This emphasizes that the contraction times of low-mass stars are much longer than the contraction time of the Sun, and conversely high-mass stars contract faster.

Exercise 2.4 What are the Kelvin–Helmholtz contraction times for stars of mass $0.5\,\mathrm{M}_\odot$, $2\,\mathrm{M}_\odot$ and $5\,\mathrm{M}_\odot$ assuming the approximate relationship above? ∎

The Kelvin–Helmholtz timescales calculated above therefore represent the length of time it has taken these stars to collapse to their present size, radiating at their current luminosity and fueled solely by gravitational collapse.

A fit to *numerical* models of the time to contract (τ_{cont}) to the main sequence give results slightly longer than the Kelvin–Helmholtz time, by a factor of a few:

$$\tau_{\text{cont}} \approx 8 \times 10^7 (M/\text{M}_\odot)(\text{L}_\odot/L_{\text{ms}}) \text{ yr}.$$

You may be surprised that the exponents in this equation do not appear to resemble those derived above. This is because the mass M and main-sequence luminosity L_{ms} are closely correlated, so one compensates for differences in the other. We can see that the dependence on mass derived from the numerical models is in fact remarkably similar, even though the coefficient is larger. Using the same approximate mass–luminosity relation for main-sequence stars as above, $L_{\text{ms}}/\text{L}_\odot = (M/\text{M}_\odot)^{3.5}$, we can rewrite the numerical result as $\tau_{\text{cont}} \approx 8 \times 10^7 (M/\text{M}_\odot)^{-2.5}$ yr which has essentially the same mass-dependence as we derived above.

2.6   Why are stars hot? Putting fusion in its place

In Chapter 1 you saw that hydrogen fusion provides the power that is radiated by the Sun and other main-sequence stars. In this Chapter, you have seen the importance of the gravitational energy released by a collapsing cloud of gas. In order to understand the roles of these two sources of energy, before we go any further, we pause to consider exactly what fusion is and is not responsible for.

In the last chapter, you saw that the average energy released for each reaction between two protons in the Sun is 15.0 MeV. In the next chapter you will calculate the timescale for the reaction between two protons and discover that there are about 1.3×10^{14} such reactions per cubic metre per second in the core of the Sun. Combining these two numbers, reactions between protons release 1.95×10^{15} MeV of energy per cubic metre per second, which is equivalent to about 300 watts per cubic metre.

● Think about this number: 300 W m^{-3}. Imagine putting three 100 W lightbulbs in a broom cupboard whose volume is about 1 cubic metre. Would that make the cupboard as bright and as hot as the Sun?

○ No, clearly it would *not*. In fact, 300 W seems like a pathetically tiny power output for a volume as large as 1 m^3, especially in something as hot as the core of the Sun! Is the calculation grossly wrong? No, the power output per cubic metre really is that small. Clearly hydrogen burning by the proton–proton chain is not much of a powerhouse! (But it is a big energy reservoir!)

So, why is the Sun (or any other main-sequence star) so hot and bright? The topic missing from this discussion so far is opacity. The solar material is very opaque. Even though the power generating rate is very small, the heat produced by it does not easily escape. Energy released from a light bulb, on the other hand, easily escapes.

● The Sun radiates with a power given by $\text{L}_\odot = 3.83 \times 10^{26}$ W. If only 300 W is generated by nuclear fusion per cubic metre, how many cubic metres of the

Sun must be involved in fusion, and what fraction of the mass of the Sun would this be? (Assume uniform density $= \rho_{\odot,c}$ from Table 1.1.)

○ The volume would be 3.83×10^{26} W/300 W m^{-3} = 1.28×10^{24} m^3. At a density of 1.48×10^5 kg m^{-3}, this corresponds to a mass of 1.28×10^{24} m$^3 \times 1.48 \times 10^5$ kg m$^{-3} = 1.89 \times 10^{29}$ kg or 1.89×10^{29} kg/1.99×10^{30} kg M$_\odot^{-1} \approx 0.1$ M$_\odot$. That is, fusion reactions occur throughout a volume of the Sun that contains about 10% of its mass (on the assumption of uniform density).

You might be tempted at this stage to think that, even though the power released in fusion is very small, the energy released heats up the core of a star, and this is why stars are hot. The energy released is indeed very important, but it is not the reason that stars heat up. In fact the opposite is true: fusion *prevents* a star from getting hotter!

In fact, a star's temperature and luminosity are not determined by nuclear burning. However, they are maintained by nuclear burning. Nuclear fusion replenishes the slow leakage of energy from the star's core and eventually from its surface. As you saw in the last Section, the release of gravitational potential energy can do exactly the same thing, but the nuclear fusion energy source has the advantage that it lasts for much longer. That is, nuclear fusion greatly delays the gravitational collapse of a star.

If someone had asked you at the beginning of this book, 'What makes stars hot?', you could have been forgiven for answering 'It is because they have thermonuclear reactions in their cores.' That answer sounds plausible, but hopefully now you can see that it is incorrect. Stars are hot because they have collapsed from large diffuse clouds, and gravitational potential energy has been converted to kinetic energy. That is why stars are hot. The role of nuclear fusion is to delay the collapse for long enough that planets, human life, and astrophysics courses could develop. Nuclear fusion in stars is also responsible for the production of most of the elements heavier than helium.

In the next chapter we shall return to the study of stellar fusion processes, but now you will be able to resist the temptation to think that fusion makes stars hot. Fusion happens because stars are hot, rather than the other way round.

Summary of Chapter 2

1. The general equation which describes the acceleration of a packet of gas in a spherical cloud, under the influence of gravitational and pressure forces is

$$-\frac{\mathrm{d}^2 r}{\mathrm{d}t^2} = g(r) + \frac{1}{\rho(r)} \frac{\mathrm{d}P(r)}{\mathrm{d}r} \qquad \text{(Eqn 2.3)}$$

   where the gravitational acceleration is

$$g(r) = \frac{Gm(r)}{r^2} \qquad \text{(Eqn 2.2)}$$

   and $\mathrm{d}P(r)/\mathrm{d}r$ must be negative in order to oppose the gravitational acceleration.

2. If there is no internal pressure support in a cloud of gas it will collapse under the influence of gravity. The free-fall time depends on the initial density of the cloud ρ, and is given by

$$\tau_{ff} = \left(\frac{3\pi}{32G\rho}\right)^{1/2}.$$ (Eqn 2.5)

3. Hydrostatic equilibrium exists when the outward force due to the pressure gradient balances the inward gravitational force at every point through the star. This is expressed mathematically as

$$\frac{dP(r)}{dr} = -\frac{G\,m(r)\,\rho(r)}{r^2}.$$ (Eqn 1.12)

The dynamical evolution of the system will be slow when this condition is approached.

4. For non-relativistic particles, the pressure P_{NR}, volume V and kinetic energy E_K are related via $P_{NR} = 2E_K/3V$. Combining this with the virial theorem gives $E_{TOT}(NR) = -E_K = E_{GR}/2$. Since E_K is positive, the total energy of a system in hydrostatic equilibrium is negative, and hence the object is bound. The equivalent expressions for ultra-relativistic particles are $P_{UR} = E_K/3V$ and $E_{TOT}(UR) = 0$. This indicates that an object whose pressure support comes from ultra-relativistic particles is unstable, because $E_{TOT} = 0$ puts the system at the boundary between bound and unbound configurations.

5. In objects with $M < 0.08\,M_\odot$, electrons become degenerate in the contracting core before hydrogen is ignited. Degeneracy sets in when the electron separation becomes comparable with the typical de Broglie wavelength of the electrons. In an object supported by degenerate electrons, the temperature ceases to rise even when it contracts further. Such objects form brown dwarfs that shine by the release of gravitational potential energy, not from thermonuclear energy as in stars.

6. Stars in which much of the pressure support comes from radiation pressure rather than gas pressure rely more on ultra-relativistic particles (photons) than on non-relativistic particles (atoms) for pressure support. The virial theorem indicates that such objects are unstable. This may help set the upper limit on the mass of main-sequence stars as $M \approx 100\,M_\odot$, because such high-mass stars have appreciable radiation pressure.

7. The timescale for the Kelvin–Helmholtz contraction of a star, in which its luminosity is provided solely by liberated gravitational potential energy as it collapses, is given by $\tau_{KH} = GM_\odot^2/R_\odot L_\odot \approx 3 \times 10^7$ yr for the Sun, or $\tau_{KH} \approx 3 \times 10^7$ yr $\times (M/M_\odot)^{-2.4}$ for solar-type stars. Numerical models suggest slightly longer contraction times $t_{cont} \approx 8 \times 10^7$ yr$(M/M_\odot)^{-2.5}$.

8. A star's temperature and luminosity are not determined by nuclear burning. However, they are *maintained* by nuclear burning. Nuclear fusion *replenishes* the slow leakage of energy from a star's core and eventually from its surface. The release of gravitational potential energy can do exactly the same thing, but the nuclear fusion energy source has the advantage that it lasts for much longer. Nuclear fusion greatly delays the gravitational collapse of a star.

Chapter 3   Nuclear fusion

Introduction

Nucleosynthesis and stellar evolution are inseparable topics: nuclear fusion provides the source of energy that keeps a star shining, while changes in the structure of a star as it ages alter the reactions that occur. In this chapter we take a more detailed look at the process of nuclear fusion. We examine the quantum properties essential for it to occur, and then study the rates at which the reactions proceed.

3.1   Quantum-mechanical properties of particles

Quantum physics is vital to understanding fusion, because of the very small sizes of nuclei. We begin this section by considering just how small a nucleus is, and then investigate the consequences.

3.1.1   Nuclear dimensions

The radius R of a nucleus containing A **nucleons**, i.e. having an atomic mass number of A, is $R \approx r_0 A^{1/3}$, where $r_0 = 1.2 \times 10^{-15}$ m. Since volume V varies as R^3, this relation tells you immediately that $V \propto A$. That is, doubling the number of nucleons in a nucleus doubles the volume occupied. In this sense, *nuclei* are *incompressible*, whereas *electrons* in an atom and *atoms* in a solid, liquid or gas can be packed together with greatly *varying density*.

● Consider for a moment just how compact the nucleus is compared with an atom. A convenient unit for measuring the radius of an atom is the Bohr radius, approximately 0.5×10^{-10} m. Atomic radii are typically a few to a few tens of Bohr radii, or of order 10^{-10} m. Nuclei are therefore of order 10^5 times smaller in radius than an atom. Think of some examples of things that differ in size by that factor, to get an impression of how compact the nucleus is compared to the atom within which it resides.

○ Here are some examples.

   ● An atom with a radius the same as the Earth's orbit (1 AU) would have a nucleus much smaller than the Sun! Since 1 AU $\approx 1.5 \times 10^{11}$ m, the scaled nucleus would have a radius 1.5×10^6 m, about the same as the radius of the Moon!

   ● The continent of Australia is roughly 4000 km across. An atom this size would have a nucleus only 40 m across, about the height of Sydney Opera House.

   ● The M25 ring road around London has a diameter of order 40 km. If an atom had a radius this size, the nucleus would be only 0.4 m across, or the width of a rubbish bin in central London.

- An athletics stadium is typically about 100 m across, so an atom on this scale would have a nucleus about 1 mm across, or the diameter of a pinhead.

- As a final example, an atom the size of a house 10 m across would have a nucleus only 0.1 mm in diameter, smaller than the full stop at the end of this sentence.

3.1.2   Wave functions and Schrödinger's equation

Very close to a nucleus (at distances of order 10^{-15} m), a force between nucleons called the **strong nuclear force** dominates over the Coulomb force and binds a nucleus together. As the distance between nucleons increases beyond 10^{-15} m, the strong force rapidly becomes negligible.

Because nuclear particles have such small separations, when they interact via the strong force it is important to consider the quantum-mechanical (wave) properties of the particles as well as their classical behaviour. In fact, if nuclear particles did not have wave properties, then fusion reactions would be exceedingly rare and we would not be here to discuss them. This is because it is the wave properties of particles that allow them to sneak through the repulsive Coulomb barrier that otherwise would hold them apart. You will study the process of barrier penetration in the following section, but first we should consider how the wave properties of particles are described.

In **quantum mechanics** a particle, whose wave properties are fundamental to its behaviour, can be described by a mathematical device called a **wave function**. A wave function ψ varies with position, r in one dimension, and time t, and is written $\psi(r, t)$. It is called a wave function because it obeys a differential equation called **Schrödinger's equation**, the solutions of which have wave-like properties.

You don't need to know the origin of Schrödinger's equation, but we will see how it can be used. For many applications, the time-dependent part $\psi_t(t)$ of the wave function can be calculated separately from the spatial part $\psi_s(r)$, and a simplified time-independent form of Schrödinger's equation can be used. We will deal with that simplified form, and consider only the spatial (time-independent) part of the wave function. Furthermore, we shall only deal with one-dimensional cases, where the distances are specified in terms of a spatial coordinate r.

For two particles A and B with **reduced mass** (see the box below for a definition) m_r, separation r, energy of approach E, and interacting with a potential energy $V(r)$, the time-independent one-dimensional Schrödinger equation is

$$\left[-\frac{\hbar^2}{2m_r} \frac{\partial^2}{\partial r^2} + V(r) \right] \psi_s(r) = E\psi_s(r). \tag{3.1}$$

In cases we shall consider, the energy of approach can be taken as the kinetic energy of one particle with respect to the other particle at rest. Note that $\hbar \equiv h/2\pi$ and notice the inclusion of a second partial derivative with respect to distance $\partial^2/\partial r^2$.

Reduced mass

When a body A is accelerated by a force F exerted by another body, B, the acceleration a_A is given by the force divided by its mass m_A, i.e. $a_A = F/m_A$.

However, an equal and opposite force $-F$ is exerted on B by A, causing B to accelerate as well: i.e. $a_B = -F/m_B$.

The acceleration a_{AB} of A *relative* to B, since *both* are being accelerated, is slightly more complicated than the case where one particle remains at rest:

$$a_{AB} = a_A - a_B = \frac{F}{m_A} - \frac{-F}{m_B} = F\left(\frac{1}{m_A} + \frac{1}{m_B}\right) = F\left(\frac{m_B + m_A}{m_A m_B}\right).$$

Fortunately, this analysis shows that the relative acceleration is still of the form: acceleration = force/mass, provided the mass of the particles is replaced by the quantity

$$m_r = \frac{m_A m_B}{m_A + m_B}, \tag{3.2}$$

which is called the *reduced mass*.

This quantity has units of mass, and if $m_B \gg m_A$ then $m_r \approx m_A m_B / m_B = m_A$, leading to the familiar result for when one particle (B) is very massive, $a_{AB} = F/m_A$.

Similarly, the reduced mass in atomic mass units, A_r, may be defined as follows. For any particle of mass m_i, the mass in atomic mass units is $A_i = m_i/u$. Hence, since $m_r = m_A m_B/(m_A + m_B)$, u can be divided into each mass separately to write

$$A_r = \frac{A_A A_B}{A_A + A_B}. \tag{3.3}$$

You may be wondering to what physical quantity the wave function ψ corresponds. Here you meet one of the conceptual hurdles in quantum mechanics, for although ψ can be calculated as a mathematical solution to the above equation, and graphs of a wave function can be plotted, a wave function *cannot be measured*! However, the *squared amplitude* of the spatial part of the wave function, $|\psi_s|^2$, corresponds to the **position probability density**, and this is something that *can* be measured. Considering a one-dimensional case, $|\psi_s(r)|^2 \, dr$ is the probability that the particle described by the wave function ψ_s is detected between r and $r + dr$. If the squared amplitude of a wave function is zero at some point A, then you can be sure that the particle will not be found at that point, whereas if the squared amplitude is greater than zero at some point B, then you know there is some probability that the particle will be found at point B. The higher the squared amplitude (= the position probability density) at any point, the more likely it is that the particle will be found there.

The preceding discussion of the *probability* that a particle will be found at some location emphasizes that on the small scales where wave properties become fundamental characteristics of objects, the word 'position' no longer has the same reliable meaning that we are accustomed to in the macroscopic (non-quantum) world. Think of any sort of wave that you are familiar with, e.g. water waves, sound waves, radio waves. A wave does not exist only at one point, but propagates in certain directions with certain amplitudes. So it is with particles on small scales when their wave properties become significant. They no longer exist just at one point, but have a potential presence in different places simultaneously. We can at best assign a probability to where the particle will be found if we try to detect its presence. The wave properties of particles have been verified experimentally through electron-diffraction experiments and similar studies of other particles.

An important application of wave functions is to use them to calculate the position probability density of an incoming particle encountering the potential barrier around a nucleus. We will do this in the next section.

3.1.3   Barrier penetration and quantum tunnelling

The existence of **quantum tunnelling**, otherwise known as **barrier penetration**, is the key to the occurrence of fusion reactions in stars. According to classical mechanics, a particle whose total energy is less than the potential energy of some barrier cannot pass that barrier. A simple example is a ball being thrown over a wall; unless the ball has sufficient kinetic energy to attain a height greater than the wall itself, there is no way it can reach the other side. That is, its kinetic energy must be enough that, once transformed to gravitational potential energy, it exceeds the gravitational potential energy at the top of the wall. In quantum mechanics, however, this restriction no longer holds. In quantum mechanics, the wave properties of the particle must also be considered, and although the wave function associated with a particle is attenuated by a potential barrier, the amplitude of its wave function goes to zero only for an *infinitely* high barrier. Where *finite* potential barriers exist in reality, the wave function *inside and beyond the barrier is non-zero*, and hence the position probability density $|\psi_s|^2$ is also *non-zero*. In this case, there is a small but non-zero probability that a particle will make its way through a barrier that appears impenetrable or insurmountable from a classical viewpoint. If a ball and wall were small enough for quantum properties to dominate over classical ones, it would be possible for a ball to reach the other side even if it did not have enough kinetic energy to overcome the gravitational potential energy at the wall's top. As bizarre as this might seem, we shall soon see that this is precisely what happens during collisions of nuclei!

The nucleons in a nucleus sit in a potential well – as they must for the nucleus to be stable – surrounded by a Coulomb potential barrier of finite height and width, shown schematically in Figure 3.1. Due to barrier penetration, there is a non-zero probability of a particle with energy less than the height of the barrier nevertheless making its way from outside the barrier to inside, and so reaching the nucleus.

Imagine that we have two nuclei of charge Z_A and Z_B with masses m_A and m_B respectively. When these nuclei are far apart they will repel each other with a potential given by Coulomb's law as $Z_A Z_B e^2 / 4\pi\varepsilon_0 r$, where ε_0 is the permittivity of free space. If we imagine what will happen as they approach one another with a certain kinetic energy E_K then, ignoring quantum mechanics, they will simply come to rest temporarily when their kinetic energy has been converted into electrical potential energy, before 'bouncing' apart again.

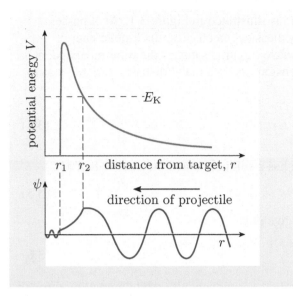

Figure 3.1 (top) A graph of potential energy V against position r for a potential barrier of finite height and thickness. Classically, a particle with kinetic energy E_K cannot get closer to the origin than $r = r_2$ because of the potential barrier of height $V > E_K$. (bottom) Quantum-mechanically, the wave function shows that there is a non-zero probability of finding the particle beyond the barrier, i.e. at $r < r_1$.

The distance of least separation (r_2 on Figure 3.1) is given by

$$E_K = \frac{Z_A Z_B e^2}{4\pi\varepsilon_0 r_2}.$$

In a non-quantum-mechanical world, the two nuclei could only undergo fusion if $r_2 < r_1$, where r_1 is the distance at which the strong nuclear force becomes important. That is to say, fusion would only be possible if the nuclei had enough kinetic energy to overcome a Coulomb barrier of height

$$V = \frac{Z_A Z_B e^2}{4\pi\varepsilon_0 r_1}.$$

● What is the height of the Coulomb barrier between two protons, in eV? (Assume $Z_A = Z_B = 1$ and $r_1 \approx 1.2 \times 10^{-15}$ m.)

○ The Coulomb barrier height is

$$V = \frac{(1.602 \times 10^{-19}\ \text{C})^2}{4\pi \times (8.854 \times 10^{-12}\ \text{C}^2\ \text{N}^{-1}\ \text{m}^{-2}) \times (1.2 \times 10^{-15}\ \text{m})}$$
$$= 1.9 \times 10^{-13}\ \text{J}.$$

Since 1 eV = 1.602×10^{-19} J, this is equivalent to $(1.9 \times 10^{-13}\ \text{J})$ / $(1.602 \times 10^{-19}\ \text{J eV}^{-1})$ or about 1.2 MeV.

● How does this energy compare to the typical thermal energy of protons in the core of the Sun?

○ For a temperature of 15.6×10^6 K, kT is $(1.381 \times 10^{-23}\ \text{J K}^{-1}) \times$ $(15.6 \times 10^6\ \text{K}) = 2.15 \times 10^{-16}$ J. Since 1 eV = 1.602×10^{-19} J, this is equivalent to $(2.15 \times 10^{-16}\ \text{J})/(1.602 \times 10^{-19}\ \text{J eV}^{-1})$ or about 1.3 keV.

Hence most protons in the core of the Sun have a kinetic energy which is 1000 times smaller than the Coulomb barrier energy between them! However, as you will have gathered by now, the solution allowing nuclear fusion to occur is provided by quantum tunnelling.

Consider a nucleus with energy E approaching a Coulomb barrier of constant height V. At large distances away from the barrier, the wave function of the nucleus has a sinusoidal behaviour, as illustrated in Figure 3.1. As it passes into the region that is forbidden by classical mechanics, the kinetic energy $E - V$ becomes negative and the wave function satisfies the time-independent Schrödinger equation. In one dimension (using a radial distance variable r as in Figure 3.1) this is:

$$\left[-\frac{\hbar^2}{2m_r} \frac{\partial^2}{\partial r^2} + V \right] \psi_s(r) = E\psi_s(r). \tag{3.4}$$

In this situation, the wave function has a solution which may be written as

$$\psi_s(r) = \exp(\chi r) \tag{3.5}$$

where χ (the Greek letter 'chi') is given by

$$\chi^2 = \frac{2m_r}{\hbar^2}(V - E). \tag{3.6}$$

Therefore the wave function $\psi_s(r)$ decays exponentially while tunnelling through the barrier towards smaller r. Beyond the barrier, the wave function has a sinusoidal behaviour once more, but its amplitude has been attenuated by the barrier, as shown in Figure 3.1.

Exercise 3.1   Verify that the wave function $\psi_s(r) = \exp(\chi r)$ is a solution of Equation 3.4.

Hint: $d\exp(ar)/dr = a\exp(ar)$ where a is a constant.   ■

Referring again to Figure 3.1, the probability of a particle successfully tunnelling through to the nucleus can be found by comparing the probability of finding a particle at the inner edge of the Coulomb barrier (at $r = r_1$) with the probability of finding it at the outer edge of the Coulomb barrier (at $r = r_2$, where $r_2 \gg r_1$). For the one-dimensional case, the probability of finding a particle in an interval from r to $r + dr$ is $|\psi_s(r)|^2 \, dr$. Therefore the penetration probability is

$$P_{\text{pen}} \approx \frac{|\psi_s(r_1)|^2}{|\psi_s(r_2)|^2} = \{\exp[-\chi(r_2 - r_1)]\}^2. \tag{3.7}$$

● With reference to an idealized square potential barrier, consider qualitatively how the penetration probability depends on (a) the height of the barrier, (b) the width of the barrier.

○ (a) The height of the barrier is given by V; increasing this increases χ, and hence reduces $P_{\text{pen}} \approx \{\exp[-\chi(r_2 - r_1)]\}^2$. A higher barrier reduces the penetration probability, as expected.

(b) A wider barrier increases the distance between r_1 and r_2, so reduces $P_{\text{pen}} \approx \{\exp[-\chi(r_2 - r_1)]\}^2$. A wider barrier reduces the penetration probability, as expected.

3.1.4   The Gamow energy

So far we have considered rectangular potential barriers whose height is uniform over a range of separations, but more realistically, barrier potentials do vary with separation, i.e. V is a function of r. For the Coulomb barrier, the penetration probability may be expressed in terms of the particle energy E, and the **Gamow energy** E_G which depends on the atomic number (and therefore charge) of the interacting nuclei, and hence the size of the Coulomb barrier:

$$P_{\text{pen}} \approx \exp\left[-\left(\frac{E_G}{E}\right)^{1/2}\right] \tag{3.8}$$

with

$$E_G = 2m_r c^2 (\pi\alpha Z_A Z_B)^2, \tag{3.9}$$

where α is the fine structure constant $\approx 1/137.0$. The rate of nuclear fusion therefore depends on the penetration probability of the Coulomb barrier, according to Equation 3.8. This penetration probability in turn is described by its Gamow energy (Equation 3.9).

● Does a higher Gamow energy increase or decrease the probability of penetration?

○ A higher E_G reduces the probability that the barrier will be penetrated. The Gamow energy measures the strength of the Coulomb repulsion, which determines the height of the Coulomb barrier.

Exercise 3.2 Calculate the Gamow energy (in both SI units and electronvolts) for the following reactions:

(a) the collision of two protons;

(b) the collision of two $^3_2$He nuclei. (For simplicity, assume the mass of a $^3_2$He nucleus, denoted by m_3, is $3m_p$.) ∎

Note from the exercise above that the one *additional charge* on each $^3_2$He nucleus compared with each $^1_1$H nucleus increases the Gamow energy by a factor of 16 compared to the proton–proton collision. The *higher masses* increase this by a further factor of 3. You can see, therefore, that the Coulomb barrier is very high for interactions involving all but the lightest nuclei.

Exercise 3.3 The core of the Sun has a temperature of 15.6×10^6 K. As you saw in an earlier bulleted question, this corresponds to an energy per particle of about $kT_c = 1.3$ keV. Using the Gamow energies computed in the previous exercise, calculate the penetration probabilities for the same two interactions in the Sun's core. ∎

From the penetration probabilities calculated above, $\sim 10^{-10} - 10^{-60}$, you can see that even with quantum tunnelling, the probability that any particular pair of nuclei will react is extremely small.

3.2 Fusion interactions

We now study fusion reactions generally, before calculating the rates of specific reactions in the next section.

Figure 3.2 The Gamow energy is named for George Gamow (1904–1968), a Russian physicist and cosmologist who escaped to the USA in 1934. His paper 'The origin of chemical elements' (1948) by Alpher, Bethe & Gamow attempted to explain much of astrophysical nucleosynthesis in the big bang and also predicted the existence of the cosmic microwave background radiation.

3.2.1   Cross-sections

Don't confuse this σ with the Stefan–Boltzmann constant, which is also denoted by σ.

A widely used concept in physics where a collision or close interaction occurs between two bodies is the cross-section, usually given the symbol σ.

The *geometric* cross-section for the collision between a small projectile and a large target is simply the cross-sectional area of the target. If some reaction may occur when they collide, the *reaction* cross-section is the geometric cross-section multiplied by the reaction probability. For instance, imagine a tennis ball colliding with a $1 \, \text{m}^2$ window and having a 10% probability of breaking it. In this case, the reaction cross-section is 10% of the geometric cross-section, or $0.1 \, \text{m}^2$. If the same window were hit by a ball made of soft foam plastic, the probability of it breaking may be only 0.1%, and the reaction cross-section would be only $0.001 \, \text{m}^2$. Note that the window is still the same physical size, $1 \, \text{m}^2$, and the ball is the same size, but the reaction cross-sections are very different depending on the nature of the projectile and the strength of the interaction. Therefore, although the reaction cross-section of a target can be considered as an effective area, and has units of area, it should not be confused with the actual physical size of the target.

As atomic nuclei are very small, with radii of order $10^{-15} \, \text{m}$, their cross-sectional areas are of order $10^{-30} \, \text{m}^2$ (since area is proportional to radius squared). Nuclear physicists have defined the area $10^{-28} \, \text{m}^2$ to be 1 barn, so nuclear cross-sections are typically measured in units of millibarns or microbarns. (The name for this unit, the barn, comes from jokes that someone who is a poor shot couldn't hit a target as big as a barn. We have already seen that nuclei are incredibly small targets for fusion reactions. The largest nuclear cross-sections are, relatively speaking, as big as a barn, whereas smaller targets have cross-sections measured only in millibarns or microbarns.)

Number density and mass fraction

We often need to express the absolute or relative amounts of a given type of particle inside a star in different ways. One of the simplest ways is to specify the **number density** of a given type of particle n_i which is simply the number of particles of that type per unit volume. Number density therefore has the unit of m^{-3}.

The **mass density** due to a given type of particle is the mass per unit volume. Usually represented by the symbol ρ (the Greek letter 'rho'). If the type of particle in question has a mass m_i per particle, then the mass density is due to particle type i is $\rho_i = n_i m_i$.

Often we are concerned with mixtures of different types of particle. One way of specifying the mix is to determine the mean molecular weight (see earlier box). This is the sum of the mass of the particles divided by the total number of particles. Alternatively we might specify the number fraction or the mass fraction of each type of particle present.

The **number fraction** is simply the number of particles of a given type per unit volume divided by the total number of particles per unit volume. If there are j different types of particles, then the number fraction of particles of type 1 is $n_1 / \sum_{i=1}^{j} n_i$.

The **mass fraction** is weighted by the mass of each type of particle. Hence if there are j different types of particles with number densities n_i and relative masses m_i, then the mass fraction of particle type 1 is $X_1 = n_1 m_1 / \sum_{i=1}^{j} n_i m_i$. In astrophysics the mass fractions of hydrogen, helium and 'all other elements' are often represented by the symbols X, Y and Z respectively. The initial mass fractions of the Sun are $X = 0.70$, $Y = 0.28$, $Z = 0.02$.

If the composition of a sample of gas is specified in terms of its mass fractions, then the number density of a particular type of particle may be expressed as

$$n_i = \frac{\rho X_i}{m_i}, \qquad (3.10)$$

where ρ is the overall density of the gas.

Imagine we have a target containing n identical particles per unit volume. As the incoming particle travels a small distance Δx through the target, the probability of it interacting is given by

$$\text{probability of reaction} = \sigma n \, \Delta x.$$

Likewise, the probability of no interaction occurring within the distance Δx (or the survival probability) is simply

$$\text{probability of no reaction} = 1 - \sigma n \, \Delta x.$$

In travelling a total distance $x = N \, \Delta x$, the overall probability of no interaction occurring is simply found by multiplying together all the individual probabilities of no interactions occurring in each small distance $\Delta x = x/N$. Hence

$$\text{probability of survival} = (1 - \sigma n x/N) \times (1 - \sigma n x/N) \times \ldots$$
$$= (1 - \sigma n x/N)^N.$$

In the limit as $N \longrightarrow \infty$ this can be re-written as

$$\text{probability of survival} = P(x) = \exp(-\sigma n x). \qquad (3.11)$$

● Consider the form of Equation 3.11. What range of values does the cross-section σ take? What range of values does the number density of target particles n take? What then is the dependence on distance of the survival probability $P(x)$ that there is no interaction over some distance x? Does this seem reasonable? Sketch the form of $P(x)$ against x.

○ The cross-section σ and the number density of target particles n must both be non-negative numbers (zero or positive), so the form of $P(x)$ against x is an exponential decay curve, taking the value 1 at $x = 0$. This is reasonable, because the survival probability of the projectile particle is 1 (i.e. survival is guaranteed) if it does not have to pass through any target particles ($x = 0$), and steadily decreases as the path through target particles lengthens (x increases to a positive number). Figure 3.3 provides a sketch of the survival probability as a function of distance.

Consider the behaviour of the projectile particle's survival probability $P(x)$ as shown in Figure 3.3. As the path length increases, the survival probability P decreases, because there is more and more chance that the particle will hit one of the targets. However, the survival probability never quite reaches zero, because we can never be absolutely sure that it will hit a target, though the survival probability becomes incredibly small (i.e. survival is unlikely) if the path length x through the target is very long.

● Consider two environments, one with target number density n_1 and one with a higher target number density n_2. What does your understanding of the physics tell you should happen to the survival probability of the particle in medium n_2 compared to n_1? What does Equation 3.11 say? Do the two agree? Treat the graph in Figure 3.3 as being for n_1, and add a sketch graph for medium n_2.

○ A medium with more targets should reduce the survival probability, because the chance of a reaction will be greater. Equation 3.11 verifies this; $\sigma n x$ is always positive, so if n increases, then P decreases. The curve for medium n_2 has the same value of P at $x = 0$ but otherwise lies below the curve for n_1. Note, however, like the curve for n_1, it does not reach the value $P = 0$, even for large values of x (see Figure 3.4).

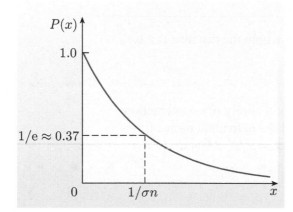

Figure 3.3   The survival probability $P(x)$ that *no* reaction occurs when the projectile travels a distance x in the target medium.

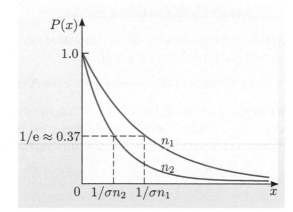

Figure 3.4   The survival probability $P(x)$ that *no* reaction occurs when the projectile travels a distance x in the target medium, for two target densities where $n_2 > n_1$.

3.2.2   Mean free path

Identical particles travelling through a medium of targets travel different distances before interacting. The average distance they travel prior to reaction is called the **mean free path** l, and is a useful indicator of the condition of the target medium and the strength of the interaction.

If the probability that a particle travels a distance between x and $x + dx$ is $P(x)\,dx$, then the mean distance travelled by a group of these particles is found by multiplying each possible distance x by the probability $P(x)$ that x is the distance travelled, summing this calculation for all possible distances, and dividing by the sum of the probabilities (= 1). Therefore the mean free path of the particle is

$$l = \int_0^\infty x\, P(x)\, \mathrm{d}x.$$

For the case of the mean free path of a particle moving through a medium containing n target particles per unit volume, $P(x)\, \mathrm{d}x$ is the probability of an interaction occurring over the interval from x to $x + \mathrm{d}x$. This is the probability of there being *no* interaction over the distance 0 to x, *and* then an interaction occurring over the interval from x to $x + \mathrm{d}x$. The probability of two independent events A and B occurring is the product of their individual probabilities, so we can write $P(x)\, \mathrm{d}x$ = (probability of survival over the distance from 0 to x) $\times$ (probability of an interaction occurring over the interval from x to $x + \mathrm{d}x$). Hence $P(x)\, \mathrm{d}x = \exp(-\sigma n x) \times \sigma n\, \mathrm{d}x$.

The mean free path is therefore

$$l = \int_0^\infty x P(x)\, \mathrm{d}x = \int_0^\infty x \exp(-\sigma n x)\, \sigma n\, \mathrm{d}x.$$

This expression is easily simplified. First, the σn term does not depend on the variable of integration x, so can be taken outside the integral. Second, it is a standard result that $\int_0^\infty x \exp(-ax)\, \mathrm{d}x = 1/a^2$, where a does not depend on x, so we can complete the simplification

$$l = \sigma n \int_0^\infty x \exp(-\sigma n x)\, \mathrm{d}x$$
$$= \sigma n \times 1/(\sigma n)^2$$
$$= 1/\sigma n. \tag{3.12}$$

To put this result into words, the mean distance — the mean free path — travelled by a class of projectiles before interacting with a target is given by 1 over the product of the number density of target particles and the reaction cross-section. You will see later that the mean free path is closely related to the average time between collisions, and hence is a very useful quantity.

3.2.3 Energy-dependence of fusion cross-sections

Having established an understanding of a reaction cross-section and the concept of mean free path, we can now proceed to consider the cross-section for the fusion of two nuclei. The cross-section for fusion is proportional to the probability of penetration of the Coulomb barrier. We therefore re-write Equation 3.8 as a fusion cross-section for nuclei with a kinetic energy between them, E, as:

$$\sigma(E) = \frac{S(E)}{E} \exp\left[-\left(\frac{E_G}{E}\right)^{1/2}\right]. \tag{3.13}$$

The factor $S(E)$ is determined by the nuclear physics of the particular fusion reaction involved.

Now, we note that the expression for the fusion cross-section has three key parts:

- Previously in this section you obtained Equation 3.8 for the probability of barrier penetration, which is exponentially dependent on the ratio of the energy of the colliding nuclei, E, to the height of the Coulomb barrier, as given by the Gamow energy, E_G: $P_{pen} \approx \exp[-(E_G/E)^{1/2}]$. The fusion cross-section reflects that penetration probability, giving rise to an energy-dependence governed by that exponential.

- At *low* energies, the fusion cross-section may be proportional to the square of the de Broglie wavelength. Since $\lambda_{dB} \equiv h/p$, where p is the relative momentum of the nuclei, and the kinetic energy of the particle is $E_K = (1/2)mv^2 = p^2/2m$, it follows that $\lambda_{dB}^2 = h^2/p^2 = h^2/2mE_K$. This gives rise to a $1/E$ term in the fusion cross-section.

- The product of these two terms with a function $S(E)$ gives rise to the standard expression for the cross-section. The **S-factor** is not strictly constant but is generally only weakly dependent on energy. (At energies close to excited states of the target nucleus, nuclear resonances arise and $S(E)$ may be strongly peaked.) S has units of energy × cross-section, which in SI units would be J m^2 but more typically is given as keV barns.

3.3   Thermonuclear fusion reaction rates

Fusion rates are crucial to our understanding of nuclear burning in stars. They are strongly dependent on temperature, and since the internal temperature of a star changes by several orders of magnitude as it evolves, different reactions occur at different stages of its life.

Let us assume we have an ionized gas containing two types of nuclei, A and B, with number densities of n_A and n_B particles per cubic metre respectively. The fusion cross-section between them is σ. To make things easier, assume that all the B nuclei are at rest, and all the A nuclei have speed v. Now, according to Equation 3.12, each A nucleus will travel an average distance $d_A = 1/n_B\sigma$ before it interacts with a B nucleus. The average time taken for a fusion reaction to occur is therefore $\tau_A = d_A/v = 1/n_B\sigma v$. So the fusion rate per unit volume for dissimilar particles is simply

$$R_{AB} = \frac{n_A}{\tau_A} = n_A n_B \sigma v. \tag{3.14}$$

If we had treated all the A nuclei as at rest, and considered all the B nuclei travelling with speed v, we would have obtained a similar result for the fusion rate R_{BA}.

There is a slight complication if both nuclei are the *same* type. Clearly if there are n_A particles of type A and n_B particles of type B, then the number of different combinations of A and B particles available to react is simply $n_A n_B$. However, if we have only one particle type, A, then clearly a given particle cannot react with itself (i.e. A1 cannot react with A1, but only with A2, A3, A4, etc.). So the number of combinations of pairs of particles from a sample of n_A particles is $n_A n_A - n_A$. We also have to avoid double counting. If we fuse particle A1 with particle A2, we cannot also count a fusion of A2 with A1. That is, we can count only half of the reaction pairs. In general therefore there are $(n_A^2 - n_A)/2$ possible reaction pairs, which is equivalent to $n_A(n_A - 1)/2$, or $\approx n_A^2/2$ since n_A

will generally be a very large number. So the fusion rate per unit volume for similar particles is simply

$$R_{AA} = (n_A^2/2)\sigma v. \tag{3.15}$$

3.3.1 The speed-averaged cross-section

In real situations, the nuclei A and B will have a range of possible speeds, rather than a single constant speed. Let us suppose that the probability of a pair of reacting nuclei having a relative speed between v_r and $v_r + dv_r$ is $P(v_r)\,dv_r$. In this case, the average value of the fusion cross-section multiplied by the relative speed is

$$\langle \sigma v_r \rangle = \int_0^\infty \sigma v_r\, P(v_r)\, dv_r. \tag{3.16}$$

This is analogous to our earlier calculation of the mean free path $l = \int_0^\infty x P(x)\, dx$, but with the independent variable changed from distance x to the speed–cross-section product σv_r, and the weighting function being the distribution of speeds rather than the distribution of penetration distances.

The necessity of averaging over the relative speed distribution alters the expressions for the fusion rates, to

$$R_{AB} = n_A n_B \langle \sigma v_r \rangle \quad \text{for fusion between dissimilar nuclei} \tag{3.17}$$

and

$$R_{AA} = (n_A^2/2)\langle \sigma v_r \rangle \quad \text{for fusion between identical nuclei.} \tag{3.18}$$

Now, we can also use the average value of the fusion cross-section multiplied by the relative speed to express the mean time it takes for a particular nucleus of type A to undergo fusion with a nucleus of type B as

$$\tau_A = \frac{1}{n_B \langle \sigma v_r \rangle}. \tag{3.19}$$

So, combining Equation 3.19 with Equation 3.17 or 3.18, the mean lifetime of a particular particle undergoing fusion is

$$\tau_A = \frac{n_A}{R_{AB}} \tag{3.20}$$

in the case of dissimilar particles, and

$$\tau_A = \frac{n_A}{2R_{AA}} \tag{3.21}$$

in the case of similar particles.

For most situations in astrophysics, the relevant speed distribution for the particles is the Maxwell–Boltzmann speed distribution, which has the form

$$P(v_r)\, dv_r = \left(\frac{m_r}{2\pi kT}\right)^{3/2} \exp\left(-\frac{m_r v_r^2}{2kT}\right) 4\pi v_r^2\, dv_r. \tag{3.22}$$

So substituting Equation 3.22 into Equation 3.16 gives

$$\langle \sigma v_r \rangle = \int_0^\infty \sigma v_r \times \left(\frac{m_r}{2\pi kT}\right)^{3/2} \exp\left(-\frac{m_r v_r^2}{2kT}\right) 4\pi v_r^2 \, dv_r$$

$$= \left(\frac{m_r}{2\pi kT}\right)^{3/2} 4\pi \int_0^\infty \sigma \exp\left(-\frac{m_r v_r^2}{2kT}\right) v_r^3 \, dv_r.$$

We now convert the above into an integration over kinetic energy $E = m_r v_r^2/2$ using the substitution that $v_r = (2E/m_r)^{1/2}$. We also note that $dE/dv_r = m_r v_r$ which may be re-written as $dE/dv_r = (2m_r E)^{1/2}$ and hence $dv_r = (2m_r E)^{-1/2} \, dE$. So in the integral above, we replace $v_r^3 \, dv_r$ by $(2E/m_r)^{3/2} \times (2m_r E)^{-1/2} \, dE$. We therefore obtain

$$\langle \sigma v_r \rangle = \left(\frac{m_r}{2\pi kT}\right)^{3/2} 4\pi \int_0^\infty \sigma(E) \exp\left(-\frac{E}{kT}\right) \frac{2E}{m_r^2} \, dE$$

$$= \left(\frac{8}{\pi m_r}\right)^{1/2} \left(\frac{1}{kT}\right)^{3/2} \int_0^\infty E \, \sigma(E) \exp\left(-\frac{E}{kT}\right) \, dE.$$

So, using Equation 3.17, we have the fusion rate for dissimilar particles as

$$R_{AB} = n_A n_B \left(\frac{8}{\pi m_r}\right)^{1/2} \left(\frac{1}{kT}\right)^{3/2} \int_0^\infty E \, \sigma(E) \exp\left(-\frac{E}{kT}\right) \, dE.$$

Then substituting in for the fusion cross-section using Equation 3.13 we get the fusion rate per unit volume as:

$$R_{AB} = n_A n_B \left(\frac{8}{\pi m_r}\right)^{1/2} \left(\frac{1}{kT}\right)^{3/2}$$

$$\times \int_0^\infty S(E) \exp\left[-\frac{E}{kT} - \left(\frac{E_G}{E}\right)^{1/2}\right] \, dE. \tag{3.23}$$

By writing the rate equation in this form, the energy-dependence of the reaction rate is made explicit. However, this is a non-trivial equation, and not many people will be able easily to picture what the function looks like.

● Give a simple physical interpretation to each term in Equation 3.23.

○ The following terms may be highlighted:

- $n_A n_B$ is the number of possible reactions;
- $(1/kT)^{3/2}$ comes from the Maxwell–Boltzmann distribution of relative speeds;
- $S(E)$ conveys the strength of the nuclear interaction;
- (E_G/E) indicates the probability of tunnelling through the Coulomb barrier;
- E/kT represents the Maxwell–Boltzmann thermal energy distribution.

3.3.2 The Gamow peak

The energy-dependence of the rate R_{AB} is all contained in the integral

$$\int_0^\infty S(E) \exp\left[-\left(\frac{E_G}{E}\right)^{1/2} - \frac{E}{kT}\right] dE.$$

Recall that $\exp(X + Y) = \exp(X) \times \exp(Y)$, so the integral can also be written

$$\int_0^\infty S(E) \exp\left[-\left(\frac{E_G}{E}\right)^{1/2}\right] \exp\left(-\frac{E}{kT}\right) dE.$$

The nuclear S-factor $S(E)$ varies only slowly with energy (apart from near low-energy resonances), so essentially all of the energy-dependence comes from the exponential terms. The term $\exp[-(E_G/E)^{1/2}]$ increases as the kinetic energy of the nuclei E increases, but the other term $\exp(-E/kT)$ decreases, so the product of the two is small at high and low energies, and becomes significant only over some range of intermediate energies. This behaviour is shown in Figure 3.5, and indicates that the distribution peaks at an energy (called the **Gamow peak**):

$$E_0 = \left(E_G \left(\frac{kT}{2}\right)^2\right)^{1/3}. \tag{3.24}$$

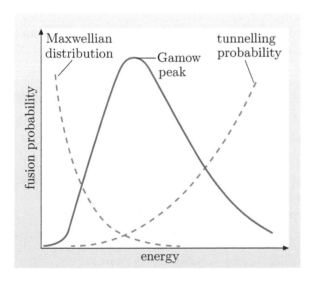

Figure 3.5 The fusion probability as a function of energy for nuclei. The two terms in the exponential of Equation 3.23 give the two dashed curves. Their product is small at both high and low energies, and is large only at some intermediate energy range, as shown. The energy at which the fusion rate is a maximum is called the Gamow peak. The region on either side of the peak in which the fusion probability is significant is called the Gamow window.

Exercise 3.4 Consider the energy integral

$$\int_0^\infty S(E) \exp\left[-\left(\frac{E_G}{E}\right)^{1/2} - \frac{E}{kT}\right] dE.$$

The integrand,

$$S(E) \exp\left[-\left(\frac{E_G}{E}\right)^{1/2} - \frac{E}{kT}\right]$$

conveys the energy-dependence of the thermonuclear fusion rate. Show that this integrand has a peak at $E_0 = [E_G(kT/2)^2]^{1/3}$.

Hint 1: A function $f(x)$ has a maximum or minimum when $df/dx = 0$.

Hint 2: By the chain rule for differentiation,

$$\frac{d}{dx}\exp(g(x)) = \frac{d\exp(g)}{dg} \times \frac{dg(x)}{dx} = \exp(g)\frac{dg(x)}{dx}.$$

Hint 3: Assume that $S(E)$ is such a weak function of energy that it can be treated as a constant. ∎

3.3.3 The width of the Gamow window

As you saw from Figure 3.5, there is a certain range of energy, or window, around the Gamow peak where fusion is most likely to occur. In order to examine the fusion rate in the vicinity of the Gamow peak, and quantify the width of this window, we first rewrite the fusion rate per unit volume as

$$R_{AB} = n_A n_B \left(\frac{8}{\pi m_r}\right)^{1/2} \left(\frac{1}{kT}\right)^{3/2} \int_0^\infty S(E)\exp[-f(E)]\ dE,$$

where the argument of the exponential function is

$$-f(E) = -\left[\left(\frac{E_G}{E}\right)^{1/2} + \left(\frac{E}{kT}\right)\right].$$

To investigate the width of the window we take a Taylor expansion of $f(E)$ about the energy of the probability peak, E_0. Such an expansion expresses the value *near* the peak in terms of the value and derivatives *at* the peak. In this case, expanding to second order and ignoring higher-order terms, we would write

$$f(E) \approx f(E_0) + (E - E_0)f'(E_0) + (1/2)(E - E_0)^2 f''(E_0) + \dots,$$

where $f(E) = (E_G/E)^{1/2} + (E/kT)$, and f' and f'' are its first and second derivatives.

Evaluating $f(E)$ at E_0 gives $f(E_0) = (E_G/E_0)^{1/2} + (E_0/kT)$. Now, from the derivation of the Gamow peak, we have $(E_G/E_0)^{1/2} = 2E_0/kT$, so we can rewrite $f(E_0)$ as $f(E_0) = 3E_0/kT$. Since $E_0 = \left(E_G(kT/2)^2\right)^{1/3}$, this yields $f(E_0) = 3(E_G/4kT)^{1/3}$.

The first derivative of $f(E)$ is $f'(E) = -(E_G^{1/2}/2E^{3/2}) + (1/kT)$, but you have already seen that, at the energy of the Gamow peak, $-(E_G^{1/2}/2E_0^{3/2}) + (1/kT) = 0$, so $f'(E_0) = 0$ (by definition).

Finally we calculate the second derivative of $f(E)$ as $f''(E) = 3E_G^{1/2}/4E^{5/2}$, so evaluating this at E_0 yields $f''(E_0) = 3E_G^{1/2}/4E_0^{5/2}$, or substituting for E_0, $f''(E_0) = (3/2^{1/3})E_G^{-1/3}(kT)^{-5/3}$.

So, the Taylor expansion is

$$f(E) \approx 3\left(\frac{E_G}{4kT}\right)^{1/3} + \frac{1}{2}(E - E_0)^2 \left(\frac{3}{2^{1/3}}\right) E_G^{-1/3}(kT)^{-5/3}.$$

This may be re-written in the form

$$f(E) \approx 3 \left(\frac{E_{\mathrm{G}}}{4kT} \right)^{1/3} + \left(\frac{E - E_0}{\Delta/2} \right)^2.$$

where Δ, the **Gamow width**, is

$$\Delta = \frac{4}{3^{1/2}2^{1/3}} E_{\mathrm{G}}^{1/6}(kT)^{5/6}, \tag{3.25}$$

which characterizes the range of energy over which fusion is likely to occur.

So the integral in the equation for the fusion rate per unit volume is then

$$\int_0^\infty S(E) \exp\left[-3\left(\frac{E_{\mathrm{G}}}{4kT} \right)^{1/3} \right] \exp\left[-\left(\frac{E - E_0}{\Delta/2} \right)^2 \right] \, \mathrm{d}E. \tag{3.26}$$

The Gamow width can also be written

$$\Delta \approx 1.8(E_{\mathrm{G}}/kT)^{1/6}kT. \tag{3.27}$$

and the half-width is therefore

$$\Delta/2 \approx (E_{\mathrm{G}}/kT)^{1/6}kT.$$

Note that the energy-dependence of the fusion rate, the last term of Equation 3.26 resembles the form

$$\exp\left(-\frac{(x - \bar{x})^2}{2\sigma^2} \right).$$

This is symmetric about E_0, and has the shape known as a **bell curve**, a **normal distribution**, or a **Gaussian distribution**.

The width of the fusion window is described by Equation 3.27. Its physical significance can be seen as follows: at $E = E_0 \pm \Delta/2$, the Gaussian term equates to

$$\exp\left[-\left(\frac{E_0 \pm \Delta/2 - E_0}{\Delta/2} \right)^2 \right] = \exp\left[-\left(\frac{\pm\Delta/2}{\Delta/2} \right)^2 \right] = \exp[-1] = 1/e.$$

That is, $\pm\Delta/2$ is the energy range over which the fusion probability falls by a factor $1/e \approx 0.37$, from its maximum at E_0. Since $\Delta/2 \approx E_{\mathrm{G}}^{1/6}(kT)^{5/6}$, we can express the Gamow window as

$$E_0 \pm \Delta/2 \approx E_0 \pm \left(\frac{E_{\mathrm{G}}}{kT} \right)^{1/6} kT. \tag{3.28}$$

The symbol σ that appears in the Gaussian is called the *standard deviation*, and is unrelated to the fusion cross-section (or the Stefan–Boltzmann constant) which use the same symbol for different quantities. The standard deviation determines the rate at which the exponential decreases as x deviates from $\bar{x}$.

3.3.4 The integrated fusion rate equation

Consider again the fusion rate equation given by Equation 3.23. Since the exponential part of this can be re-written in terms of a second-order Taylor expansion as Equation 3.26, we can re-write the fusion rate as

$$R_{\mathrm{AB}} = n_{\mathrm{A}}n_{\mathrm{B}} \left(\frac{8}{\pi m_{\mathrm{r}}} \right)^{1/2} \left(\frac{1}{kT} \right)^{3/2}$$

$$\times \int_0^\infty S(E) \exp\left[-3\left(\frac{E_{\mathrm{G}}}{4kT} \right)^{1/3} \right] \exp\left[-\left(\frac{E - E_0}{\Delta/2} \right)^2 \right] \, \mathrm{d}E. \tag{3.29}$$

● The revised form of the fusion rate equation still looks daunting, but in fact just a few physical variables determine its value. What are they?

○ Note that:

- the energy of the Gamow peak, $E_0 = [E_G(kT/2)^2]^{1/3}$, and
- the Gamow width $\Delta \approx 1.8(E_G/kT)^{1/6}kT$

are primarily determined by two variables. These variables are:

1. the temperature T which governs the energy of the nuclei, and
2. the Gamow energy E_G which characterizes the Coulomb barrier to be penetrated. Since $E_G = 2m_r c^2(\pi\alpha Z_A Z_B)^2$, this energy depends on the charges of the nuclei and their reduced mass.

Another factor, which we have assumed is only weakly dependent on energy is the S-factor, which encapsulates the nuclear physics of the strength of the fusion interaction. The remaining variables are the particle densities and their reduced mass.

These few quantities summarize the physics underlying thermonuclear fusion; don't lose sight of this when confronted by the algebra and calculus.

The first exponential term in Equation 3.29 is independent of energy E, so it can be taken out of the integral, and $S(E)$ usually varies so little over the width of the Gaussian that it can safely be replaced by its value at the peak, $S(E_0)$. The rate equation then becomes

$$R_{AB} = n_A n_B \left(\frac{8}{\pi m_r}\right)^{1/2} \left(\frac{1}{kT}\right)^{3/2}$$

$$\times S(E_0)\exp\left[-3\left(\frac{E_G}{4kT}\right)^{1/3}\right] \int_0^\infty \exp\left[-\left(\frac{E-E_0}{\Delta/2}\right)^2\right] dE$$

where the integration is now only over the Gaussian distribution. That integration is straightforward, but is a mathematical distraction outside the scope of this book. The result of performing it is our final equation for the integrated fusion rate per unit volume:

$$R_{AB} = \frac{6.48\times10^{-24}}{A_r Z_A Z_B} \times \frac{n_A n_B}{[m^{-6}]} \times \frac{S(E_0)}{[\text{keV barns}]} \times \left(\frac{E_G}{4kT}\right)^{2/3}$$

$$\times \exp\left[-3\left(\frac{E_G}{4kT}\right)^{1/3}\right] \text{m}^{-3}\text{s}^{-1} \tag{3.30}$$

where the numerical constant has been evaluated assuming that $S(E_0)$ is specified in keV barns and the particle densities are in m^{-3}. A_r is the reduced mass in atomic mass units: $A_r = m_r/u$. As noted earlier, since

$m_r = m_A m_B/(m_A + m_B)$, so u can be divided into each mass separately to write $A_r = A_A A_B/(A_A + A_B)$.

Note again that Equation 3.30 is determined by:

- the temperature T which governs the thermal energy of the nuclei,
- the Gamow energy E_G which characterizes the Coulomb barrier to be penetrated, and
- the S-factor, which encapsulates the strength of the fusion interaction.

As you saw in the previous Chapter, the timescale of hydrogen burning in the p–p chain is set by the slowest reaction in that chain, the $p + p \longrightarrow d + e^+ + \nu_e$ reaction. The following example examines this reaction in the light of the equations developed in this chapter.

Worked Example 3.1

Calculate the reaction rate R_{pp} and mean lifetime of a proton in the first step of the p–p chain at the solar core.

Hint 1: Use Table 1.1 for solar-core values, and assume that the mass fraction of hydrogen is 0.5. You will also need the value $S_{pp}(E_0) = 3.8 \times 10^{-22}$ keV barns.

Hint 2: The number density (number of atoms or nuclei per unit volume) of some isotope Z is given by Equation 3.10 as $n_Z = \rho X_Z/m_Z$, where ρ is the matter density (typically in kg m^{-3}), X_Z is the mass fraction of the isotope Z, and m_Z is the mass of each atom of the isotope (typically in kg).

Solution

The first step of the p–p chain is $p + p \longrightarrow d + e^+ + \nu_e$.

The fusion rate (Equation 3.30), with the number-density term appropriate for *identical* particles, is

$$R_{pp} = \frac{6.48 \times 10^{-24}}{A_r Z_p Z_p} \times \frac{n_p^2}{2[m^{-6}]} \times \frac{S(E_0)}{[\text{keV barns}]}$$
$$\times \left(\frac{E_G}{4kT}\right)^{2/3} \times \exp\left[-3\left(\frac{E_G}{4kT}\right)^{1/3}\right] m^{-3} s^{-1}.$$

Step 1: Compute the energy factor $E_G/4kT$.

$$\frac{E_G}{4kT} = \frac{2m_r c^2 \times (\pi \alpha Z_p Z_p)^2}{4kT}$$
$$= \frac{2c^2 \times (\pi \alpha Z_p Z_p)^2}{4kT} \times \frac{m_p m_p}{m_p + m_p}$$
$$= \frac{2c^2 \times (\pi \alpha Z_p Z_p)^2}{4kT} \times \frac{m_p}{2}$$
$$= \frac{(2.998 \times 10^8 \text{ m s}^{-1})^2 \times (\pi \times \frac{1}{137} \times 1 \times 1)^2 \times 1.673 \times 10^{-27} \text{ kg}}{4 \times 1.381 \times 10^{-23} \text{ J K}^{-1} \times 15.6 \times 10^6 \text{ K}}$$
$$= 91.8 \frac{\text{kg m}^2 \text{ s}^{-2}}{\text{J}} = 91.8.$$

Step 2: Compute the fusion rate.

The proton density in the solar core is given by $n_p = \rho_{\odot,c} X_p / m_p$, where X_p is the mass fraction of hydrogen.

The reduced atomic mass (in atomic mass units) is

$$A_r = \frac{m_r}{u} = \frac{1}{u}\left(\frac{m_p m_p}{m_p + m_p}\right) = \frac{m_p}{2u}$$

so we can write

$$R_{pp} = \frac{6.48 \times 10^{-24}}{Z_p Z_p} \times \frac{2u}{m_p} \times \frac{(\rho_{\odot,c} X_p / m_p)^2}{2\,[\text{m}^{-6}]} \times \frac{S(E_0)}{[\text{keV barns}]}$$
$$\times \left(\frac{E_G}{4kT}\right)^{2/3} \times \exp\left[-3\left(\frac{E_G}{4kT}\right)^{1/3}\right] \text{m}^{-3}\,\text{s}^{-1}$$

$$= \frac{6.48 \times 10^{-24}}{Z_p Z_p} \times \frac{(\rho_{\odot,c} X_p)^2 u}{m_p^3\,[\text{m}^{-6}]} \times \frac{S(E_0)}{[\text{keV barns}]}$$
$$\times \left(\frac{E_G}{4kT}\right)^{2/3} \times \exp\left[-3\left(\frac{E_G}{4kT}\right)^{1/3}\right] \text{m}^{-3}\,\text{s}^{-1}$$

$$= \frac{6.48 \times 10^{-24}}{1 \times 1} \times \frac{(1.48 \times 10^5\,\text{kg m}^{-3} \times 0.5)^2 \times 1.661 \times 10^{-27}\,\text{kg}}{(1.673 \times 10^{-27}\,\text{kg})^3\,[\text{m}^{-6}]}$$
$$\times \frac{3.8 \times 10^{-22}\,\text{keV barns}}{[\text{keV barns}]} \times (91.8)^{2/3} \times \exp\left[-3 \times (91.8)^{1/3}\right] \text{m}^{-3}\,\text{s}^{-1}$$

$$= 1.3 \times 10^{14}\,\text{m}^{-3}\,\text{s}^{-1}.$$

That is, there are 130 million million fusion reactions per cubic metre per second!

Step 3: Compute the proton lifetime.

From Equation 3.21 the lifetime of a single proton before it fuses with another proton is $\tau_p = n_p / 2 R_{pp}$. Noting again that $n_p = \rho_{\odot,c} X_p / m_p$ gives

$$\tau_p = \frac{\rho_{\odot,c} X_p}{2 m_p R_{pp}}$$
$$= \frac{1.48 \times 10^5\,\text{kg m}^{-3} \times 0.5}{2 \times 1.673 \times 10^{-27}\,\text{kg} \times 1.3 \times 10^{14}\,\text{m}^{-3}\,\text{s}^{-1}}$$
$$= 1.7 \times 10^{17}\,\text{s} \approx 5.4 \times 10^9\,\text{years}.$$

This is, the average proton lifetime τ_p is more than 5 billion years.

3.3.5   The CN cycle revisited

You can now use some of what you have learnt about nuclear fusion in this Chapter to investigate the CN cycle. We examine the rates of the reactions in the next example, and the exercise that follows it, and then look at the isotope ratios that the CN cycle produces.

Worked Example 3.2

Work out the fusion rate R_{p12C} per unit mass fraction X_{12C}, given as R_{p12C}/X_{12C}, for the fusion reaction

$$p + {}^{12}_{6}C \longrightarrow {}^{13}_{7}N + \gamma$$

for conditions in the core of the Sun. Assume a hydrogen mass fraction $X_p = 0.5$.

Hint 1: Use the solar-core values from Table 1.1.

Hint 2: Use the S-factor, $S(E_0) = 1.5$ keV barns.

Hint 3: Use the approximation that the mass of a nucleus Z is $m_Z = A_Z u$ where A_Z is the atomic mass number and $u = 1$ amu.

Solution

The fusion rate for *dissimilar* particles is

$$R_{AB} = \frac{6.48 \times 10^{-24}}{A_r Z_A Z_B} \times \frac{n_A n_B}{[m^{-6}]} \times \frac{S(E_0)}{[\text{keV barns}]}$$
$$\times \left(\frac{E_G}{4kT}\right)^{2/3} \times \exp\left[-3\left(\frac{E_G}{4kT}\right)^{1/3}\right] m^{-3} s^{-1}.$$

From Equation 3.10, $n_A = \rho X_A / m_A$.

To analyse the reaction $p + {}^{12}_{6}C$, we consider particle A to be a proton p, and particle B to be ${}^{12}_{6}C$. Writing $n_p = \rho X_p / m_p$ and $n_{12C} = \rho X_{12C} / m_{12C}$, we can rewrite the fusion rate equation as

$$R_{p12C} = \frac{6.48 \times 10^{-24}}{A_r Z_p Z_{12C}} \times \frac{n_p n_{12C}}{[m^{-6}]} \times \frac{S(E_0)}{[\text{keV barns}]}$$
$$\times \left(\frac{E_G}{4kT}\right)^{2/3} \times \exp\left[-3\left(\frac{E_G}{4kT}\right)^{1/3}\right] m^{-3} s^{-1}$$
$$= \frac{6.48 \times 10^{-24}}{A_r Z_p Z_{12C}} \times \frac{\rho_c X_p}{m_p [m^{-3}]} \frac{\rho_c X_{12C}}{m_{12C} [m^{-3}]} \times \frac{S(E_0)}{[\text{keV barns}]}$$
$$\times \left(\frac{E_G}{4kT}\right)^{2/3} \times \exp\left[-3\left(\frac{E_G}{4kT}\right)^{1/3}\right] m^{-3} s^{-1}.$$

We can divide both sides by X_{12C} to give the fusion rate per unit mass fraction of ${}^{12}_{6}C$, R_{p12C}/X_{12C}. At the same time, we recall that the reduced atomic mass is $A_r = \frac{A_p A_{12C}}{A_p + A_{12C}}$, so

$$\frac{R_{p12C}}{X_{12C}} = \frac{6.48 \times 10^{-24}}{\frac{A_p A_{12C}}{A_p + A_{12C}} Z_p Z_{12C}} \times \frac{\rho_c^2 X_p}{m_p m_{12C} [m^{-6}]} \times \frac{S(E_0)}{[\text{keV barns}]}$$
$$\times \left(\frac{E_G}{4kT}\right)^{2/3} \times \exp\left[-3\left(\frac{E_G}{4kT}\right)^{1/3}\right] m^{-3} s^{-1}.$$

Using $m_Z = A_Z u$, we can substitute for m_p and m_{12C} and write

$$\frac{R_{p12C}}{X_{12C}} = 6.48 \times 10^{-24} \times \frac{(A_p + A_{12C})\rho_c^2 X_p}{(A_p A_{12C} u)^2 [m^{-6}] Z_p Z_{12C}}$$

$$\times \frac{S(E_0)}{[\text{keV barns}]} \left(\frac{E_G}{4kT}\right)^{2/3} \exp\left[-3\left(\frac{E_G}{4kT}\right)^{1/3}\right] m^{-3} s^{-1}.$$

We evaluate $E_G/4kT$ separately for convenience: recall that $E_G = 2m_r c^2 (\pi \alpha Z_A Z_B)^2$, so

$$\frac{E_G}{4kT} = \frac{2m_p m_{12C}}{m_p + m_{12C}} \times \frac{c^2 (\pi \alpha Z_p Z_{12C})^2}{4kT}$$

$$= \frac{2 \times 1u \times 12u}{1u + 12u} \times \frac{(2.998 \times 10^8 \text{ m s}^{-1})^2 \times (\pi \times \frac{1}{137} \times 1 \times 6)^2}{4 \times 1.381 \times 10^{-23} \text{ J K}^{-1} \times 15.6 \times 10^6 \text{ K}}$$

$$= 3.645 \times 10^{30} u \frac{(\text{m s}^{-1})^2}{\text{J}}$$

$$= 3.645 \times 10^{30} \times 1.661 \times 10^{-27} \text{ kg} \frac{\text{m}^2 \text{ s}^{-2}}{\text{kg m}^2 \text{ s}^{-2}}$$

$$= 6054.$$

Since both E_G and kT have energy units, the ratio is dimensionless. Four significant figures have been retained to minimize rounding errors in the next computation. Hence

$$\frac{R_{p12C}}{X_{12C}} = 6.48 \times 10^{-24} \times \frac{(1 + 12) \times (1.48 \times 10^5 \text{ kg m}^{-3})^2 \times 0.5}{(1 \times 12 \times 1.661 \times 10^{-27} \text{ kg})^2 \times [m^{-6}] \times 1 \times 6}$$

$$\times \frac{1.5 \text{ keV barns}}{[\text{keV barns}]} \times (6054)^{2/3} \times \exp\left[-3 \times (6054)^{1/3}\right] m^{-3} s^{-1}$$

$$= 3.5 \times 10^{17} \text{ m}^{-3} \text{ s}^{-1}$$

That is, there are $3.5 \times 10^{17} \times X_{12C}$ proton-capture $p + {}^{12}_{6}C$ reactions every second in one cubic metre at the core of the Sun.

Exercise 3.5   Following the example above, compute the fusion rate R_{pZ} per unit mass fraction X_Z, for the following $p + Z$ reactions:

(a) R_{p13C}/X_{13C} for $p + {}^{13}_{6}C \longrightarrow {}^{14}_{7}N + \gamma$

(b) R_{p14N}/X_{14N} for $p + {}^{14}_{7}N \longrightarrow {}^{15}_{8}O + \gamma$

(c) R_{p15N}/X_{15N} for $p + {}^{15}_{7}N \longrightarrow {}^{12}_{6}C + \alpha$

for conditions in the core of the Sun. Assume a hydrogen mass fraction $X_p = 0.5$.

Hint 1: Use the solar-core values from Table 1.1.

Hint 2: Use the S-factors: $S(E_0) = 5.5, 3.3, 78$ keV barns respectively for the three reactions.

Hint 3: Use the approximation that the mass of a nucleus Z is $m_Z = A_Z u$ where A_Z is the atomic mass number and $u = 1$ amu. ∎

The reaction rates per unit mass fraction, $R_{\mathrm{pZ}}/X_{\mathrm{Z}}$, for the four proton-capture reactions in the CN cycle, which you have calculated in the previous Example and Exercise are summarized in Table 3.1. They indicate that the $^{14}_{7}\mathrm{N} + \mathrm{p} \longrightarrow ^{15}_{8}\mathrm{O} + \gamma$ step is the slowest and hence is the bottleneck in the CN cycle.

Since the $^{14}_{7}\mathrm{N} + \mathrm{p} \longrightarrow ^{15}_{8}\mathrm{O} + \gamma$ rate is so much slower than the others, you might expect that the bottleneck causes a build-up of $^{14}_{7}\mathrm{N}$ as the cycle proceeds. You would be right! In fact, the CN cycle is believed to be a major source of nitrogen in the Universe.

When the CN cycle reaches equilibrium, the abundances of the isotopes cease changing. From then on, the rates of all reactions in the cycle are equal:

$$R_{\mathrm{p12C}} = R_{\mathrm{p13C}} = R_{\mathrm{p14N}} = R_{\mathrm{p15N}},$$

and the equilibrium isotope ratios are readily found from Table 3.1 as the following Worked Example and Exercise 3.6 demonstrate.

Table 3.1 Reaction rates per unit mass fraction, $R_{\mathrm{pZ}}/X_{\mathrm{Z}}$, for the four proton-capture reactions in the CN cycle at the solar core temperature.

Z	$R_{\mathrm{pZ}}/X_{\mathrm{Z}}$
$^{12}_{6}\mathrm{C}$	3.5×10^{17} m^{-3} s^{-1}
$^{13}_{6}\mathrm{C}$	10×10^{17} m^{-3} s^{-1}
$^{14}_{7}\mathrm{N}$	0.015×10^{17} m^{-3} s^{-1}
$^{15}_{7}\mathrm{N}$	0.30×10^{17} m^{-3} s^{-1}

Worked Example 3.3

What is the isotope ratio by mass of $^{12}_{6}\mathrm{C}$ to $^{13}_{6}\mathrm{C}$ produced by the CN cycle? What is the isotope ratio by number?

Solution

$$R_{\mathrm{p12C}} = 3.5 \times 10^{17} X_{\mathrm{12C}} \text{ m}^{-3}\text{ s}^{-1}$$

and

$$R_{\mathrm{p13C}} = 10 \times 10^{17} X_{\mathrm{13C}} \text{ m}^{-3}\text{ s}^{-1}$$

but since in equilibrium $R_{\mathrm{p12C}} = R_{\mathrm{p13C}}$, it follows that $3.5 \times 10^{17} X_{\mathrm{12C}} = 10 \times 10^{17} X_{\mathrm{13C}}$, and hence $X_{\mathrm{12C}}/X_{\mathrm{13C}} = 10 \times 10^{17}/3.5 \times 10^{17} = 2.9$.

That is, the isotope ratio *by mass* of $^{12}_{6}\mathrm{C}$ to $^{13}_{6}\mathrm{C}$ is 2.9.

The isotope ratio *by number* of nuclei, $n_{\mathrm{12C}}/n_{\mathrm{13C}}$, which is more commonly written $^{12}_{6}\mathrm{C}/^{13}_{6}\mathrm{C}$, differs by the ratio of their atomic masses, a factor 13/12. From Equation 3.10

$$n_{\mathrm{12C}} = \rho X_{\mathrm{12C}}/m_{\mathrm{12C}}$$

and

$$n_{\mathrm{13C}} = \rho X_{\mathrm{13C}}/m_{\mathrm{13C}}$$

so

$$\frac{^{12}_{6}\mathrm{C}}{^{13}_{6}\mathrm{C}} = \frac{n_{\mathrm{12C}}}{n_{\mathrm{13C}}} = \frac{\rho X_{\mathrm{12C}}/m_{\mathrm{12C}}}{\rho X_{\mathrm{13C}}/m_{\mathrm{13C}}} = \frac{X_{\mathrm{12C}}}{X_{\mathrm{13C}}}\frac{m_{\mathrm{13C}}}{m_{\mathrm{12C}}} = \frac{X_{\mathrm{12C}}}{X_{\mathrm{13C}}}\frac{13u}{12u} = \frac{X_{\mathrm{12C}}}{X_{\mathrm{13C}}}\frac{13}{12}$$

i.e. $^{12}_{6}\mathrm{C}/^{13}_{6}\mathrm{C} = 2.9 \times 13/12 = 3.1$.

Exercise 3.6 Compute the equilibrium isotope ratios by mass and by number for the CN cycle at the conditions of the solar core, for

(a) $^{14}_{7}\mathrm{N}$ relative to $^{12}_{6}\mathrm{C}$,

(b) $^{14}_{7}\mathrm{N}$ relative to $^{15}_{7}\mathrm{N}$. ∎

Table 3.2   Equilibrium isotope ratios from the CN cycle for $T = 15.6 \times 10^6$ K, the temperature of the solar core.

Isotope ratio	Ratio by number
$^{12}_{6}$C/$^{13}_{6}$C	3.1
$^{14}_{7}$N/$^{12}_{6}$C	200
$^{14}_{7}$N/$^{15}_{7}$N	21

Most $^{12}_{6}$C is produced by helium burning, whereas $^{13}_{6}$C is produced only by hydrogen burning when $^{12}_{6}$C already exists and where temperatures are high enough to overcome (via quantum tunnelling) the substantial Coulomb barrier. In general, then, $^{12}_{6}$C is abundant in the Universe, but $^{13}_{6}$C is produced copiously only where the CN cycle has operated. For this reason, measurements of the $^{12}_{6}$C/$^{13}_{6}$C isotope ratio are important diagnostics of CN-cycle activation. We will return to this later, when we discuss giant star evolution.

To conclude this subsection, the results of Worked Example 3.3 and Exercise 3.6 are summarized in Table 3.2.

● The first step in the proton–proton chain was the bottleneck. Which step is the bottleneck in the CN cycle, and what is the consequence of this?

○ The bottleneck of the CN cycle is the reaction, p + $^{14}_{7}$N $\longrightarrow$ $^{15}_{8}$O + γ. The consequence of this being in the middle of the cycle is that some of the original $^{12}_{6}$C is burnt but doesn't progress beyond $^{14}_{7}$N, so the cycle sees the net conversion of $^{12}_{6}$C to $^{14}_{7}$N.

3.4   The temperature-dependence of the fusion rate

As you have seen, the fusion rate is expressed by Equation 3.30. The Gamow energy appears in two $E_G/4kT$ terms, of which the exponential rather than the (2/3)-power term dominates. An increase in the temperature increases the kinetic energy of the nuclei, increasing the probability that they will penetrate the Coulomb barrier, thus increasing the fusion rate. This physics is captured in the exponential term, where an increase in T results in a decrease in the *magnitude* of the argument to the exponential function, $-3(E_G/4kT)^{1/3}$, but since the argument is negative, the exponential function gives a higher value.

3.4.1   The temperature derivative of the fusion rate

A more complete view of the temperature-dependence is found by examining the derivative of the rate R_{AB} with respect to temperature T.

Exercise 3.7   (a) Compute the temperature derivative of the fusion rate:

$$R_{AB} = \frac{6.48 \times 10^{-24}}{A_r Z_A Z_B} \times \frac{n_A n_B}{[\text{m}^{-6}]} \times \frac{S(E_0)}{[\text{keV barns}]}$$
$$\times \left(\frac{E_G}{4kT}\right)^{2/3} \times \exp\left[-3\left(\frac{E_G}{4kT}\right)^{1/3}\right] \text{m}^{-3}\,\text{s}^{-1}$$

to obtain $\mathrm{d}R_{AB}/\mathrm{d}T$ in terms of R_{AB}, the Gamow energy, the temperature and some constants.

Hint 1: You may find the algebra is tidier if you rewrite the non-temperature-dependent term as a single constant a such that

$$a = \frac{6.48 \times 10^{-24}}{A_r Z_A Z_B} \times \frac{n_A n_B}{[\text{m}^{-6}]} \times \frac{S(E_0)}{[\text{keV barns}]}.$$

Hint 2: Recall, from the product rule, that $d(uv)/dx = u\,dv/dx + v\,du/dx$. To exploit this, make the definitions

$$u = a\left(\frac{E_G}{4kT}\right)^{2/3} \quad \text{and} \quad v = \exp\left[-3\left(\frac{E_G}{4kT}\right)^{1/3}\right]$$

so $R_{AB} = uv$.

Hint 3: Note, from the chain rule, that

$$\frac{d\exp(y)}{dx} = \frac{d\exp(y)}{dy} \times \frac{dy}{dx} = \exp(y)\frac{dy}{dx}.$$

(b) Use the chain rule to find an expression for $d\log_e R_{AB}/d\log_e T$ in terms of dR_{AB}/dT, then combine this with your answer from (a) to express $d\log_e R_{AB}/d\log_e T$ in terms of the Gamow energy and kT. ∎

In the previous exercise you computed $d\log_e R_{AB}/d\log_e T$. The significance is that if $d\log_e R_{AB}/d\log_e T = \nu$, then $R_{AB} \propto T^\nu$. That is, ν indicates the sensitivity of the fusion rate to temperature. The proof is straightforward.

If $R_{AB} \propto T^\nu$, with some unknown constant of proportionality, we can write R_{AB} = const $\times T^\nu$. We can take natural logarithms of both sides, to give

$$\log_e R_{AB} = \log_e \text{const} + \log_e T^\nu = \log_e \text{const} + \nu\log_e T.$$

Differentiating with respect to $\log_e T$ gives

$$\frac{d\log_e R_{AB}}{d\log_e T} = \nu.$$

That is, the statement

$$\frac{d\log_e R_{AB}}{d\log_e T} = \nu$$

implies that $R_{AB} \propto T^\nu$.

Note that since ν is itself a function of T, this differential equation is an approximation. However since ν only varies as $T^{-1/3}$, the approximation is adequate in most circumstances.

The logarithmic derivative derived at the end of Exercise 3.7 therefore indicates that a variation in temperature leads to a rate change of the form

$$R_{AB} \propto T^{\left[(E_G/4kT)^{1/3} - (2/3)\right]}.$$

Note that the exponent (power) of the temperature, commonly given the symbol ν, depends on the ratio of the Gamow energy to the temperature. It can be expanded as

$$\nu \equiv \frac{d\log_e R_{AB}}{d\log_e T} = \left(\frac{E_G}{4kT}\right)^{1/3} - \frac{2}{3}$$

$$= \left(\frac{2m_r c^2(\pi\alpha Z_A Z_B)^2}{4kT}\right)^{1/3} - \frac{2}{3}, \tag{3.31}$$

which shows the dependence of the temperature-exponent ν on the mass and charge of the nuclei. Clearly, reactions involving heavier particles with greater atomic number depend much more strongly on temperature.

Exercise 3.8   Calculate the power of the temperature-dependence of the fusion rate for the following reactions at the core of the Sun:

(a) $p + p \longrightarrow d + e^+ + \nu_e$

(b) $p + {}^{14}_7N \longrightarrow {}^{15}_8O + \gamma$.

Hint 1: Use the solar-core values from Table 1.1.

Hint 2: For part (b), assume the mass of a proton is $1u$ and the mass of ${}^{14}_7N$ is $14u$.   ∎

The previous exercise shows that $R_{pp} \propto T^{3.8}$ and $R_{p14N} \propto T^{19.6}$, i.e. the $p + {}^{14}_7N$ reaction is far more sensitive to temperature.

3.4.2   The dependence of hydrogen burning on stellar mass

You have learnt that there are two hydrogen-burning reactions, the proton–proton chain and the CNO cycle. Which is more important, and in what circumstances? In this section you will discover that the difference in their sensitivity to temperature has major consequences.

In the previous section you calculated the temperature-dependence of the fusion rate, showing that

$$\nu \equiv \frac{\mathrm{d}\log_e R_{AB}}{\mathrm{d}\log_e T} = \left(\frac{E_G}{4kT}\right)^{1/3} - \frac{2}{3},$$

where the fusion rate is $R_{AB} \propto T^\nu$. That is, the temperature-dependence increases as the Coulomb barrier, characterized by the Gamow energy, becomes more impenetrable. In the last exercise you showed that whereas the reaction rate R_{pp} of the p–p chain reaction $p + p \longrightarrow d + e^+ + \nu_e$ varies as $R_{pp} \propto T^{3.8}$, the rate R_{p14N} of the CNO cycle stage $p + {}^{14}_7N \longrightarrow {}^{15}_8O + \gamma$ varies as $R_{p14N} \propto T^{19.6}$. Strictly, those calculations were for solar conditions. Calculations for other stars lead to typical values for the p–p chain and CNO cycle rates $R_{pp} \propto T^4$ and $R_{CNO} \propto T^{16 \text{ to } 20}$ being quoted in textbooks.

The cores of more-massive stars have higher temperatures. Low-mass stars have core temperatures too low to initiate the CNO cycle to any significant degree, so for them energy generation increases slowly with stellar mass, according to T^4. However, once a mass is reached at which CNO processing is initiated, it rapidly becomes important due to the much stronger temperature-dependence of this reaction. Consequently:

> The proton–proton chain dominates in low-mass main-sequence stars ($M \sim M_\odot$), and the CNO cycle dominates in higher-mass stars ($M \geq$ a few $M_\odot$).

Because of the *very* steep temperature-dependence of the CNO cycle, energy generation in that cycle is highly concentrated, leading to a steep temperature gradient and the onset of convection (the reason for this is discussed in the next chapter). Higher-mass stars with the $p + {}^{14}_7N$ reaction activated therefore have convective cores, whereas low-mass stars, in which the $p + p$ reaction dominates, do not.

3.4.3 The energy generation rate

The nuclear energy generation rate per unit volume, $\varepsilon_{\mathrm{nuc}}$, measures the amount of energy liberated per second in one cubic metre of material, and hence has units of joules per second per cubic metre, or watts per cubic metre (W m^{-3}). As you already know, R_{AB} is the fusion rate, i.e. the number of reactions per cubic metre per second. Therefore $\varepsilon_{\mathrm{nuc}}$ is just $R_{\mathrm{AB}} \times$ the energy released per reaction, which you can calculate from the mass defect Δm using $E = \Delta m\, c^2$:

$$\varepsilon_{\mathrm{nuc}} = \Delta m\, c^2 \times R_{\mathrm{AB}}. \qquad (3.32)$$

Therefore, just as we have written the *temperature*-dependence of the rate as $R_{\mathrm{AB}} \propto T^{\nu}$, we can also write the *temperature*-dependence of $\varepsilon_{\mathrm{nuc}} \propto T^{\nu}$. Figure 3.6 shows the nuclear energy generation rate as a function of core temperature, for the proton–proton chain and the CNO cycle. The cross-over point is in stars slightly more massive than the Sun.

A final comment needs to be made about the expression $R_{\mathrm{AB}} \propto T^{\nu}$. In deriving this in the previous section we assumed that the number density of particles was constant, but in later work we relax that constraint and permit variations in the density of particles. Particle densities appear in R_{AB} as the product $n_{\mathrm{A}} n_{\mathrm{B}}$. Since $n_{\mathrm{A}} = \rho_{\mathrm{A}}/m_{\mathrm{A}} = \rho X_{\mathrm{A}}/m_{\mathrm{A}}$ and $n_{\mathrm{B}} = \rho_{\mathrm{B}}/m_{\mathrm{B}} = \rho X_{\mathrm{B}}/m_{\mathrm{B}}$, we can write more generally that $R_{\mathrm{AB}} \propto \rho^2 T^{\nu}$, and hence

$$\varepsilon_{\mathrm{nuc}} \propto \rho^2 T^{\nu}$$

or

$$\varepsilon(r) = \varepsilon_0 \rho^2(r) T^{\nu}(r), \qquad (3.33)$$

where ε_0 is a constant of proportionality. This is therefore the final linking equation which can be used to solve the differential equations of stellar structure referred to in Chapter 1.

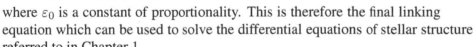

Figure 3.6 Stellar energy generation rate $\varepsilon_{\mathrm{nuc}}$ as a function of core temperature, for the proton–proton chain and the CNO cycle. Three stars are shown: a cool M0 dwarf, the Sun (a G2 dwarf), and Sirius A (an A0 dwarf).

Exercise 3.9 You have shown that the rate R_{pp} of the p–p chain reaction p + p $\longrightarrow$ d + e^{+} + ν_{e} varies as $R_{\mathrm{pp}} \propto T^{3.8}$ (for solar conditions), while the rate R_{p14N} in the CNO cycle varies as $R_{\mathrm{p14N}} \propto T^{19.6}$. On Figure 3.6, draw lines corresponding to $\varepsilon \propto T^{3.8}$ and $\varepsilon \propto T^{19.6}$ through the point for the Sun. You will see that your lines are a good match to the slopes of the curves shown in the figure.

Hint: Figure 3.6 plots ε versus T on log–log scales, i.e. it plots $\log_{10} \varepsilon$ against $\log_{10} T$. If you have a function $y = x^{\alpha}$, then taking logs of each side gives $\log_{10} y = \alpha \log_{10} x$, i.e. a straight line of slope α and zero intercept. So, a function $\varepsilon \propto T^{\nu}$ will appear on the log–log graph as a straight line with slope ν. ∎

Summary of Chapter 3

1. Nuclei have radii of order $R \approx 10^{-15}$ m (typically $R = r_0 A^{1/3}$, where $r_0 = 1.2 \times 10^{-15}$ m), a factor 10^5 times smaller than atoms (i.e. including the electrons) which have radii around 10^{-10} m.

2. In quantum mechanics, particles are described (in one dimension) by wave functions $\psi_s(r)$ which are solutions to the time-independent Schrödinger equation,

$$\left[-\frac{\hbar^2}{2m_r}\frac{\partial^2}{\partial r^2} + V(r)\right]\psi_s(r) = E\psi_s(r). \qquad \text{(Eqn 3.4)}$$

The wave function can be calculated but not measured, but the squared amplitude $|\psi_s|^2$ can be measured and is the probability that the particle described by the wave function ψ_s is located between r and $r + dr$.

3. The quantum properties of particles lead to barrier penetration and quantum tunnelling. The amplitude of a wave function encountering a potential barrier goes to zero only for an *infinite* potential; in realistic *finite* barriers, there is a finite probability that a particle represented by the wave function will pass through the barrier even though this would be forbidden under classical (non-quantum) mechanics.

4. For fusion reactions, the barrier results from the Coulomb repulsion of (positively charged) nuclei. The height of the barrier is characterized by the Gamow energy $E_G = 2m_r c^2 (\pi\alpha Z_A Z_B)^2$, where α is the fine structure constant $\approx 1/137.0$.

5. The reaction cross-section σ is the product of the geometrical cross-section with the reaction probability, and thus has units of area. For nuclear reactions, the standard unit is the barn $= 10^{-28}$ m^2. The reaction probability, and therefore also the cross-section, depends on the projectile energy and the strength of the interaction. The mean free path of a projectile travelling through a target with n particles per unit volume is $l = 1/n\sigma$.

6. The fusion cross-section may be written

$$\sigma(E) = \frac{S(E)}{E}\exp\left[-\left(\frac{E_G}{E}\right)^{1/2}\right] \qquad \text{(Eqn 3.13)}$$

where the exponential indicates that the barrier-penetration probability is higher at higher energies, and $S(E)$ is the S-factor which indicates the strength of the interaction.

7. The speed-averaged cross-section leads to the following formulae for the fusion rate:

$R_{AB} = n_A n_B \langle\sigma v_r\rangle$ for fusion of dissimilar nuclei and

$R_{AA} = (n^2/2)\langle\sigma v_r\rangle$ for fusion of identical nuclei.

For a Maxwellian speed distribution, the former becomes

$$R_{AB} = n_A n_B \left(\frac{8}{\pi m_r}\right)^{1/2}\left(\frac{1}{kT}\right)^{3/2}\int_0^\infty S(E)\exp\left[-\left(\frac{E_G}{E}\right)^{1/2} - \frac{E}{kT}\right]dE.$$

$$\text{(Eqn 3.23)}$$

The exponential function in the integrand has one term increasing with energy – the Coulomb barrier-penetration probability – and one term decreasing with energy – reflecting the fact that (for the given temperature) there are fewer particles possessing higher energy. This gives the rate R_{AB} an approximately Gaussian form, with a local maximum at an energy $E_0 = (E_G(kT/2)^2)^{1/3}$ called the Gamow peak.

The rate falls to a value $1/e \approx 0.37$ of the peak at an energy

$$E_0 \pm \Delta/2 = E_0 \pm \frac{2}{3^{1/2}2^{1/3}} E_G^{1/6}(kT)^{5/6} \approx E_0 \pm \left(\frac{E_G}{kT}\right)^{1/6} kT \quad \text{(Eqn 3.28)}$$

where $\Delta \approx 1.8(E_G/kT)^{1/6}kT$ is known as the Gamow width.

8. The integrated reaction rate

$$R_{AB} = \frac{6.48 \times 10^{-24}}{A_r Z_A Z_B} \times \frac{n_A n_B}{[\text{m}^{-6}]} \times \frac{S(E_0)}{[\text{keV barns}]}$$
$$\times \left(\frac{E_G}{4kT}\right)^{2/3} \times \exp\left[-3\left(\frac{E_G}{4kT}\right)^{1/3}\right] \text{m}^{-3}\,\text{s}^{-1}$$

$$\text{(Eqn 3.30)}$$

can be differentiated with respect to temperature to obtain the temperature-sensitivity of the reactions, dR_{AB}/dT.

The *logarithmic* derivative

$$\nu \equiv \frac{d \log_e R_{AB}}{d \log_e T} = \left(\frac{E_G}{4kT}\right)^{1/3} - \frac{2}{3} = \left(\frac{2m_r c^2(\pi\alpha Z_A Z_B)^2}{4kT}\right)^{1/3} - \frac{2}{3}$$

$$\text{(Eqn 3.31)}$$

where $R_{AB} \propto T^\nu$ indicates that the temperature-sensitivity increases for reactions involving nuclei of higher charge and higher mass.

9. The reaction p + p depends on $\sim T^4$, and the reaction p + $^{14}_{7}$N depends on $\sim T^{20}$ (at the solar core). The p–p chain predominates in stars with $M \sim M_\odot$, whereas the CNO cycle predominates in stars with $M \geq$ a few $M_\odot$.

10. Because of the very steep temperature-dependence of the CNO cycle, the energy-generating region is highly concentrated. This leads to a very steep temperature gradient, and hence the onset of convection. High-mass main-sequence stars therefore have convective cores.

11. In the CN cycle, initially, not all steps of the reaction proceed at the same rate, and the cycle converts $^{12}_{6}$C into $^{13}_{6}$C and $^{14}_{7}$N. When the CN cycle reaches equilibrium, the abundance of the isotopes ceases to change, and the rates of all reactions in the cycle will be equal. The equilibrium isotope ratios can readily be found. Two important (easily observable) isotopes ratios are, by *number*, $^{12}_{6}$C/$^{13}_{6}$C ≈ 3 and $^{14}_{7}$N/$^{12}_{6}$C ≈ 200.

12. The energy generation rate is given by the product of the energy associated with the mass defect multiplied by the reaction rate, $\varepsilon_{\text{nuc}} = \Delta m\, c^2 \times R_{AB}$. Hence the energy generation rate as a function of density and temperature may be written as $\varepsilon(r) = \varepsilon_0 \rho^2(r)\, T^\nu(r)$.

Chapter 4 From main sequence to red-giant branch

Introduction

Hydrogen burning accounts for the two most obvious features of the Hertzsprung–Russell diagram: the main sequence and the giant branch. Stars spend most of their lives burning hydrogen on the main sequence, but once core hydrogen burning ceases, they expand and cool to become red giants. Nevertheless, hydrogen burning continues in a shell outside the inert core during this phase. In this section, we examine the timescales for evolution and the changes that take place when a star exhausts the hydrogen in its core. We also see what we can learn from observations of stars nearing the end of this stage of their evolution.

4.1 Hydrogen-burning timescales

So far you have looked at hydrogen-burning reactions and their energy contribution to a star. Your calculations in the last chapter showed the lifetime of a *proton* in the core of the Sun during hydrogen burning by the proton–proton chain is more than $\approx 5 \times 10^9$ yr, i.e. 5 billion years. In this section we will go from studying *particle* lifetimes to *stellar* lifetimes.

A star uses its fusion energy to replace the radiation losses at its surface, so its **nuclear lifetime** τ_{nuc} is given by the energy content of its fusion fuel supply E_{fusion} divided by the rate at which energy is radiated away from the star, its luminosity L:

$$\tau_{\mathrm{nuc}} = E_{\mathrm{fusion}}/L. \tag{4.1}$$

Exercise 4.1   Assume that at birth, the Sun is 70% hydrogen by mass. (The Big Bang made a primordial helium fraction close to 24% by mass which was increased to about 28% by nucleosynthesis in stars, and heavier elements account for about 2%.) Using the mass defect for hydrogen burning (Section 1.5), calculate how long the Sun could continue to radiate at its current rate, if it converted *all* of its hydrogen into helium.   ■

Exercise 4.1 above shows that if the Sun could convert all of its hydrogen into helium, it would have a lifetime around 70 billion years (70×10^9 yr). In reality, we find that stars finish hydrogen burning after they have used of order 10% of their hydrogen. We therefore find a hydrogen-burning lifetime for the Sun of order 10 billion years.

The fuel supply depends on the mass of a star ($\propto M$). You know already from Section 1.3 that for high-mass stars, the mass–luminosity relationship is $L \propto M^3$ and this is borne out by observations. For lower-mass stars (below one solar mass or so), observations indicate that $L \propto M^{3.5}$ is a better approximation to reality. So the hydrogen-burning lifetimes vary as $M/M^{3.0\,\text{to}\,3.5} = M^{-2.0\,\text{to}\,-2.5}$, and are therefore *longer* for *lower-mass* stars.

Hydrogen burning is the longest phase of evolution for any star. You will find out why in Chapter 5 on helium burning. Hydrogen burning is such a slow and stable process that a star evolves little during this period. Apart from gradually brightening by about 0.5 magnitude, stars maintain much the same conditions until they have burnt about 10% of their hydrogen.

Exercise 4.2   Use the mass-dependence of stellar lifetimes, and the approximate hydrogen-burning lifetime of the Sun (assume 10 billion years), to estimate the hydrogen-burning lifetimes of $0.5\,M_\odot$ and $10\,M_\odot$ stars. For simplicity, assume that the observed mass–luminosity relation has an exponent of 3.5 (i.e. $L \propto M^{3.5}$) across the whole mass range. ■

4.2   Assigning ages from hydrogen-burning timescales

One of the most important and difficult tasks in astronomy is finding the ages of objects. In this section we see how to derive ages for stars from their hydrogen-burning timescales.

There are essentially two types of ages measurable in astronomy: radioactive decay ages, which require the measurement of elements or isotopes that undergo natural decay, and evolutionary ages, which require that we have an accurate model of stellar evolution and that we can assign an age to an object depending on how advanced its evolution is. When trying to measure ages for stars, we almost always have to rely on the latter, and the evolutionary process we measure is hydrogen burning.

- Carbon dating, which measures the radioactive decay of $^{14}_{6}\mathrm{C}$ with a half-life of 5730 yr, is often used for assigning ages of archaeological discoveries. Could $^{14}_{6}\mathrm{C}$ dating be used to measure the ages of stars? Justify your answer.

○ The half-life of $^{14}_{6}\mathrm{C}$ is very short compared with stars' lives of millions or billions of years. The probability of seeing *any* short-lived isotope is extremely low; isotopes with half-lives of millions or billions of years must be measured instead. Suitable elements include thorium (Th) and uranium (U). (Besides the short half-life, there are other reasons why $^{14}_{6}\mathrm{C}$ is unsuitable.)

The evolutionary dating method is as follows. Hydrogen-burning lifetimes depend strongly on stellar mass, as you have seen from Exercise 4.2. You can tell the mass of a main-sequence star from its temperature and luminosity, but you can't tell easily how long ago it formed. However, you know that a star *just leaving the main sequence* to become a red giant has just finished burning its core hydrogen, and you know from its mass how long that takes. This way, you *can* assign ages to stars just leaving the main sequence. This point in a star's life is called the *main-sequence turnoff*, or m-s turnoff.

The H–R diagram of a star cluster, a group of many stars of *different masses* that formed from the *same gas cloud* at the *same time*, clearly shows which stars are just finishing their main-sequence evolution. The age of stars at the main-sequence turnoff therefore dates the whole cluster. Cluster dating is of major importance for astrophysics, and is usually the *only* way of obtaining ages for stellar systems. Because of its importance, much effort goes into producing more realistic theoretical models of stellar evolution.

● If you observe a star cluster with 1 M$_\odot$ stars at the main-sequence turnoff, approximately how old must the cluster be? Recall that the Sun is 4.5 billion years old.

○ The *age* of the Sun sets only a lower limit on the age of a cluster with 1 M$_\odot$ stars at the main-sequence turnoff. Moreover, you know that the hydrogen-burning lifetime of the Sun is ≈ 10 billion years, so the cluster age must be ≈ 10 billion years.

Figure 4.1 compares the observed H–R diagrams of several **open clusters** of different ages, which have vastly different main-sequence turnoffs. A more detailed comparison between the observed H–R diagram of the old globular cluster M92 and theoretical models for a range of ages is shown in Figure 4.2. Images of these or similar clusters, which also emphasize the different *appearances* of clusters of *vastly different ages*, are shown in Figure 4.3 (on page 82).

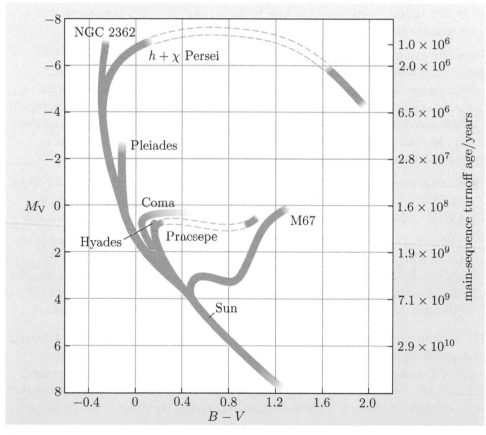

Figure 4.1   H–R diagram of several open clusters of different ages. The youngest clusters, e.g. NGC 2362 and the h and χ Persei clusters, have main sequences populated by bright, blue stars, whereas older clusters, e.g. M67, have only cooler dwarfs and a well-developed red-giant branch. The main-sequence turnoffs (m-s turnoffs) of older clusters are found at lower luminosities and temperatures; the ages corresponding to these turnoff luminosities are shown on the right-hand vertical axis.

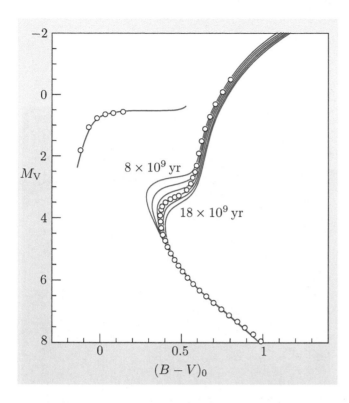

Figure 4.2 H–R diagram of the old globular cluster M92, showing the main sequence, subgiant branch, red-giant branch, and horizontal branch of the cluster (circles), and theoretical H–R diagrams for ages 8, 10, 12, 14, 16 and 18 billion years. The m-s turnoff indicates an age of 14 billion years for this cluster.

● Referring to Figure 4.3 overleaf, what difference do you see in the colours of the stars in a young open cluster like the Pleiades and a globular cluster like NGC 6093? How would you explain that difference from what you know about the different H–R diagrams of these types of objects?

○ Young open clusters are dominated by hot blue or white stars, whereas globular clusters have mostly yellow–red stars. Hot, blue main-sequence stars must be massive and hence very young, so open clusters must be young. Globular clusters, on the other hand, are dominated by red giants and yellow solar-temperature stars, so they must be much older than open clusters.

4.3 Rescaling the stellar structure equations

In this and the next few sections, we examine the changes in a star that drive it away from the main sequence onto the red-giant branch. Of prime importance are changes in its chemical composition resulting from the conversion of hydrogen into helium.

In Chapter 1 you saw how a simple analysis of the equations of stellar structure led to approximate relationships between the luminosity and mass of a main-sequence star. In this section we consider how the luminosity, radius and temperature of a star vary as a function of its composition and mass. These relationships will be developed using the assumptions of **homology**. This simply states that as a star undergoes changes in luminosity, density, temperature, composition, etc. its structure changes by the *same factor at all radii*. That is, if the surface luminosity changes by some factor A, then the luminosity everywhere in the star (at all radii) $L(r)$ changes by the same factor. Similarly, if the core

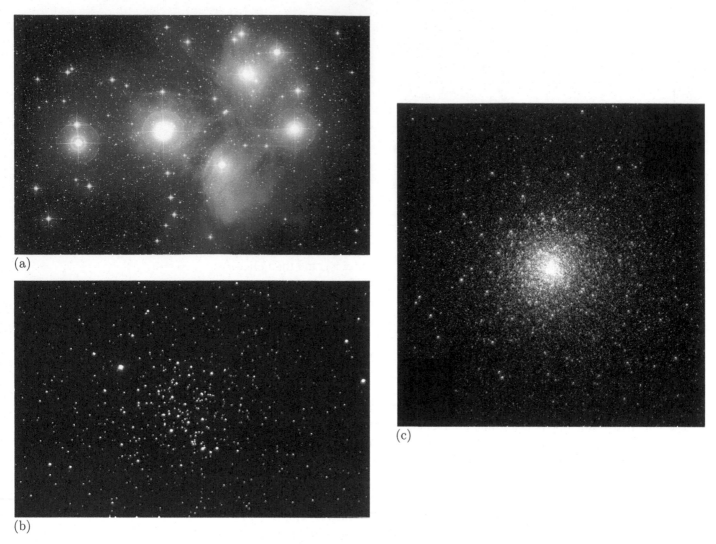

(a)

(b)

(c)

Figure 4.3   Images of three star clusters of vastly different ages: (a) the Pleiades, one of the youngest open clusters $(100 \times 10^6$ yr), (b) M67, one of the oldest open clusters $(4 \times 10^9$ yr), and (c) NGC 6093 (M80), a globular cluster similar to M92 $(14 \times 10^9$ yr).

temperature T_c changes by some factor B, then the temperature everywhere in the star (at all radii) $T(r)$ changes by the same factor. The same can be said of the density and pressure profiles throughout the star. They would each also scale in a similar fashion at all radii, although by different factors (C and D say) relative to the core density and pressure. The purpose of homology analysis, as shown in the next few subsections, is to derive these scaling factors.

In the following subsections, we assume a star has mass M, radius R, surface luminosity L and mean molecular mass μ. Its core temperature and core pressure are T_c and P_c respectively, and its mean density is $\overline{\rho}$.

4.3.1   Mass distribution

The first equation for which we will write a scaling relation is the density equation. The mean density of a star is $\overline{\rho} = 3M/4\pi R^3$, which clearly scales with the mass and radius as

$$\overline{\rho} \propto \frac{M}{R^3}. \tag{4.2}$$

Under homologous conditions, the density $\rho(r)$ at some radius r scales in the same way as the mean density.

4.3.2 Ideal gas law

We now use the familiar ideal gas law $P(r) = \rho(r)\, kT(r)/\overline{m}$ (Equation 1.11), where $\overline{m}$ is the mean mass of the particles, given by $\overline{m} = \mu u$. We can therefore also express the ideal gas law as

$$P(r) = \frac{k}{u} \frac{\rho(r)\, T(r)}{\mu}.$$

According to the homology assumption, the pressure $P(r)$ scales in the same way as the core pressure P_c; the density $\rho(r)$ scales in the same way as the mean density $\overline{\rho}$; and the temperature $T(r)$ scales in the same way as the core temperature T_c. Consequently we can write the scaling relation for the ideal gas law as

$$P_c \propto \frac{\overline{\rho} T_c}{\mu}. \tag{4.3}$$

4.3.3 Hydrostatic equilibrium

We take the equation for hydrostatic equilibrium from Equation 1.12,

$$\frac{\mathrm{d}P(r)}{\mathrm{d}r} = \frac{G\, m(r)\, \rho(r)}{r^2}$$

and write the scaling relations for that. As before, for a homologous change: the pressure interval $\mathrm{d}P(r)$ scales in the same way as the core pressure P_c; a radial interval $\mathrm{d}r$ and some radius r both scale in the same way as the outer radius R; the mass $m(r)$ enclosed by some radius r is proportional to the total mass M; and the density $\rho(r)$ scales in the same way as the mean density $\overline{\rho}$. Consequently we can write the scaling relation for hydrostatic equilibrium as

$$\frac{P_c}{R} \propto \frac{M\overline{\rho}}{R^2}.$$

So

$$P_c \propto \frac{M\overline{\rho}}{R}. \tag{4.4}$$

We can now equate the two scaling relations for the core pressure, derived from the ideal gas law (Equation 4.3) and hydrostatic equilibrium (Equation 4.4), to obtain

$$\frac{\overline{\rho} T_c}{\mu} \propto \frac{M\overline{\rho}}{R},$$

which gives a relation between the core temperature, chemical composition, mass and radius of a star as:

$$T_c \propto \frac{M}{R}\mu. \tag{4.5}$$

● Consider a given star whose chemical composition is unchanging. What relation do you expect between core temperature and radius?

○ For a given star, M is constant, and if μ is also constant then Equation 4.5 gives $T_c \propto 1/R$.

4.3.4   Radiative diffusion

In Equation 1.14 we stated the equation for luminosity when energy transport is dominated by radiative diffusion. This may be rearranged as

$$L(r) = -4\pi r^2 \frac{16\sigma}{3} \frac{T^3(r)}{\rho(r)\,\kappa(r)} \frac{\mathrm{d}T(r)}{\mathrm{d}r}.$$

For a homologous change: the luminosity $L(r)$ at some radius r scales in the same way as the surface luminosity L; a radial interval $\mathrm{d}r$ and some radius r both scale in the same way as the outer radius R; the temperature $T(r)$ and the temperature interval $\mathrm{d}T(r)$ scale in the same way as the core temperature T_c; and the density $\rho(r)$ scales in the same way as the mean density $\overline{\rho} \propto M/R^3$.

We will assume that the opacity in the stars is largely due to electron scattering, for which $\kappa(r) = \text{constant}$. This means our analysis will be more appropriate for stars with $M \geq$ a few $M_\odot$ than for those with $M \sim M_\odot$, but the latter are in any case affected by convection, making the adoption of the radiative diffusion equation rather dubious anyway.

Substituting these scaling relations into the radiative diffusion equation gives

$$L \propto R^2 \frac{T_c^3}{\overline{\rho}} \frac{T_c}{R} \propto R \frac{T_c^4}{\overline{\rho}}.$$

Substituting in for T_c using Equation 4.5 and for $\overline{\rho}$ using Equation 4.2 gives

$$L \propto R \frac{[(M/R)\mu]^4}{M/R^3} \propto R^{1-4-(-3)} M^{4-1} \mu^4$$

and finally

$$L \propto M^3 \mu^4. \tag{4.6}$$

● What does the luminosity relation say about stars that have the same chemical composition?

○ The mean molecular masses of stars that have the same chemical composition are constant, and hence the term μ^4 is constant, so $L \propto M^3$. This is, of course, the observed mass–luminosity relation for stars with $M \geq$ a few $M_\odot$, as we already derived in Chapter 1 using a simple scaling analysis.

We have now obtained relations for:

- T_c in terms of M, R and μ, and

- L in terms of M, R and μ (actually, R has cancelled out).

but we still need a relation between M, R and μ. For this we turn to the energy generation equation introduced in Chapter 1, and seek to express the radius as a function of the mass and composition.

4.3.5 Energy generation

The luminosity increases outwards from the core of a star, as new sources of energy (or more accurately, sites of energy liberation) are encountered. The energy generation equation describes the increase in luminosity $dL(r)$ per step in radius dr as:

$$\frac{dL(r)}{dr} = 4\pi r^2 \, \varepsilon(r) \qquad \text{(Eqn 1.15)}$$

where $\varepsilon(r)$ is the energy generation rate per unit volume. During the stages of stellar evolution that we consider here, the luminosity is supplied mostly by fusion. We know already from Equation 3.33 that

$$\varepsilon(r) = \varepsilon_0 \, \rho^2(r) \, T^\nu(r).$$

So combining these two equations we have

$$\frac{dL(r)}{dr} = 4\pi r^2 \, \varepsilon_0 \, \rho^2(r) \, T^\nu(r).$$

Under the homology condition, a luminosity increment $dL(r)$ scales in the same way as the surface luminosity L, a radial increment dr and a radius r both scale in the same way as the stellar radius R, the density $\rho(r)$ scales in the same way as the mean density $\bar{\rho}$ and the temperature $T(r)$ scales in the same way as the core temperature T_c. We can therefore write the scaling relation for the energy generation equation as

$$\frac{L}{R} \propto R^2 \bar{\rho}^2 T_c^\nu \quad \text{or simply} \quad L \propto R^3 \bar{\rho}^2 T_c^\nu.$$

Now we can use the results $\bar{\rho} \propto M/R^3$ (Equation 4.2) and $T_c \propto (M/R)\mu$ (Equation 4.5) to write

$$L \propto R^3 \left(\frac{M}{R^3}\right)^2 \left(\frac{M}{R}\mu\right)^\nu = \frac{M^{2+\nu}}{R^{3+\nu}}\mu^\nu.$$

We can then equate this with the result from homologous scaling of the radiative equilibrium equation, $L \propto M^3 \mu^4$ (Equation 4.6) to write

$$M^{2+\nu} R^{-(3+\nu)} \mu^\nu \propto M^3 \mu^4,$$

which we simplify as $R^{3+\nu} \propto M^{\nu-1} \mu^{\nu-4}$.

Taking the $(3+\nu)$-root of both sides gives

$$R \propto M^{(\nu-1)/(\nu+3)} \mu^{(\nu-4)/(\nu+3)}. \qquad (4.7)$$

We can now use Equation 4.7 to eliminate the dependence on R in T_c (Equation 4.5), giving it as a function of M and μ only:

$$T_c \propto \frac{M}{R}\mu \propto \frac{M\mu}{M^{(\nu-1)/(\nu+3)}\mu^{(\nu-4)/(\nu+3)}}$$

$$\propto M^{1-\frac{\nu-1}{\nu+3}} \mu^{1-\frac{\nu-4}{\nu+3}}$$

$$\propto M^{\frac{\nu+3-(\nu-1)}{\nu+3}} \mu^{\frac{\nu+3-(\nu-4)}{\nu+3}},$$

which finally yields

$$T_c \propto M^{4/(\nu+3)} \mu^{7/(\nu+3)}. \qquad (4.8)$$

4.3.6 Taking stock

We have now developed homologous scaling relations for the equations of stellar structure, the ideal gas law, and the energy generation rate, assuming the opacity to be constant. The three scaling laws we use below are Equations 4.6, 4.7 and 4.8:

$$L \propto M^3 \mu^4$$
$$R \propto M^{(\nu-1)/(\nu+3)} \mu^{(\nu-4)/(\nu+3)}$$
$$T_{\rm c} \propto M^{4/(\nu+3)} \mu^{7/(\nu+3)}.$$

We can use these relationships in two ways:

- to ask how the luminosity, radius, and core temperature of stars having the same composition depend on their mass M, and
- to ask how a single star (i.e. of fixed mass) evolves as its chemical composition μ changes, for example as a result of burning hydrogen to helium.

It can be instructive to consider two cases:

- Low-mass stars ($M \sim {\rm M_\odot}$) where the proton–proton chain dominates energy production. For this case, the temperature-sensitivity of energy generation suggests $\nu \approx 4$.
- Higher-mass stars ($M \geq$ a few ${\rm M_\odot}$) where the CNO cycle dominates energy production and $\nu \approx 17$.

In the next two sections we therefore examine how the properties of stars depend on their mass and on their chemical composition, in the case of low-mass stars and high-mass stars.

Warning note: Scaling relations involving the temperature need to be written in terms of the core temperature $T_{\rm c}$, rather than the effective surface temperature $T_{\rm eff}$ which is characteristic of just one layer in the star's atmosphere. If scaling relations involving $T_{\rm eff}$ are required, then they must be derived from the scaling relations for luminosity and radius using $L \propto R^2 T_{\rm eff}^4$ which comes from the Stefan–Boltzmann law.

4.4 Relations between stars of different mass along the main sequence

For stars covering a range of masses but having the same composition, we can use Equations 4.6, 4.7 and 4.8 to derive relationships between the radius, core temperature, luminosity and mass.

Worked Example 4.1

For low-mass stars ($M \sim {\rm M_\odot}$), use Equations 4.6, 4.7 and 4.8 to find expressions for the following quantities as a function of mass, assuming uniform chemical composition: (a) luminosity (i.e. the mass–luminosity relationship), (b) core temperature, (c) radius.

Solution

For low-mass stars, the proton–proton chain dominates energy production and $\nu \approx 4$. For stars of uniform chemical composition, μ is constant, so the μ-term can be absorbed into the unknown constant of proportionality.

(a) Equation 4.6 becomes $L \propto M^3 \mu^4$, i.e. $L \propto M^3$.

(b) Equation 4.8 becomes

$$T_c \propto M^{4/(\nu+3)} \, \mu^{7/(\nu+3)} \propto M^{4/(4+3)} \propto M^{4/7}$$

i.e. $T_c \propto M^{0.6}$.

(c) Equation 4.7 becomes

$$R \propto M^{(\nu-1)/(\nu+3)} \, \mu^{(\nu-4)/(\nu+3)} \propto M^{(4-1)/(4+3)} \propto M^{3/7}$$

i.e. $R \propto M^{0.4}$.

That is, the results suggest that the luminosity increases steeply with mass, the core temperature increases somewhat with mass, and the radius barely increases with mass.

Exercise 4.3   For high-mass stars ($M \geq$ a few $M_\odot$), use Equations 4.6, 4.7 and 4.8 to find expressions for the following quantities as a function of mass, assuming uniform chemical composition: (a) luminosity, (b) core temperature, (c) radius.   ■

Table 4.1 summarizes the differences for these two mass ranges of stars having the same composition, using the results of Worked Example 4.1 and Exercise 4.3. Since the main sequence is one of the few places where we can see a wide range of stellar masses at the same time, these results are useful as giving an understanding of the physics that shapes the main sequence. The derived equation for the high-mass stars, $L \propto M^3$, matches the observed relation very well, but for the low-mass stars these scaling results differ from observations, which suggest $L \propto M^{3.5}$. Nevertheless, the scaling relations are useful for seeing the overall behaviour of different stars, and indicate how the energy generation mechanism (ν) affects the outcome.

Table 4.1   Mass-dependence of luminosity, core temperature and radius for stars having the same composition.

Mass range	low mass $M \sim M_\odot$	high mass $M \geq$ few $M_\odot$
Energy source	proton–proton chain $\nu \approx 4$	CNO cycle $\nu \approx 17$
Results from homologous scaling of stellar structure equations	$L \propto M^3$ $T_c \propto M^{0.6}$ $R \propto M^{0.4}$	$L \propto M^3$ $T_c \propto M^{0.2}$ $R \propto M^{0.8}$

4.5 The effect on a star of its changing composition

The second area of use for the re-scaled equations is to consider the case where M is fixed, i.e. where a single star is being considered, but now its chemical composition, characterized by the mean molecular mass, μ, is allowed to vary. We can express the luminosity, core temperature, and radius as a function of μ.

Worked Example 4.2

For low-mass stars ($M \sim M_\odot$), use Equations 4.6, 4.7 and 4.8 to find expressions for the following quantities as a function of chemical composition, assuming constant mass: (a) luminosity, (b) core temperature, (c) radius.

Solution

For a low-mass star, the proton–proton chain dominates energy production and $\nu \approx 4$. For a star of constant mass, the M-term can be absorbed into the unknown constant of proportionality.

(a) Equation 4.6 becomes $L \propto M^3 \mu^4$, i.e. $L \propto \mu^4$.

(b) Equation 4.8 becomes

$$T_c \propto M^{4/(\nu+3)}\, \mu^{7/(\nu+3)} \propto \mu^{7/(4+3)} \propto \mu^{7/7}$$

i.e. $T_c \propto \mu$.

(c) Equation 4.7 becomes

$$R \propto M^{(\nu-1)/(\nu+3)}\, \mu^{(\nu-4)/(\nu+3)} \propto \mu^{(4-4)/(4+3)} \propto \mu^{0/7}$$

i.e. $R \propto \mu^0$, which means that R is constant.

Note that according to these calculations for low-mass stars, the luminosity increases steeply with increasing mean molecular mass, the temperature increases linearly with mean molecular mass, and the radius remains constant.

Exercise 4.4   For a high-mass star ($M \geq$ a few $M_\odot$), use Equations 4.6, 4.7 and 4.8 to find expressions for the following quantities as a function of mean molecular mass, assuming constant mass: (a) luminosity, (b) core temperature, (c) radius. ■

Table 4.2 summarizes the results of Worked Example 4.2 and Exercise 4.4 for these two mass ranges of stars. These differences indicate that the different physics at work in each case produces different evolutionary effects.

You would not be expected to remember the contents of Table 4.1 and Table 4.2, though you might be expected to remember how you calculated them. They are tabulated here because we will make use of them shortly, and to remind you that it is possible for you to compute the impact of evolution on a star, albeit subject to some simplifying assumptions.

Table 4.2 Dependence of luminosity, core temperature and radius on chemical composition.

Mass range	low mass $M \sim \mathrm{M}_\odot$	high mass $M \geq$ few $\mathrm{M}_\odot$
Energy source	proton–proton chain $\nu \approx 4$	CNO cycle $\nu \approx 17$
Results from homologous scaling of stellar structure equations	$L \propto \mu^4$ $T_\mathrm{c} \propto \mu$ R is constant	$L \propto \mu^4$ $T_\mathrm{c} \propto \mu^{0.4}$ $R \propto \mu^{0.7}$

Before moving on to the evolution of a hydrogen-burning star away from the main sequence, complete one last exercise on the structure proportionalities.

Exercise 4.5 (a) Calculate the mean molecular mass μ of the following samples of neutral gas: (i) fully ionized hydrogen and helium in the Big-Bang proportions, 93% H and 7% helium by number (which is equivalent to 76% and 24% by mass); (ii) fully ionized helium. Calculate the ratio of μ in case (ii) to μ in case (i).

(b) Using the result of (a) and the proportionalities summarized in Table 4.2, calculate how much larger the radius, temperature, and luminosity of a star would be after a primordial sample of hydrogen and helium has been burnt *completely* to helium via (i) proton–proton chain, (ii) the CNO cycle. ∎

The main point of this subsection, as summarized by Exercise 4.5, is that objects which convert hydrogen into helium become hotter and very much more luminous. It is important to remember, however, that these changes are taking place in the *core* of the star; the *surface* temperature of a star also depends on its radius, so the *core and surface temperatures need not evolve in unison.*

4.6 The subgiant transition to shell-hydrogen burning

Having studied the physical process occurring in the core of hydrogen-burning stars, we now study the transition from the main sequence to the **red-giant branch** in the H–R diagram.

4.6.1 Expansion of the envelope, contraction of the core

When stars have burnt a fraction of their hydrogen to helium, around 10% in solar-mass objects, they undergo significant changes that send them from the main sequence to the red-giant branch. Numerical evolutionary models that incorporate all of the known contributing physics reproduce the observations very well, so astronomers have confirmed that they understand the process sufficiently well to be able to reproduce it on computers. However, despite this triumph, one regrettable problem persists: it has not yet proven possible to reduce those processes to just a few simple statements that encapsulate the major physics driving this phase of evolution. It is possible to point out *parts* of the contributing physics, but these always fail to provide a robust explanation of what takes place. This makes it difficult to grasp the elusive key explanation of subgiant

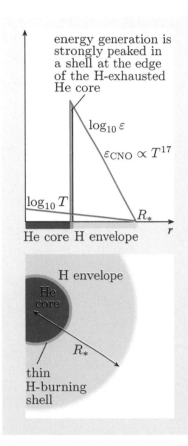

energy generation is strongly peaked in a shell at the edge of the H-exhausted He core

$\log_{10} \varepsilon$

$\varepsilon_{CNO} \propto T^{17}$

$\log_{10} T$

R_*

r

He core H envelope

H envelope

He core

R_*

thin H-burning shell

Figure 4.4   Schematics of the CNO cycle: (top) the steep dependence of energy generation on temperature, $\varepsilon_{CNO} \propto T^{17}$, leading to strongly peaked energy generation on the edge of the hydrogen-exhausted helium core, and (bottom) cross-section of a star with this structure.

evolution. Fortunately the problem is more pedagogical (to do with teaching) than physical, since when all of the physics *is* incorporated in the numerical models, the predictions *do* agree well with the observations. In this subsection we can present *some* of the physical processes involved, but unfortunately we cannot provide a definitive and simplified explanation of this phase of evolution.

Low-mass ($M \sim M_\odot$), initially p–p-chain dominated stars are not well-mixed, which means that the build-up of helium produced from hydrogen burning in the **core** is confined there. Material in their outer layers – the **envelope** – does not experience the same progressive increase in the mean molecular mass that occurs with the conversion of hydrogen to helium. (You will learn later about specific evolutionary phases during which mixing does occur.) So, while the core becomes helium-rich as the star sits on the main sequence, the envelope retains its original chemical composition. Nevertheless, the changing mean molecular mass in the core does alter the structure of the star, and this probably contributes to the expansion of the envelope, as we will now see.

As hydrogen is converted into helium, the mean molecular mass and core temperature increase. The higher temperature increases the relative importance of the CNO cycle compared to the p–p chain for energy generation. As the CNO cycle has such a strong dependence on temperature ($\varepsilon \propto T^{17}$), this has the effect of concentrating energy generation in a narrow shell of hot material sandwiched between the already-burnt helium interior and the yet-to-be-burnt, cooler hydrogen envelope (Figure 4.4). This condition is called **shell-hydrogen burning**. Under this condition, burning by the CNO cycle becomes more important than the initial p–p chain.

That is all very well for the helium-rich core, but the mean molecular mass of the H-rich envelope is unchanged from before. However, it now has a much hotter core embedded within it than previously. This increased pressure at the base of the envelope may be one factor contributing to the envelope's subsequent expansion. The expansion is gradual throughout the main-sequence life, causing the star to expand slowly.

Detailed numerical calculations show that the expansion becomes particularly noticeable when about 10% of the star's mass is in its helium core. By this stage, the expansion of the envelope is sufficient to drive the star off the main sequence (Figure 4.5). (It is for this reason that, in previous main-sequence lifetime calculations, we have assumed that only 10% of the hydrogen mass would be converted into helium during the star's main-sequence lifetime.) The star maintains its luminosity during this envelope expansion, and hence moves essentially *horizontally* across the H–R diagram towards larger radii. Recall from Chapter 1 that since $L = 4\pi R^2 \sigma T_{eff}^4$, an expansion in R at constant L corresponds to a decrease in the surface temperature. The star therefore becomes cooler and larger: a red giant.

Although material from the helium-rich core is not mixed into the H-rich envelope, the envelope is not static. In particular, low-mass main-sequence stars have outer **convection zones** that circulate material through the stars' outer layers. As the star becomes cooler, it also becomes more convective as it approaches the so-called **Hayashi limit** in the H–R diagram. As the expanding envelope causes the outer layers to cool, the outer convection zone extends deeper into the star. For a discussion of convection, see the Box on page 92.

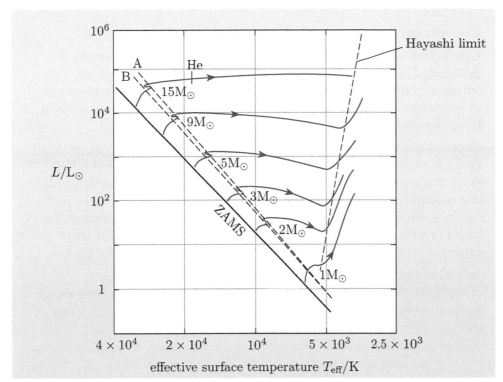

Figure 4.5 Evolutionary tracks of stars with masses in the range $M_\odot \leq M \leq 9\,M_\odot$ during hydrogen burning, i.e. from hydrogen ignition on the main sequence – the zero-age main sequence or ZAMS – to helium ignition on the red-giant branch. The slow evolution from ZAMS to line A is due to small structural changes during core hydrogen burning. The zigzag in stars with $M \geq 2\,M_\odot$ marks (first, along line A, the cooler turning point) the contraction of the core as gravitational energy release dominates hydrogen burning, and (second, along line B, the hotter turning point) the transition to shell-hydrogen burning. A $15\,M_\odot$ star is also shown; this ignites helium (point marked 'He') while still close to the main sequence.

In stars more massive than a few $M_\odot$, core temperatures are high enough on the main sequence for the CNO cycle to be the major hydrogen-burning process. The high temperature-dependence of the CNO cycle, $\varepsilon \propto T^{17}$, causes energy generation to be so centrally concentrated that a steep temperature profile is established. Consequently convection occurs in the cores of these stars, which keeps their cores well mixed.

Although *energy* production is strongly centrally concentrated, the CNO cycle is nevertheless active over a large radial distance within the star. This affects the *abundances* of the CNO isotopes at different depths. In particular, the CN cycle converts $^{12}_{6}C$ into $^{13}_{6}C$ and $^{14}_{7}N$, and in the core of the star, the higher temperature activates the ON cycle which converts $^{16}_{8}O$ into $^{14}_{7}N$ (Figure 4.6). The surface of the star will not reveal this process at this stage, but it is important to keep in mind that it is happening below the surface.

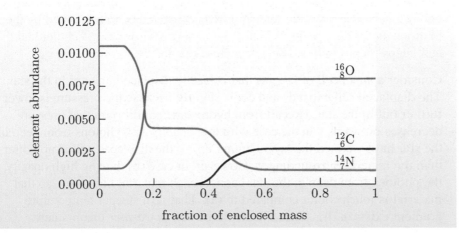

Figure 4.6 CNO abundances as a function of enclosed mass within a $2.5\,M_\odot$ star, around the end of its main-sequence evolution. When the star reaches the giant branch, the outer convection zone will deepen and may reach layers in which the CN-cycled isotopes are dredged up to the surface where they can be observed.

Convection

Convection is motion within a fluid in a gravitational field, driven by differences in density from place to place. Less-dense material rises, whereas more-dense material falls. Often density differences are the result of temperature differences, and in this case, fluid motion also transfers heat from one place to another.

Convection can be a very important mechanism of transporting energy (heat) throughout the star, especially when the opacity of stellar material is high since that prevents radiation from transporting energy. High opacity means that radiation does not travel easily through the stellar material, and a steep radiative temperature gradient is set up. The reason is straightforward: if a hotter layer is prevented by high opacity from shining on a cooler layer further out, then radiation is prevented from heating up the cooler layer.

When convection is about to begin, we say that a star is *unstable to convection*. Steep radiative temperature gradients are notorious for making stars unstable to convection. Figure 4.7 helps illustrate this by comparing two cases for (a) low-opacity material and (b) high-opacity material.

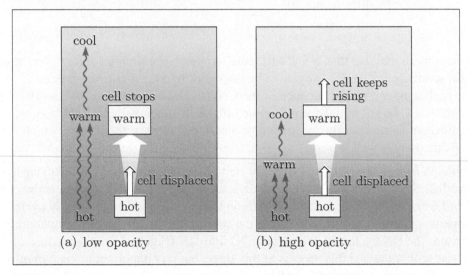

Figure 4.7   A cell of material is displaced upwards in a star, and expands because of the decreasing pressure. If the opacity is low, as in (a), its new environment is at a similar temperature because it has been warmed by the photons, but if the opacity is high (b), the new environment is cooler and the cell keeps rising. In this case, convection sets in.

Consider a small cell of hot gas that is briefly displaced upward in the star. The displaced cell expands and cools slightly because the pressure is lower further out in the star. (Recall from hydrostatic equilibrium that pressure decreases outwards.) In the case with low opacity, (a), photons from deep in the star have warmed the layers higher up, so the displaced cell is no hotter than the material surrounding it. However, in case (b), having high opacity, the photons from deep in the star cannot reach material higher up, so that material is much cooler compared to (a). That is, a steeper temperature gradient exists in (b). Since the displaced cell is *warmer* than its new (relatively cool) surroundings, it follows the old adage that hot air rises, and *continues* to rise.

Convection, however is a cyclic process. The rising cell of material eventually deposits its energy in the surrounding gas and so cools. It is then denser than its surroundings, and so will begin to fall back to a region of comparable density and temperature. From this point, the cycle can of course repeat, and consequently, large-scale motions of material result. This motion is convection.

4.6.2 The Hertzsprung gap

The interval during which the star moves horizontally to the right across the H–R diagram is called the **subgiant phase**. As the envelope expansion is quite rapid, stars are in this phase for only a relatively short period of time compared to main-sequence hydrogen burning. Consequently very few stars are seen in the subgiant stage of evolution. This gives rise to a gap in the H–R diagram between the end of the main sequence – called the main-sequence turnoff – and the base of the red-giant branch. This gap is called the **Hertzsprung gap**.

Although there are relatively *few* stars in the Hertzsprung gap compared to the number on the main sequence or giant branch, some that are there are highly conspicuous. There is a roughly vertical region of the H–R diagram, extending upward from near the main-sequence turnoff of a star slightly more massive than the Sun, in which the outer layers of the star pulsate; that is, they repeatedly expand and contract. Not surprisingly, this region of the H–R diagram is called the **instability strip** (Figure 4.8 overleaf). The changing radius of the star brings about a change in its brightness and its effective temperature, so it draws attention to itself in the same way as a flashing light on an emergency vehicle, though with a period of a few hours rather than a second. As subgiants with $M \approx 2\,M_\odot$ evolve off the main sequence to make the transition from the main-sequence turnoff to the red-giant branch, they go through the instability strip, and hence undergo a short episode of pulsation. These are called δ **Scuti stars** (after the first such star discovered) or **dwarf Cepheids** (because of their similarity to some giant stars also in the instability strip which were named after their prototype δ Cephei). These **pulsating variable stars** should be distinguished from stars with companions in a **binary system**, which may be variable due to eclipses, distortions to the shape of the stars, or the presence of matter streaming from one star to the other.

4.7 The red-giant phase: shell-hydrogen burning

Even low-mass stars like the Sun rely on convective energy transport throughout some portion of their interior. In the case of the Sun, the convection zone extends throughout the upper 30% or so of its radius. However, once a subgiant reaches the Hayashi limit, its convection zone deepens substantially, to such an extent that the star can be considered fully convective. The increasing convection does two things: it affects the energy transport through the envelope, and also transports matter from deep in the star to the surface. The results of these processes will be discussed in this section.

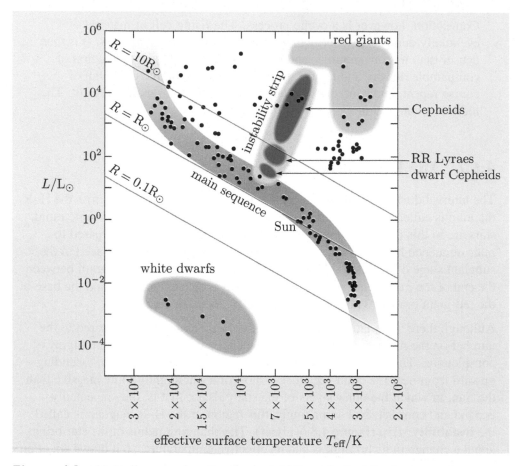

Figure 4.8   H–R diagram showing the instability strip that lies between the giant branch and the upper main sequence. The locations of three classes of pulsating variable stars are indicated.

4.7.1   Convective energy transport

Before studying the effects of convection in red giants, it is important to understand why convection sets in as a subgiant cools. A packet of gas in a star will rise if it is less dense than its surroundings. As long as it finds itself in a cooler and more dense environment, it will continue to rise. Conversely, a packet of gas in a star will fall (due to gravity) if it is more dense than its surroundings. As long as it finds itself in a hotter and less dense environment, it will continue to fall. Complex convection currents can therefore be set up within a star, and their action depends on the variation in temperature gradient from place to place. We examine the condition for convection below.

The ideal gas law (Equation 1.11) may be rearranged as

$$\rho = \frac{\overline{m}}{k}\frac{P}{T}$$

so, if we take natural logarithms of this equation, we obtain

$$\log_e \rho = \log_e \left(\frac{\overline{m}}{k}\right) + \log_e P - \log_e T.$$

Differentiating this equation with respect to P gives

$$\frac{d \log_e \rho}{dP} = \frac{d \log_e (\overline{m}/k)}{dP} + \frac{d \log_e P}{dP} - \frac{d \log_e T}{dP}.$$

Using the chain rule and setting the first right-hand term to zero (as $\overline{m}$ and k are constant) gives

$$\frac{d \log_e \rho}{d\rho} \frac{d\rho}{dP} = \frac{d \log_e P}{dP} - \frac{d \log_e T}{dT} \frac{dT}{dP}.$$

Now we note that $d \log_e x/dx = 1/x$, so multiplying through by dP gives

$$\frac{d\rho}{\rho} = \frac{dP}{P} - \frac{dT}{T}.$$

Considering *small* finite increments (rather than infinitesimals) gives

$$\frac{\Delta\rho}{\rho} = \frac{\Delta P}{P} - \frac{\Delta T}{T}. \tag{4.9}$$

This equation describes how the temperature, pressure and density of a small packet of gas will adjust, relative to its initial properties, when subject to change.

> In the following argument we will refer to changes in a packet of rising gas by δ, e.g. $\delta\rho$, whereas changes in the surroundings will be referred to by Δ, e.g. $\Delta\rho$.

Now consider a packet of gas at a certain height in a stellar envelope, whose density ρ, temperature T and pressure P match those of its surroundings. If the packet of gas is displaced upwards to a slightly different height where the density, temperature and pressure are $\rho + \Delta\rho$, $T + \Delta T$ and $P + \Delta P$, in general its properties will no longer match those of its surroundings. (Remember that $\Delta\rho$, ΔT and ΔP can be positive or negative.) Let us suppose the density, temperature and pressure of the packet of gas are now $\rho + \delta\rho$, $T + \delta T$ and $P + \delta P$ respectively. We can further suppose that:

- the pressure in the packet will rapidly change to match that of the surroundings where the pressure is $P + \Delta P$, so $\delta P = \Delta P$;

- the packet will expand or contract adiabatically, i.e. without input or extraction of energy, in order to achieve this matched pressure.

In an adiabatic process, the pressure P, density ρ and volume V of an ideal gas obey the relationship

$$PV^\gamma = \text{constant} \quad \text{or} \quad P \propto \rho^\gamma, \tag{4.10}$$

where γ (the **adiabatic index**) of a classical ideal gas is related to the number of **degrees of freedom** s of the particles in the gas (see box below) by

$$\gamma = \frac{1 + s/2}{s/2}. \tag{4.11}$$

If each gas particle has just the three translational degrees of freedom, then $s = 3$ and $\gamma = 5/3$. However, if the number of degrees of freedom is larger, then γ is smaller. As s becomes large, so γ approaches 1. This situation will occur in a molecular gas where the particles have additional degrees of freedom due to rotation or vibration. It will also be the case if heat can be absorbed by ionizing atoms.

Degrees of freedom

The total energy of a gas particle is said to arise from s degrees of freedom when there are s independent terms in the expression for that energy, each term arising from the square of a component of translational, vibrational or angular velocity. An ideal, monatomic gas has just three degrees of freedom, corresponding to independent velocities in each of the x-, y- and z-directions. However, molecular ideal gases have a few additional degrees of freedom associated with molecular rotation and vibration, and non-ideal gases have even more degrees of freedom associated with ionization. Figure 4.9 shows that a diatomic molecule has three forms of energy: translational, rotational and vibrational. These give rise, respectively, to three degrees of freedom due to velocities in three independent directions, two degrees of freedom due to there being two independent axes about which the molecule can rotate (a rolling motion about the axis joining the atoms is not counted), and two degrees of freedom due to the kinetic and potential energies of vibration, giving a total of seven degrees of freedom.

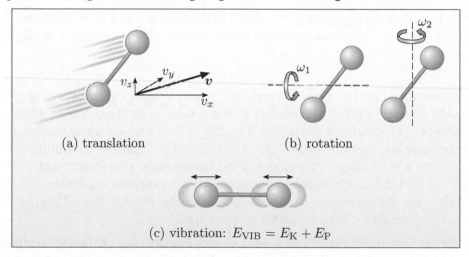

(a) translation

(b) rotation

(c) vibration: $E_{\text{VIB}} = E_{\text{K}} + E_{\text{P}}$

Figure 4.9   Three forms of energy of a diatomic molecule.

Now, let the constant of proportionality in $P \propto \rho^{\gamma}$ be C, so $P = C\rho^{\gamma}$. Differentiating with respect to density gives $\mathrm{d}P/\mathrm{d}\rho = C\gamma\rho^{\gamma-1}$, i.e. $(1/\gamma)\,\mathrm{d}P = C\rho^{\gamma-1}\,\mathrm{d}\rho$. Dividing through by P gives $(1/\gamma)\,\mathrm{d}P/P = C\rho^{\gamma-1}\,\mathrm{d}\rho/P$. On the right-hand side, we then use the fact that $P = C\rho^{\gamma}$ to obtain $C\rho^{\gamma-1}\,\mathrm{d}\rho/C\rho^{\gamma} = \rho^{-1}\,\mathrm{d}\rho = \mathrm{d}\rho/\rho$. Substituting this into the previous expression therefore gives

$$\frac{1}{\gamma}\frac{\mathrm{d}P}{P} = \frac{\mathrm{d}\rho}{\rho}.$$

Provided we consider very small but finite increments $\delta P \ll P$ and $\delta\rho \ll \rho$, it is permissible to replace the infinitesimals by finite increments: $dP \to \delta P$ and $d\rho \to \delta\rho$. Doing so gives:

$$\frac{1}{\gamma}\frac{\delta P}{P} = \frac{\delta\rho}{\rho}. \tag{4.12}$$

Now, convection will be possible if the packet of gas is buoyant and continues to rise, that is if its density is less than that of its surroundings, i.e. $(\rho + \delta\rho) < (\rho + \Delta\rho)$ or simply $\delta\rho < \Delta\rho$. So substituting Equation 4.12 into Equation 4.9, we have the condition for convection as

$$\frac{1}{\gamma}\frac{\delta P}{P} < \frac{\Delta P}{P} - \frac{\Delta T}{T},$$

where δP is the pressure change in the packet and ΔP and ΔT are the pressure and temperature changes in the surrounding gas. As noted above, we can assume that $\delta P = \Delta P$, since the pressure of the packet will respond rapidly to its new environment, and rewrite this condition as

$$\frac{\Delta T}{T} < \frac{(\gamma - 1)}{\gamma}\frac{\Delta P}{P}. \tag{4.13}$$

So the critical temperature gradient for convection to occur is

$$\frac{dT}{dr} < \frac{(\gamma - 1)}{\gamma}\frac{T}{P}\frac{dP}{dr}. \tag{4.14}$$

Note that since r is measured outwards in the star, both dT/dr and dP/dr are negative (T and P decrease outwards). Therefore, the statement that dT/dr is *less* than another negative number actually means that the temperature gradient is *steeper* than the pressure gradient term. Taking absolute magnitudes of the gradients, we get

$$\left|\frac{dT}{dr}\right| > \frac{(\gamma - 1)}{\gamma}\frac{T}{P}\left|\frac{dP}{dr}\right|, \tag{4.15}$$

i.e. the inequality switches from < to >. You may use either form, but you must be unambiguous; merely saying that the temperature gradient must be less than the pressure gradient, without adding that both are negative, could be misleading.

Exercise 4.6 Consider the critical temperature gradient for convection, and evaluate the coefficient $(\gamma - 1)/\gamma$ for two values of s: (a) $s = 3$, and (b) $s \to \infty$. In case (b), under what circumstances is the gas unstable to convection (assuming $dT/dr < 0$)? ∎

The critical temperature gradient for convection can be written in several different ways. Since $dx/x = d\log_e x$, we can also write the critical condition for convection in terms of the logarithmic temperature and pressure gradients

$$\frac{d\log_e T}{dr} < \frac{(\gamma - 1)}{\gamma}\frac{d\log_e P}{dr}.$$

One can even go a step further, and make use of the chain rule to remove all reference to the radius r, and consider the change in log (temperature) with log (pressure):

$$\frac{d\log_e T}{d\log_e P} > \frac{\gamma - 1}{\gamma}. \tag{4.16}$$

(Note the change of sense of the inequality from < to > because we have divided both sides by a negative quantity, $d\log_e P/dr$.)

4.7.2   The increase in luminosity on the red-giant branch

When a star's envelope is in radiative mode, energy escapes from the core to the surface by a process of random walks, in which radiative energy (i.e. photons) is constantly absorbed and re-emitted. As the light is re-emitted in a random direction, it takes a long time for it to make its way to the surface. If the opacity of the material were quite low, photons could at least travel a decent distance in one step. However, stellar material is generally very opaque, and a photon is absorbed and re-emitted many times before reaching the surface. In the Sun, for example, it is estimated that it takes $\sim 50\,000$ years for photons to diffuse from the core to the surface.

Physical transport of hot gas by convection has the potential to be far more efficient at transporting energy to a star's surface. As the stellar envelope in a subgiant becomes fully convective, this in fact happens.

During the red-giant phase, energy generation, and hence the star's luminosity, is closely related to the mass of the stellar core. The core mass steadily increases during shell-hydrogen burning, as the shell progressively burns hydrogen-rich material surrounding the core. Hence the luminosity of the star increases. Convection, being a far more effective energy-transport mechanism than radiative diffusion, allows the liberated energy to escape from the star far more easily than before. In evolving to higher luminosity, the star moves up the H–R diagram and makes the transition from the horizontally evolving subgiant branch to become a more vertically evolving red-giant branch (RGB) star (see Figure 4.5).

4.7.3   First dredge-up

As the convection zone deepens when the star evolves towards the Hayashi limit, it may eventually reach a depth where material which has undergone hydrogen burning by the CN cycle is dredged up to the surface. This is called the **first dredge-up**. (Two more dredge-up episodes may occur during the **asymptotic-giant branch (AGB)** phase of a star's evolution, which we discuss in Chapter 5). Recall from Figure 4.6 that CN cycling may be active enough to modify isotopic abundances even in the intermediate zones of a star where the *energy* contribution from the CN cycle is negligible. When first dredge-up happens, the star's surface $^{12}_{6}C/^{13}_{6}C$ ratio decreases, due to the greater proportion of $^{13}_{6}C$ in CN-cycled material. (Recall from Worked Example 3.3 that the CN cycle $^{12}_{6}C/^{13}_{6}C$ equilibrium ratio is ~ 3; the corresponding interstellar medium value, in contrast, is ~ 70.) The $^{12}_{6}C/^{13}_{6}C$ ratio observed in stars is therefore a vital diagnostic of hydrogen burning via the CN cycle and the degree of deep mixing in stars. An increase in the nitrogen abundance, due to $^{14}_{7}N$ production, may also be observed.

Summary of Chapter 4

1. The observed mass–luminosity relation gives $L \propto M^{3.0}$ for stars with $M \geq$ a few $M_{\odot}$ and $L \propto M^{3.5}$ for $M \sim M_{\odot}$.

2. The luminosity reflects the rate of fuel consumption, and the mass reflects the fuel supply. Hence the hydrogen-burning lifetimes vary as

$\tau_{\text{nuc}} = E_{\text{fusion}}/L \propto M/L \propto 1/M^{2.0 \text{ to } 2.5}$. Core hydrogen burning ends when of order 10% of the hydrogen has been burnt. Using the Sun as a reference, a star of mass M will have a main-sequence lifetime of $\sim 10^{10}(M/M_\odot)^{-2.5}$ yr. Hence a $10\,M_\odot$ star has a main-sequence lifetime of only $\sim 30 \times 10^6$ yr, whereas a $0.5\,M_\odot$ star lasts almost 60×10^9 yr!

3. The main-sequence lifetimes of stars are used to assign ages to stellar populations. Young star clusters have large numbers of massive stars lying on the hot (blue), luminous portion of the main sequence. Older clusters have a main sequence that terminates (the main-sequence turnoff) at lower luminosity and also have a well-developed giant branch.

4. Scaling relations can be written for the stellar structure equations and the ideal gas law under the assumption of homology. These describe the dependence of the luminosity, core temperature and radius of stars on stellar mass and mean molecular mass. The equations are parameterized by ν, the temperature exponent of the energy generation rate.

   - $\nu \approx 4$ for the p–p chain; this is applicable for solar-mass stars.
   - $\nu \approx 17$ for the CNO cycle; this is applicable for $M \geq$ a few $M_\odot$.

5. For stars of the same composition, μ is constant, so the equations give the dependence of L, T_c and R on M.

6. For a given star, M is fixed, so the equations give the dependence of L, T_c and R on μ.

7. The luminosity increases *extremely* rapidly with increasing mean molecular mass, with $L \propto \mu^4$. The core temperature also increases. The *cores* of stars converting hydrogen to helium therefore become hotter and more luminous.

8. The increased temperature of the helium-enriched core eventually causes even p–p chain dominated stars to initiate the CNO cycle and develop a H-burning shell around the helium-rich core. More-massive stars ($M \geq$ a few $M_\odot$) are dominated by the CNO cycle, and have well-mixed cores burning hydrogen in the centre.

9. The increased temperature of the helium-enriched core ($T_c \propto \mu^{0.4 \text{ to } 1}$) increases the pressure inside the unenriched envelope. This increased internal pressure contributes to the expansion of the envelope.

10. Once $\sim$10% of the hydrogen has been burnt to helium, expansion of the envelope becomes very significant and drives the star at roughly constant luminosity to the right across the H–R diagram, along the subgiant branch through the Hertzsprung gap. The cooling of the outer layers causes the outer convection zone to deepen as the star approaches its Hayashi limit.

11. In an adiabatic process, the pressure and density are related by $P \propto \rho^\gamma$ where γ is the adiabatic index. For a classical ideal gas, $\gamma = (1 + s/2)/(s/2)$ where s is the number of degrees of freedom of the particles. For three translational degrees of freedom, $\gamma = 5/3$.

12. A stellar layer is unstable to convection if the density of a cell of gas displaced adiabatically is less than that of its new surroundings. By

considering an adiabatic displacement of an ideal gas, it can be shown that convection occurs if

$$\frac{dT}{dr} < \frac{(\gamma - 1)}{\gamma} \frac{T}{P} \frac{dP}{dr}. \qquad \text{(Eqn 4.14)}$$

Both dT/dr and dP/dr are *negative*, so convection occurs if the radiative temperature gradient is *steeper* than the pressure gradient term, i.e. if

$$\left| \frac{dT}{dr} \right| > \frac{(\gamma - 1)}{\gamma} \frac{T}{P} \left| \frac{dP}{dr} \right|. \qquad \text{(Eqn 4.15)}$$

The condition is often written in terms of the logarithmic gradients

$$\frac{d \log_e T}{d \log_e P} > \frac{\gamma - 1}{\gamma}. \qquad \text{(Eqn 4.16)}$$

13. The critical gradient for convection becomes less steep, i.e. is more easily exceeded, as γ is reduced below its maximum value (5/3), such as by the appearance of additional degrees of freedom (s) in zones where hydrogen or helium is being ionized. The radiative gradient will be very steep if the material has high opacity or if the energy source is very concentrated, as when it is strongly temperature-dependent (e.g. the CNO cycle). Both effects favour the onset of convection.

14. As the star becomes more fully convective, bulk energy transport provides a much more efficient mechanism for moving energy from the core to the surface of the star. As the core mass increases, so does the energy generation rate. Efficient convection transports this energy to the surface, and the luminosity of the star increases, which drives it up the red-giant branch, almost vertically on the H–R diagram.

15. The deepening convection zone may carry CNO-cycled material to the surface. This is called first dredge-up. Observable signatures of first dredge-up include a reduction in the $^{12}_{6}C/^{13}_{6}C$ ratio *towards* the equilibrium value of the CNO cycle (≈ 3) and an increase in the nitrogen abundance due to $^{14}_{7}N$ production.

Chapter 5 Helium-burning stars

Introduction

Stars spend the majority of their lives burning hydrogen, first as core-burning main-sequence stars and later as shell-burning red-giant-branch stars. Burning phases involving heavier nuclei such as helium, carbon, oxygen and silicon are no less important than those involving hydrogen. In this chapter we study helium burning, which is the source of most of the carbon, oxygen and (indirectly) nitrogen in the Universe, and is the final phase of nuclear fusion for many stars. We begin, however, by reviewing how ineffective nucleosynthesis would be without helium burning.

5.1 Nucleosynthesis without helium burning

Hydrogen burning produces new helium nuclei from protons, but it does not contribute to the net production of heavier elements. In the CNO cycle, for example, the nitrogen abundance increases dramatically, but only at the expense of the carbon abundance as $^{12}_{6}$C is converted into $^{14}_{7}$N. Likewise, the NeNa and MgAl cycles are two further hydrogen-burning reactions that occur in sufficiently massive stars, and they too do not produce heavier nuclei; they merely convert pre-existing nuclei, neon into sodium, and magnesium into aluminium. The net production of elements heavier than helium, which astronomers refer to collectively as **metals**, is due to stellar nucleosynthesis beyond hydrogen burning. We will examine some of the more important reactions in this chapter.

Nucleosynthesis in the first few minutes of the Big Bang produced hydrogen as $^{1}_{1}$H and $^{2}_{1}$H, helium as $^{3}_{2}$He and $^{4}_{2}$He, and a tiny amount of lithium as $^{7}_{3}$Li, but no significant quantity of any other isotope. In fact the vast majority was in the form of $^{1}_{1}$H and $^{4}_{2}$He in proportions of about 76% and 24% (by mass) respectively. Primordial nucleosynthesis of heavier elements was prevented by three factors:

1. the greater Coulomb barriers of elements with higher atomic numbers; recall that the electrostatic repulsive force between two nuclei having charges Z_A and Z_B scales as the product $Z_A Z_B$, and that the Gamow energy scales as the square of this: $E_G = (\pi \alpha Z_A Z_B)^2 2 m_r c^2$;

2. the lack of stable isotopes having mass numbers 5 and 8;

3. the decreasing density of matter as the Universe expanded.

So, although nucleosynthesis in the Big Bang produced the raw materials from which the first stars formed, all other elements are produced in stars. Stellar evolution has therefore enriched the interstellar medium with elements heavier than hydrogen and helium. If we examine the relative abundances of the elements in the interstellar medium, a few features become apparent:

1. Hydrogen and helium are still the dominant elements, each having at least 1000 times the abundance of any other element.

2. Elements with atomic number 3, 4 and 5 (lithium, beryllium and boron) are virtually absent (with abundances typically 10^{-8} to 10^{-10} that of hydrogen).

These elements are largely prevented from forming because there are no stable isotopes with mass numbers 5 or 8.

3. The elements resulting from nuclear fusion in stars (i.e. carbon, oxygen, neon, silicon and elements around iron) have relatively large abundances (typically 10^{-3} to 10^{-5} that of hydrogen).

4. Elements more massive than iron have very low abundances (typically less than 10^{-7} that of hydrogen). These elements arise from neutron capture reactions in the later stages of a star's evolution.

● Helium makes up 28% of the Sun. Where did that helium come from?

○ 24% was made in the Big Bang, and most of the remaining 4% was made in stars that lived and died before the Sun formed. Both the Big Bang and the stars synthesized helium from hydrogen (protons).

5.2   Equilibrium and the chemical potential

The thermodynamic properties of a gas are described by its temperature T, pressure P and **chemical potential** μ'. In the next section we will use the chemical potential to calculate the equilibrium ratios of nuclei involved in helium-burning reactions, so we begin this subsection by defining μ'.

Note: Some texts use the symbol μ for chemical potential, but we add the prime $'$ so you won't mistake it for the mean molecular mass.

The chemical potential is a measure of the internal energy that each particle in a gas brings to the total, and for non-relativistic particles it may be written

$$\mu' = mc^2 - kT \log_e(g_s n_{\mathrm{QNR}}/n). \tag{5.1}$$

The first term is the mass–energy (or **rest energy**) $E = mc^2$ of each particle, where we recognize that each particle is an energy reservoir by virtue of its mass. It is possible to tap that reservoir in nuclear reactions and convert some fraction of the mass into thermal energy, reducing the mass of the system but increasing the kinetic energy. Of course, this is precisely what happens in nuclear fusion, where the small mass difference between the initial and final nuclei is realized as a change in the energy of the particles. For example, recall from Chapter 1 that the p–p chain converts $4\mathrm{p} \longrightarrow {}_2^4\mathrm{He} + 2\mathrm{e}^+ + 2\nu_\mathrm{e}$. The sum of the masses of the particles on the right-hand side is lower than the sum of the masses of the particles on the left; the decrease in mass goes hand in hand with the release of 26.74 MeV of energy (for the ppI branch).

The second term of the chemical potential equation relates the particle energy to the quantum properties discussed in more detail later in this chapter. It comprises a kT term, which you will by now recognize as a thermal energy, multiplied by a factor (recall that logarithms are dimensionless) which depends on:

- g_s, the number of polarizations possible for each particle (see box below), and
- the ratio between n, the number density of gas particles (i.e. the number per unit volume), and the (non-relativistic) **quantum concentration**:

$$n_{\mathrm{QNR}} = (2\pi mkT/h^2)^{3/2}, \tag{5.2}$$

where h is Planck's constant. The value of n_{QNR} reflects the number of quantum states available to the particle.

Spin and polarizations

The **spin** of a particle may be thought of as the angular momentum it possesses about its own centre of mass, analogous to the spin of the Earth. This distinguishes it from the angular momentum a body may possess because of the motion of its centre of mass, analogous to the motion of the Earth around the Sun. Spin is quantized with **quantum number** s giving the magnitude of the spin vector s, i.e. $|s| = s$.

The *number* of **polarizations** of a particle, g_s, is the number of independent spins it can have. The values of these quantum numbers differ for massive and massless particles.

Electrons, protons and neutrons have spin $s = 1/2$. The number of polarizations g_s is given by $g_s = (2 \times \text{spin}) + 1$, and thus takes the value 2. The two polarizations are sometimes referred to as spin-up and spin-down, with corresponding spin vectors $s = +1/2$ and $s = -1/2$.

The spin of a nucleus, usually given the symbol I, depends on the numbers of protons and neutrons.

- Where an even number of protons is present, the protons with spin vector $s = +1/2$ pair with those of opposing spin vector $s = -1/2$, and the net proton spin is zero. Similar pairing occurs for an even number of neutrons. Nuclei with an *even* number of protons and an *even* number of neutrons – called even–even nuclei – therefore have zero net spin $I = 0$, and hence $g_s = (2 \times \text{spin}) + 1 = 1$. That is, there is only one polarization when the spin is zero.

- Nuclei with an even number of protons but an odd number of neutrons (or vice versa) – odd nuclei – must have one unpaired particle and hence non-zero spin. The spin I for such nuclei can be found from data tables.

- Odd–odd nuclei, for which both the neutron number and the proton number are odd, generally have non-zero spins, which likewise can be found from data tables.

For *massless* particles, e.g. photons and neutrinos, the relation between spin and the number of polarizations g_s *differs*. Photons have spin $= 1$ but *two* possible polarizations i.e. $g_s = 2$, corresponding to their **circular polarization** axis being aligned either forwards or backwards along the photon's direction of propagation (these are sometimes referred as clockwise and anticlockwise circular polarizations). Neutrinos have spin $= 1/2$, but only one polarization, i.e. $g_s = 1$; the opposite polarization is possessed by the antineutrino. These details are summarized in Table 5.1.

Table 5.1 Spins and polarizations of various particles.

Particle	s	s	g_s
e^-, e^+, p, n	$\frac{1}{2}$	$\pm\frac{1}{2}$	2
even–even nuclei	0	0	1
odd, odd–odd nuclei	> 0	various	> 1
γ (photon)	1	± 1	2
ν (neutrino)	$\frac{1}{2}$	$\frac{1}{2}$	1

We began this section with the statement that the thermodynamic properties of a gas are described by its temperature T, pressure P, and chemical potential μ'. When a state of thermodynamic equilibrium exists between various particles, they have the same temperature, pressure, and also the same chemical potential. For this reason, the chemical potential allows you to compute the relative abundances of the particles in *equilibrium* reactions. For the triple-alpha process (which you will study in the next section), the stage: $^4_2\text{He} + ^4_2\text{He} \longleftrightarrow ^8_4\text{Be}$ is one such equilibrium reaction. (The double-headed arrow indicates that the reaction can go in both directions.)

> The sum of chemical potentials on one side of the reaction must equal the sum of chemical potentials on the other side.

If they were not equal, the reaction would not be in equilibrium. This allows us to write: $\mu'(^4_2\text{He}) + \mu'(^4_2\text{He}) = \mu'(^8_4\text{Be})$. We will see below how to make use of this result.

- How many polarization states, g_s, do the nuclei (i) ^4_2He and (ii) ^8_4Be possess?
- (i) ^4_2He contains $Z = 2$ protons and $A - Z = 4 - 2 = 2$ neutrons, so is even–even, which signifies zero spin and hence that $g_s = 1$.

  (ii) ^8_4Be contains $Z = 4$ protons and $A - Z = 8 - 4 = 4$ neutrons, so is also even–even, which signifies zero spin and hence that $g_s = 1$.

5.3   Helium burning

In this section we examine helium-burning reactions to see how much energy helium fusion can liberate, and the implications for the lifetime of a helium-burning star.

Helium burning produces both carbon and oxygen, which are two of the elements that are vital for life as we know it. In fact, the elements carbon and oxygen comprise 0.39% and 0.85% respectively of the ordinary, baryonic matter in the Universe and are the most abundant elements after hydrogen and helium.

As noted in the last chapter, hydrogen burning in the core of a star stops when most of the hydrogen is converted into helium. The core then contracts, converting gravitational energy into thermal energy. The temperature and pressure of the core consequently increase, leading to a large expansion of the outer envelope of the star. A shell of hydrogen burning ignites surrounding the core and the star is then a red giant.

In stars whose mass is greater than about $0.5\,\text{M}_\odot$, the core becomes hot and dense enough (i.e. $T_c \sim (1\text{–}2) \times 10^8$ K and $\rho_c \sim (10^5\text{–}10^8)$ kg m^{-3}) for helium burning to begin, via the **triple-alpha process**. There are three steps in this process, as follows:

$$\text{Step 1: } ^4_2\text{He} + ^4_2\text{He} \longleftrightarrow ^8_4\text{Be}$$
$$\text{Step 2: } ^4_2\text{He} + ^8_4\text{Be} \longleftrightarrow ^{12}_6\text{C}^*$$
$$\text{Step 3: } ^{12}_6\text{C}^* \longrightarrow ^{12}_6\text{C} + 2\gamma.$$

Notice that $^{12}_{6}\text{C}^*$ indicates a nucleus of carbon-12 in an excited state. Note also that Steps 1 and 2 of this sequence can proceed in either direction. If they are in equilibrium, we can use the chemical potentials of the reactants to calculate their proportions and the rate of the reaction. We begin by examining Step 1.

5.3.1 The triple-alpha process: Step 1

The important point to note is that the $^{8}_{4}\text{Be}$ nucleus is unstable, since the mass of $^{8}_{4}\text{Be}$ is slightly *greater* than the mass of two $^{4}_{2}\text{He}$ nuclei. In fact the mass–energy difference between a nucleus of beryllium-8 and two helium-4 nuclei is just 91.8 keV and we may write

$$(2m_4 - m_8)c^2 = -91.8 \text{ keV},$$

where m_8 and m_4 are the masses of the beryllium-8 nucleus and the helium-4 nucleus respectively. Therefore 91.8 keV of energy must be *put in* so that this reaction may proceed in the forward direction. Since the beryllium-8 nuclei are unstable, they rapidly break down into helium-4 nuclei with a timescale of $\tau = 2.6 \times 10^{-16}$ s.

However, if two helium-4 nuclei approach each other with a combined kinetic energy of close to 91.8 keV, there is an enhanced probability of them interacting to form an intermediate excited state (a so-called *resonance*) which is close to the ground state of beryllium-8. Under these circumstances, some beryllium-8 nuclei can form and exist for a very brief time.

Exercise 5.1 (a) Write the chemical potential equation for the equilibrium reaction $^{4}_{2}\text{He} + {}^{4}_{2}\text{He} \longleftrightarrow {}^{8}_{4}\text{Be}$.

(b) Using the expression for the chemical potential (Equation 5.1), find the ratio of $^{8}_{4}\text{Be}$ nuclei to $^{4}_{2}\text{He}$ nuclei at $T = 2 \times 10^8$ K and $\rho = 10^8$ kg m^{-3}.
(For this reaction, the energy required to go in the forward direction is $\Delta Q = \Delta m\, c^2 = -91.8$ keV where Δm is the mass difference between $^{8}_{4}\text{Be}$ and $^{4}_{2}\text{He}$.)

Hint 1: Recall that $\log_e a - \log_e b = \log_e(a/b)$, and that $\exp(\log_e x) = x$.

Hint 2: Recall that the number density of helium-4 nuclei, $n_4 = \rho X_4/m_4$, and further assume that the gas is primarily $^{4}_{2}\text{He}$. ∎

Since the relative number of beryllium-8 nuclei is so small, the number density of helium-4 nuclei can be calculated simply as $n_4 \sim \rho/m_4$ where $m_4 \sim 4u$. So at the density considered in Exercise 5.1 we have $n_4 \sim 10^8$ kg m$^{-3}/(4 \times 1.661 \times 10^{-27}$ kg$) \sim 1.5 \times 10^{34}$ m^{-3}. Then in equilibrium at 2×10^8 K, we have the number density of beryllium-8 nuclei as $n_8 \sim 1.5 \times 10^{34}$ m$^{-3}/(2.1 \times 10^7) \sim 7.1 \times 10^{26}$ m^{-3}.

Exercise 5.2 (a) Calculate the Gamow energy for the fusion of two $^{4}_{2}\text{He}$ nuclei, expressing it in both joules and eV.

(b) At what temperature does the energy of the Gamow peak coincide with the 91.8 keV energy deficit of Step 1? ∎

Exercise 5.2 shows that the energy of the Gamow peak for the interaction of two helium-4 nuclei is close to the value of ΔQ in Step 1 in the triple-alpha process when the temperature is just over 10^8 K. Hence, this is roughly the temperature that is required for helium burning. In fact, evaluating the energy of the Gamow peak, and the width of the Gamow window at 10^8 K and at 2×10^8 K yields energies of 83 ± 31 keV and 132 ± 55 keV respectively. These energy ranges both overlap the value of ΔQ for Step 1 in the triple-alpha process, and indicate that temperatures in the approximate range $(1\text{--}2) \times 10^8$ K are required for helium burning to occur.

Moreover, at a temperature of 2×10^8 K, the ratio of helium-4 nuclei to beryllium-8 nuclei was shown to be about 21 million to one. At the lower temperature of 10^8 K, this ratio can be calculated as about 2 billion to one. Hence, although the beryllium-8 nuclei only exist in small numbers, and they only exist for a very short time before decaying into helium-4 nuclei again, they do provide the raw material for Step 2 of the triple-alpha process, which we examine next.

5.3.2   The triple-alpha process: Step 2

The next step in the triple-alpha process requires the existence of an excited state of the carbon-12 nucleus ($^{12}_{6}\text{C}^*$). Although such a state was not known on Earth at the time, in 1946 the British astronomer Sir Fred Hoyle predicted that such a state must exist in order to explain how carbon could be produced in stars. In fact the mass–energy of this excited state of carbon is just 287.7 keV above the combined mass–energy of a helium-4 nucleus plus a beryllium-8 nucleus, and we may write

$$(m_4 + m_8 - m_{12*})c^2 = -287.7 \text{ keV},$$

where m_{12*} is the mass of the excited state of the carbon-12 nucleus. Note that similarly to Step 1, 287.7 keV of energy must be *put in* so that this reaction may proceed in the forward direction. Hence excited carbon-12 nuclei are formed very occasionally and will have a brief existence enabling the final stage of the triple-alpha process to occur.

In an earlier exercise, we saw that by equating the chemical potentials of the first step in the triple-alpha process, the ratio of helium-4 nuclei to beryllium-8 nuclei in equilibrium could be calculated as about 21 million to 1 at a temperature of 2×10^8 K. By following a similar procedure the ratio of beryllium-8 nuclei to excited carbon-12 nuclei in equilibrium at this temperature can be calculated as about 2.2 trillion (2.2×10^{12}) to 1.

● In equilibrium at 2×10^8 K what is the approximate ratio of helium-4 nuclei to excited carbon-12 nuclei?

○ Since there are about 21 million (2.1×10^7) helium-4 nuclei for every beryllium-8 nucleus, and about 2.2 trillion (2.2×10^{12}) beryllium-8 nuclei for every excited carbon-12 nucleus, there are roughly ($2.1 \times 10^7 \times 2.2 \times 10^{12}$) = 4.6×10^{19} helium-4 nuclei for every excited carbon-12 nucleus in equilibrium at this temperature.

● What is the number density of excited carbon-12 nuclei in equilibrium at 2×10^8 K?

○ Since the number density of beryllium-8 nuclei is $n_8 \sim 7.1 \times 10^{26}$ m^{-3}, the number density of excited carbon-12 nuclei is $n_{12*} \sim 7.1 \times 10^{26}$ m$^{-3}/(2.2 \times 10^{12}) \sim 3.2 \times 10^{14}$ m^{-3}.

Exercise 5.3 (a) Calculate the Gamow energy for the fusion of a $^4_2$He nucleus and a $^8_4$Be nucleus.

(b) At what temperature does the energy of the Gamow peak coincide with the 287.8 keV energy deficit of Step 2. ■

The results of Exercises 5.2 and 5.3 show that the temperatures for the fusion of two $^4_2$He nuclei, and for the fusion of a $^4_2$He nucleus and $^8_4$Be nucleus, are around 1×10^8 K and 3×10^8 K respectively. This indicates that helium burning in stars occurs when they attain temperatures around 10^8 K. Recall that the core temperature of the Sun is currently 1.56×10^7 K, so the Sun is much too cool for helium burning to occur at present.

5.3.3 The triple-alpha process: Step 3

Nearly all of the excited carbon-12 nuclei produced in Step 2 of the triple-alpha process almost immediately break down again to beryllium-8 and helium-4. However, about one in 2500 of these nuclei decay to the ground state of carbon-12 via Step 3 of the triple-alpha process. In this decay, an energy of 7.65 MeV is released, carried away by gamma-rays. The timescale for Step 3 is $\tau^* \sim 1.8 \times 10^{-16}$ s. This reaction does not significantly change the two equilibrium reactions discussed above, as the number of nuclei involved is relatively small.

The rate of production of carbon-12 nuclei in the ground state is given by

$$\frac{\mathrm{d}n_{12\mathrm{C}}}{\mathrm{d}t} = \frac{n_{12*}}{\tau^*}. \tag{5.3}$$

Since the number density of excited carbon-12 nuclei at a temperature of $T = 2 \times 10^8$ K and a density of $\rho \sim 10^8$ kg m^{-3} is $n_{12*} \sim 3.2 \times 10^{14}$ m^{-3}, the rate of production for carbon-12 nuclei under these conditions is

$$\frac{\mathrm{d}n_{12\mathrm{C}}}{\mathrm{d}t} \sim \frac{3.2 \times 10^{14} \text{ m}^{-3}}{1.8 \times 10^{-16} \text{ s}} \sim 1.8 \times 10^{30} \text{ m}^{-3} \text{ s}^{-1}.$$

The energy production rate of this process is simply

$$\varepsilon = \Delta Q^* \times \frac{\mathrm{d}n_{12\mathrm{C}}}{\mathrm{d}t}.$$

Since the energy released in Step 3 is $\Delta Q^* = 7.65$ MeV or $(7.65 \times 10^6 \text{ eV}) \times (1.602 \times 10^{-19} \text{ J eV}^{-1}) = 1.23 \times 10^{-12}$ J, the energy production rate is

$$\varepsilon = 1.23 \times 10^{-12} \text{ J} \times 1.8 \times 10^{30} \text{ m}^{-3} \text{ s}^{-1} = 2.2 \times 10^{18} \text{ W m}^{-3}.$$

● How does this energy production rate compare to the value in the core of the Sun, due to the p–p chain?

○ The energy production rate in the Sun's core was calculated as 300 W m^{-3} at the end of Chapter 2. The energy production rate due to helium burning is sixteen orders of magnitude greater!

5.3.4   The energy released by the triple-alpha process

We can calculate the overall mass defect for the triple-alpha process in the same way as for the p–p chain.

Exercise 5.4   (a) Calculate the mass defect for the triple-alpha process in atomic mass units, ignoring the electron contributions to the mass (since, unlike the p–p chain, there are no beta-decays which change the number of electrons present).

(b) Convert the mass defect into an energy value in MeV, and show that this is in agreement with the values of ΔQ for each step in the triple-alpha process mentioned in the previous subsections.

(c) Calculate the mass defect as a fraction of the initial atomic mass. Compare the fractional mass defect of the triple-alpha process to that for hydrogen burning.

Hint: The *atomic* mass of $^4_2$He is 4.002 60 amu, and for $^{12}_6$C it is exactly 12 by definition. ■

Hydrogen burning converts 0.0066 of a star's burnable (core) hydrogen into energy. Exercise 5.4 above shows that for the triple-alpha process, however, the mass fraction converted into energy is a factor of ten lower, 0.000 65. One immediate implication of this is that helium burning liberates at most one-tenth of the energy that hydrogen burning does. If a star had the same luminosity during both phases of evolution, its helium-burning lifetime would therefore be a factor of ten shorter (since it would have to burn ten times as much fuel in the same time interval in order to have the same power output). In fact, the luminosity during helium burning is even higher, which shortens the helium-burning lifetime further. This indicates why the hydrogen-burning phase, on the main sequence, is longer lasting than the helium-burning phase of stellar evolution. In fact, it is longer than any other phase of stellar evolution.

The temperature-dependence of the energy generation rate for the triple-alpha process turns out to be roughly $\varepsilon \propto T^{30}$! This creates an extremely steep temperature gradient within the helium-burning core.

● How does this temperature-dependence compare with those of the p–p chain and the CNO cycle?

○ Their temperature-dependences are roughly $\varepsilon \propto T^4$ and $\varepsilon \propto T^{20}$ respectively,

5.3.5   Further helium-burning reactions

When stars undergo helium burning, the reaction

$$^4_2\text{He} + {}^{12}_6\text{C} \longrightarrow {}^{16}_8\text{O} + \gamma \tag{5.4}$$

is also possible. This converts pre-existing and freshly synthesized carbon-12 to oxygen-16. However, there are no resonances near to this fusion window to aid in the production of oxygen, and moreover the nuclear S-factor for this reaction is not well known, although it is small (probably around 0.3 MeV barns). Nevertheless it determines how much of the carbon-12 produced by the triple-alpha process remains as carbon and how much is burnt to oxygen. The C/O ratio that results in the core of a helium-burning star greatly influences its future evolution. It determines the luminosity of runaway thermonuclear burning in the white dwarfs involved in type Ia supernovae, which are used as standard candles by cosmologists. Consequently the uncertainty in this rate is one of the most important long-standing uncertainties in stellar astrophysics.

Following Equation 5.4, there is a further reaction possible to create neon-20, namely

$$\ce{^{4}_{2}He} + \ce{^{16}_{8}O} \longrightarrow \ce{^{20}_{10}Ne} + \gamma. \tag{5.5}$$

However, this reaction happens very rarely, due to its very high Coulomb barrier. So, helium burning essentially consists of the triple-alpha process and the production of oxygen-16 only. The production of further nuclei by the addition of even more helium-4 nuclei (such as neon-20, magnesium-24, silicon-28, etc) has an extremely low rate of occurrence during helium burning.

Note also that helium burning bypasses the stable nuclei with atomic numbers 3, 4 and 5 between helium and carbon (i.e. lithium-6 or 7, beryllium-9 and boron-10 or 11). These elements are therefore *not* produced as a result of stellar nucleosynthesis. Any nuclei of these elements found in the interstellar medium are likely to have been produced by the interaction of cosmic rays (high energy protons probably generated in supernovae) with other nuclei.

5.4 Quantum states and degeneracy

As a star on the red-giant branch burns hydrogen in a shell, its contracting core continues to increase in density and pressure. As you have already seen, once the internal temperature reaches $\sim 10^8$ K, helium fusion becomes possible. However, before we discuss helium burning further, we need to understand the condition of the stellar material when it reaches that temperature.

We begin by discussing the quantum-mechanical wave properties of the particles that constitute a gas. The volume of space containing the gas may be considered to be occupied by particles whose wave properties give rise to standing waves fitting neatly within that volume. They therefore have quantized values of wavelength, and therefore also of energy and momentum. This quantization means that there is a discrete rather than continuous distribution of possible energies. The number of possible quantum states available to particles below some energy is finite; often it is very large, but nevertheless finite. In the very dense interiors of some astronomical objects, the finite number of available quantum states, though large, can be insufficient for the huge number of particles squeezed into the confined volume. Under these circumstances physics then takes new twists, giving rise to some remarkable properties.

As you saw in Chapter 2, when we derived the minimum mass at which fusion could occur in the core of a star, the critical point of having too few available quantum states for the number of particles present is reached when the separations of the gas particles become smaller than their de Broglie wavelengths. At that point, the wave properties dominate over the particle properties of the gas, and a quantum gas is said to exist.

The de Broglie wavelength of a particle of momentum p is $\lambda_{\mathrm{dB}} = h/p$. For non-relativistic particles, the (thermal) kinetic energy is $3kT/2$. We can therefore write $mv^2/2 = 3kT/2$, i.e. $p^2/(2m) = 3kT/2$, where $p = mv$. Hence $p = (3mkT)^{1/2}$. The de Broglie wavelength of non-relativistic particles is therefore $\lambda_{dB} = h/(3mkT)^{1/2}$.

Exercise 5.5 (a) Calculate the de Broglie wavelength of (i) a proton and (ii) an electron, in the core of the Sun. (Use values from Table 1.1.)

(b) Which of (i) and (ii) has the larger de Broglie wavelength, and by what factor?

(c) How does this ratio vary from star to star? ∎

Since the de Broglie wavelength of an electron is ≈ 40 times longer than that of a nucleon, electrons run out of space and hence reach the quantum limit corresponding to degeneracy sooner than nucleons as the stellar core contracts and its density increases.

Degeneracy can be described in two equivalent ways:

- a condition in which the separation between identical particles is less than their de Broglie wavelength;

- a condition in which the number of particles per unit volume, n, is higher than the number of available quantum states n_Q at their energy, where n_Q is the quantum concentration. Under non-relativistic (NR) conditions, $n_{QNR} = (2\pi mkT/h^2)^{3/2}$.

Before going further, complete the following exercise to convince yourself that these first two conditions *are* equivalent.

Exercise 5.6 Beginning with the second statement of the condition for degeneracy, $n \gg n_Q$, use the definition of the quantum concentration $n_Q = (2\pi mkT/h^2)^{3/2}$ to derive an expression for the mean separation l of the particles in terms of the de Broglie wavelength $\lambda_{dB} = h/p \approx h/(3mkT)^{1/2}$. Show that this leads to the first statement for the condition of degeneracy.

Hint: If the mean separation of particles is l, then the number of particles in a volume $l^3 = 1$. That is, the number of particles per unit volume is $n = 1/l^3$. ∎

You have now shown that there are two expressions for degeneracy that are actually equivalent. That is, they are merely alternative expressions of the same physics. In the following exercise you will derive a third equivalent expression.

Exercise 5.7 (a) Rewrite the second description of degeneracy, $n \gg n_Q$, using the definition of the quantum concentration $n_Q = (2\pi mkT/h^2)^{3/2}$, to find a limit for the thermal energy kT. State a third equivalent condition for degeneracy as a limit on the temperature.

(b) Calculate values of the quantity $h^2 n^{2/3}/(2\pi mk)$ for (i) protons and (ii) electrons in the solar core. Assume the mass fraction of hydrogen $X_H = 0.5$, and the mass fraction of helium $X_{He} = 0.5$. For other values in the solar core, use Table 1.1.

Hint 1: In the solar core, all atoms are fully ionized.

Hint 2: Recall that $n_H = \rho X_H / m_H$.

(c) Consider the results to part (b), and state whether (i) protons and (ii) electrons are degenerate in the core of the Sun. (Use values as required from Table 1.1.) ■

You will have found in the previous Exercise that a third equivalent expression for the condition of degeneracy can be written:

- a condition in which the temperature T of the particles is much less than the value $h^2 n^{2/3} / (2\pi m k)$.

Sometimes this third condition is used to describe a degenerate gas as *cold*, because its temperature falls below some limit. However, this can be misleading, because electrons may become degenerate at temperatures of millions of kelvin, which is not cold in the common usage of the word.

5.5 Electron degeneracy

The state of the electrons in the core of the star, in particular whether they are degenerate or non-degenerate, plays a major role in determining what happens to a star when helium ignites. For this reason, the current section considers degeneracy in the context of stellar cores.

When hydrogen burning ends, interior pressure support tends to diminish so gravity makes the core contract. Hence the particle density in the core increases, since $n \propto R^{-3}$. The temperature in the core also increases since $T \propto 1/R$. This in turn drives up the quantum concentration, which depends on the temperature as $n_{QNR} \propto T^{3/2} \propto R^{-3/2}$. However, this is not fast enough to keep pace with the increasing particle density. Eventually the density of particles catches up with the quantum concentration, and hence the matter becomes electron-degenerate.

Electrons, along with protons and neutrons, are **fermions**, which means they have spin $= \pm 1/2$. The occupation of quantum states by fermions is restricted by the **Pauli exclusion principle**, which says that no more than one identical fermion can occupy a given quantum state. The *only* way to distinguish one fermion from another particle of the same type (e.g. to distinguish one electron from another) is its spin, and since this has only two possible values, $+1/2$ or $-1/2$, at most two fermions of opposite spin can occupy a given quantum state.

The *average* number of *identical* fermions in a state of energy E_p is given by

$$f(E_p) = \frac{1}{\exp\left[(E_p - \mu')/kT\right] + 1} \tag{5.6}$$

where μ' is the chemical potential. By this definition, $0 \leq f(E_p) \leq 1$. If the gas had a temperature of 0 K, the term $\left[(E_p - \mu')/kT\right]$ would take values $+\infty$ if $E_p > \mu'$ or $-\infty$ if $E_p < \mu'$, and the term $f(E_p)$ would take values 0 and 1 respectively. That is, at 0 K, all quantum states with energy $E_p < \mu'$ would be filled, and all quantum states with energy $E_p > \mu'$ would be empty. In a cold electron gas, the energy of the most energetic, degenerate electron is called the

Fermi energy, E_F, and the momentum of particles with this energy is called the Fermi momentum, p_F.

In a degenerate gas, all quantum states up to those with momentum p_F are occupied, and all those above this momentum are unoccupied. The density of states is the number of quantum states dN whose momentum lies within a certain range between p and $p + dp$, which is given by:

$$dN = g_s \frac{V}{h^3} 4\pi p^2 \, dp.$$

We can find the total number of degenerate electrons in the gas by adding up (i.e. by integrating over) these occupation numbers for all degenerate momentum values. Since no degenerate electrons have a momentum exceeding the Fermi momentum, it is sufficient to integrate from $p = 0$ to $p = p_F$, which gives:

$$N_e = \int_{p=0}^{p=p_F} dN_e = \int_{p=0}^{p=p_F} g_s \frac{V}{h^3} 4\pi p^2 \, dp.$$

This is evaluated simply as

$$N_e = \frac{4\pi g_s V}{3h^3} p_F^3. \tag{5.7}$$

Since the number density of degenerate electrons is simply $n_e = N_e/V$, and there are two spin states for electrons ($g_s = 2$) this equation can be rearranged as:

$$p_F = \left(\frac{3n_e}{8\pi} \right)^{1/3} h. \tag{5.8}$$

Similarly, the total internal energy of the gas is the sum of the energies E_p of each particle:

$$E = \int_{p=0}^{p=p_F} E_p \, dN = \int_{p=0}^{p=p_F} E_p \, g_s \frac{V}{h^3} 4\pi p^2 \, dp. \tag{5.9}$$

(The particle energy E_p appears *inside* the integral because it depends on momentum, which is the variable of integration.)

5.5.1   Non-relativistic degenerate electrons

The total energy of a particle is given by the sum of its kinetic energy and rest-mass energy. In the case of non-relativistic electrons we can write $E_p \approx \frac{1}{2}m_e v^2 + m_e c^2$ and since in this case the momentum is $p \approx m_e v$, we have $E_p \approx p^2/2m_e + m_e c^2$. Substituting this into the equation for the total internal energy of the gas (Equation 5.9) we have,

$$E = \int_{p=0}^{p=p_F} \left(\frac{p^2}{2m_e} + m_e c^2 \right) g_s \frac{V}{h^3} 4\pi p^2 \, dp$$

$$= \frac{4\pi g_s V}{h^3} \int_{p=0}^{p=p_F} \left(\frac{p^4}{2m_e} + m_e c^2 p^2 \right) dp$$

which can be integrated to give

$$E = \frac{4\pi g_s V}{h^3} \left[\frac{p_F^5}{10 m_e} + \frac{m_e c^2 p_F^3}{3} \right]$$

$$= \frac{4\pi g_s V p_F^3}{h^3} \left[\frac{p_F^2}{10 m_e} + \frac{m_e c^2}{3} \right].$$

Now, since the total number of degenerate electrons is given by Equation 5.7, this may be re-written as

$$E = 3N_e \left[\frac{p_F^2}{10 m_e} + \frac{m_e c^2}{3} \right]$$

$$= N_e \left[\frac{3 p_F^2}{10 m_e} + m_e c^2 \right]. \tag{5.10}$$

The second term in the bracket above is simply the rest-mass energy per particle, so the first term must be the kinetic energy per particle: $3 p_F^2 / 10 m_e$. Furthermore, the kinetic energy per unit volume is the kinetic energy per particle multiplied by the number of particles per unit volume. So the kinetic energy per unit volume is $3 p_F^2 n_e / 10 m_e$.

Now, in Chapter 2 you saw that the pressure P_{NR} provided by non-relativistic particles is 2/3 of the kinetic energy per unit volume (Equation 2.7), so

$$P_{NR} = \frac{2}{3} \times \frac{3 p_F^2 n_e}{10 m_e} = \frac{n_e p_F^2}{5 m_e}.$$

Then using Equation 5.8, we have

$$P_{NR} = \frac{n_e}{5 m_e} \times \left(\frac{3 n_e}{8\pi} \right)^{2/3} h^2 = \frac{h^2}{5 m_e} \left[\frac{3}{8\pi} \right]^{2/3} n_e^{5/3}.$$

So the equation of state for a gas of non-relativistic degenerate electrons has the form

$$P_{NR} = K_{NR} n_e^{5/3}, \tag{5.11}$$

where K_{NR} is a constant given by the previous expression.

5.5.2 Ultra-relativistic degenerate electrons

We can also develop a similar expression for the equation of state for ultra-relativistic particles. In the case of ultra-relativistic electrons, the energy per particle is given by the relativistic energy equation as $E_p^2 = p^2 c^2 + (mc^2)^2$. Since the kinetic energy will be much greater than the rest-mass energy, $E_p^2 \approx p^2 c^2$ and so $E_p \approx pc$. Substituting this into the equation for the total internal energy of the gas (Equation 5.9) we have,

$$E = \int_{p=0}^{p=p_F} pc\, g_s \frac{V}{h^3} 4\pi p^2 \,\mathrm{d}p$$

$$= \frac{4\pi\, g_s\, cV}{h^3} \int_{p=0}^{p=p_F} p^3 \,\mathrm{d}p,$$

which can be integrated to give

$$E = \frac{4\pi g_s c V}{h^3} \frac{p_F^4}{4}.$$

Now, since the total number of degenerate electrons is given by Equation 5.7, this may be re-written as

$$E = \frac{3}{4} N_e p_F c. \tag{5.12}$$

Since we have already noted that in this case the rest-mass energy is negligible, the kinetic energy per particle is $3p_F c/4$. Furthermore, the kinetic energy per unit volume is the kinetic energy per particle multiplied by the number of particles per unit volume. So the kinetic energy per unit volume is $3p_F n_e c/4$.

Now, in Chapter 2 you saw that the pressure P_{UR} provided by ultra-relativistic particles is 1/3 of the kinetic energy per unit volume (Equation 2.8), so

$$P_{UR} = \frac{1}{3} \times \frac{3p_F n_e c}{4} = \frac{n_e p_F c}{4}.$$

Then using Equation 5.8, we have

$$P_{UR} = \frac{n_e c}{4} \times \left(\frac{3n_e}{8\pi}\right)^{1/3} h = \frac{hc}{4} \left[\frac{3}{8\pi}\right]^{1/3} n_e^{4/3}.$$

So the equation of state for a gas of ultra-relativistic degenerate electrons has the form

$$P_{UR} = K_{UR} n_e^{4/3}, \tag{5.13}$$

where K_{UR} is a constant given by the previous expression.

Notice that the equations of state for degenerate electrons, whether non-relativistic or ultra-relativistic (or indeed somewhere in between), show that the pressure is *independent* of temperature and depends only on the number density of electrons in each case. In both non-relativistic and ultra-relativistic cases, it is the independence of the pressure from the temperature that helps to give a degenerate gas its interesting properties.

● What is the key difference between the equation of state for a *non*-relativistic electron-degenerate gas compared to an *ultra*-relativistic electron-degenerate gas?

○ The pressure of an ultra-relativistic electron-degenerate gas has a weaker dependence on density, so as the density increases, the pressure in an ultra-relativistic gas increases less markedly.

5.6   The helium flash

Recall from earlier that the minimum mass for the onset of hydrogen burning in the p–p chain is about $0.08 \, M_\odot$; the cores of lower-mass objects do not reach temperatures required to initiate the reaction. Helium burning also has a mass threshold, linked to the higher temperature (10^8 K) required to initiate it. Only stars with $M \geq 0.5 \, M_\odot$ achieve this threshold. Stars below the $0.5 \, M_\odot$ mass limit end their lives with inert helium cores.

- What is the number density of electrons in the core of the Sun?
- According to Exercise 5.7(b), the number density in the core of the Sun is $n_e = 6.65 \times 10^{31}$ m^{-3}.
- What is the quantum concentration in the core of the Sun?
- The quantum concentration is $n_{QNR} = (2\pi m_e kT/h^2)^{3/2}$. At the solar core temperature of 15.6×10^6 K,

$$n_{QNR} = \left(\frac{2\pi \times 9.109 \times 10^{-31}\,\text{kg} \times 1.381 \times 10^{-23}\,\text{J K}^{-1} \times 15.6 \times 10^6\,\text{K}}{(6.626 \times 10^{-34}\,\text{J s})^2} \right)^{3/2}$$
$$= 1.49 \times 10^{32}\,\text{m}^{-3}.$$

In the Sun, the number density of electrons in the core (n_e) is less than the quantum concentration (n_{QNR}), but only by a factor of about 2. So the core of the Sun behaves essentially as a classical (non-degenerate) gas, although it is not that far from being degenerate.

However, during hydrogen burning as the helium core grows, the number density of electrons will eventually exceed the quantum concentration when the Sun becomes a red giant. When the core is at a temperature of about 10^8 K, it will behave as a cold degenerate electron gas, even though it is hot enough to ignite helium fusion.

When helium fusion begins, it will release an enormous amount of energy into the core. This energy goes partly into increasing the temperature of the He-burning region further, and partly into driving an expansion of the core. If the core were a classical gas, this huge dump of energy into the core would cause the gas to expand greatly, and the increasing volume would lower both the density and the temperature, so reducing the fusion rate. However, the decoupling of temperature from pressure in an electron-degenerate gas means that increasing the temperature of the electrons will not on its own drive an expansion. Of course, the core is also composed of non-degenerate protons and neutrons. (Recall that the de Broglie wavelength of a particle depends inversely on its mass, so protons and neutrons require much higher densities before they become degenerate.) Furthermore, the condition for degeneracy may not be met in full by all electrons throughout the core. The temperature increase of the non-degenerate atomic nuclei and the fraction of electrons that are not fully degenerate does result in an increase in pressure, but this is a relatively small contribution to the total pressure, which is dominated by the pressure of degenerate electrons. Consequently, the expansion of the core is not sufficient (initially at least) to compensate for the temperature increase, and the fusion rate increases further. This consequently raises the temperature further still, and in turn increases the fusion rate. The runaway condition this produces is called the **helium flash**. The core luminosity increases tremendously, to perhaps 1 million times the luminosity of the hydrogen-burning shell (which lies further out in the star). Conditions in the core change significantly on timescales of days when the flash peaks.

Calculations of the evolution of the star during the flash are quite sensitive to assumptions about the role of convection and other processes in removing energy from the fusion zone. Were it not for energy transport by convection, the runaway reactions in the core would be complete and result in a thermonuclear explosion. Several textbooks suggest that the rise in temperature ultimately removes the degeneracy of the electrons — see Section 5.4 to understand why this is logical, if perhaps not entirely factual. Stellar evolution calculations seem to show that indeed much of the energy released in the helium flash goes into the mechanical work of expansion of the core. The surface of the star does not brighten as a result of helium ignition. Ultimately, the core density falls sufficiently that the electrons are no longer degenerate, and the star settles down to a state in which helium fusion continues in a more controlled way in a less dense core that behaves like a classical gas once more, possibly after it goes through four or five He flashes of diminishing luminosity.

More-massive stars have a lower density for a given temperature, so the temperature for helium ignition will be reached in more-massive stars at lower density. Stars more massive than about $2.5\,M_\odot$ ignite helium *before* they reach the density required for electron degeneracy. For these stars, in contrast to the lower mass ones like the Sun, helium ignition is controlled rather than explosive.

● To recap, what is the essential difference between the conditions under which helium ignites in a $1\,M_\odot$ star and a $5\,M_\odot$ star? What is the consequence?

○ A $1\,M_\odot$ star will be electron-degenerate in its core when T_c reaches $\sim 10^8\,\mathrm{K}$, whereas the core of a $5\,M_\odot$ star will not be degenerate when this temperature is reached. When nuclear fusion begins in a degenerate gas, where the pressure no longer depends on the temperature, the temperature will increase without triggering an expansion of the core to cool it, so a thermonuclear runaway occurs. This is the *helium flash*.

5.7   Core-helium burning stars

Once helium ignites, stars experience a new period of steady burning. We consider the phase of core-helium burning stars in this section.

The helium flash of a low-mass star *terminates* its evolution up the red-giant branch, and it begins a new stage of life as a low-mass, core-helium burning star. The electron degeneracy is removed at the same time, due to the expansion and heating of the core, so the new phase is again stable, with pressure and temperature recoupled through the equation of state of an ideal gas. This recoupling leads to an expansion of the core and also of the hydrogen-burning shell further out in the star. The expansion of the hydrogen-burning shell reduces its temperature, density, and hence its energy generation. In some mass ranges, hydrogen fusion ceases. The outer hydrogen-rich envelope of the star reacts to this temperature decrease by contracting. Since $L = 4\pi R^2 \sigma T_{\mathrm{eff}}^4$, the contraction would lead to an increase in the effective surface temperature if the luminosity were constant. However, the luminosity of the star is also lower, so the increase in temperature is less pronounced. Changes in luminosity and effective surface temperature do, of course, shift the star in the H–R diagram. The star's new stable location is partway down and slightly towards the left of the giant branch. Because the new configuration of stable core-helium burning and shell-hydrogen burning is reasonably long lived, stars remain in this region of the H–R diagram for some time. The location is recognized observationally as a clump of stars in the H–R diagram, due to the slow evolution through this phase of life, and they are called **red-giant clump stars** (Figure 5.1).

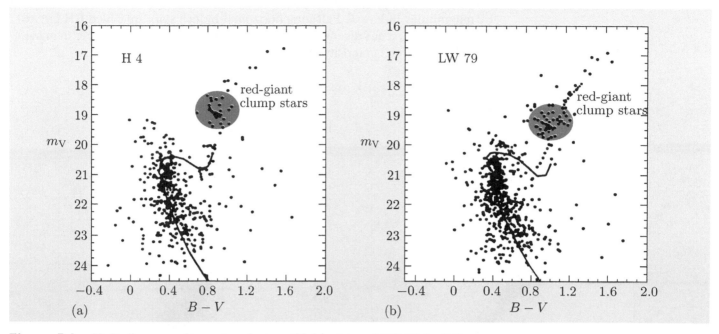

Figure 5.1 H–R diagram of two star clusters (H 4 in (a) and LW 79 in (b)) showing a well-populated main sequence and a red-giant branch with distinct core-helium burning red-giant clump stars at $m_V \approx 19$ and $B - V \approx 0.8 - 1.0$.

It is important to note that the location of low-mass, core-helium burning, shell-hydrogen burning stars depends a great deal on their heavy-element content. (Until now, we have discussed the impact of a growing helium abundance in hydrogen-burning stars, but have ignored the trace ($\leq 2\%$ by mass) elements heavier than helium, which astronomers refer to collectively as *metals*.) Stars with a similar metal composition to the Sun will sit close to the red-giant branch as the red-giant clump stars. However, stars with *lower* metal content have higher temperatures so their location in the H–R diagram is called the **horizontal branch** (HB in Figure 5.2 overleaf).

The position of stars on the horizontal branch also depends on their mass, lower-mass stars being hotter. Stellar evolution models show that horizontal-branch stars must have lost a significant fraction ($\approx 25\%$) of their initial mass; stars of low metallicity which had main-sequence masses $M_{ms} \approx 0.8\,\mathrm{M_\odot}$ apparently lose up to $\approx 0.2\,\mathrm{M_\odot}$ of their envelope soon after helium ignition. Moreover, they must lose *different* amounts from one star to the next, as horizontal-branch stars in a metal-poor cluster (where all stars have the *same* metallicity) span a range of temperatures, the lowest mass stars being hottest. This mass loss process is poorly understood.

Recall from the discussion of $M \approx 2\,\mathrm{M_\odot}$ hydrogen-burning stars leaving the main sequence and going through the δ Scuti phase, that the instability strip for stellar pulsation cuts roughly vertically downwards through the H–R diagram (see Figure 4.8). The instability strip therefore cuts across the horizontal branch, with the result that a significant fraction of horizontal-branch stars pulsate. In globular clusters, the intersection of the instability strip with the horizontal branch is at about the same colour (effective surface temperature) as the main-sequence turnoff, and causes brightness variations of order $\Delta V \approx \pm 0.5$ magnitudes (as well as colour variations $\Delta V - I \approx \pm 0.10$), for stars over an interval of perhaps

0.2 magnitudes in $V - I$. Pulsating horizontal-branch stars are called **RR Lyrae** stars after the first one discovered. Their periods, ≈ 0.5 days, are longer than the δ Scuti stars, whose luminosity is lower.

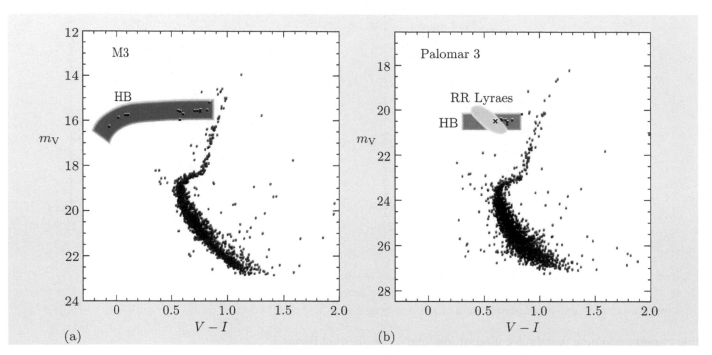

Figure 5.2   H–R diagrams of the globular clusters (a) M3 and (b) Palomar 3, showing the horizontal branches (HB) at $m_V \approx 15.5$ and 20.5 respectively, which are populated by core-helium burning stars with a low metal content. These clusters have ages around 13×10^9 years, and their metal content is only $\approx 1/30$ of the solar value. Note that whereas the HB of Palomar 3 is quite short, that of M3 has a blue extension to $V - I \approx 0$. The instability strip intersects the HB almost directly above the main-sequence turnoff, giving rise to pulsating stars called RR Lyrae variables. Their approximate location, and typical brightness and colour variations are indicated for Palomar 3 along with one confirmed RR Lyrae star (marked by a very small cross). These H–R diagrams are for only a small field within each cluster; many more cluster stars could be observed and added to these diagrams.

RR Lyrae stars are of immense importance in observational astronomy. Because horizontal-branch stars have the same intrinsic luminosity once any light variation is allowed for – they are, after all, *horizontal*-branch stars – they are excellent standard candles for field (i.e. non-cluster) stars. RR Lyrae stars are instantly identifiable from their light curves, which have distinctive periods, amplitudes, and shapes (Figure 5.3). Once identified, their inferred *intrinsic* luminosities can be compared with their apparent brightness, and their distances computed.

- ● Why are RR Lyrae variables used as standard candles while non-pulsating HB stars are not?

- ○ RR Lyrae variables are recognizable from their light curves, whereas other HB stars could be mistaken for nearby main-sequence stars or distant supergiants.

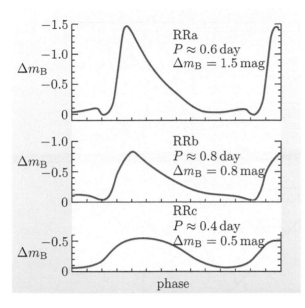

Figure 5.3 Typical light-curves (in blue light) for three classes of RR Lyrae variable star.

The terms *clump red giant* and *red-giant clump star* are interchangeable.

Whereas low-mass core-helium burning stars occupy the region of clump red giants or the horizontal branch depending on their metal content, higher-mass stars start performing **blue loops** in the H–R diagram once helium burning begins. Because these stars are non-degenerate in their cores, helium ignition leads to an expansion of the core and the hydrogen-burning shell. The temperature at the base of the outer (hydrogen) envelope therefore decreases, and the envelope contracts, decreasing the radius and driving up the effective surface temperature. That is, the star becomes bluer, the extent of the blueward variation being greater for higher-mass stars; see Figure 5.4 overleaf. (This is the reverse of the mass distribution along the horizontal branch for low-mass, low-metallicity stars.) The extent of the blue loop also depends on the chemical composition of the star.

● Summarize the main similarities and differences between clump red giants, horizontal-branch stars, and RR Lyrae variable stars.

○ All three classes are burning helium in their cores. Horizontal-branch stars have a lower metal content than clump red giants, and so are hotter. RR Lyraes are a subset of horizontal-branch stars, where the instability strip intersects the horizontal branch. Since horizontal-branch stars have a known luminosity, and RR Lyraes are easily recognized, they are valuable standard candles for measuring distances.

5.8 Helium-burning in a shell

In the same way that core-hydrogen burning is followed by shell-hydrogen burning, so core-helium burning is followed by shell-helium burning. This second stage of giant-star evolution is explained in this section.

Whereas the temperature-dependence of energy generation for the p–p chain is $\varepsilon \propto T^4$ and for the CNO cycle goes as roughly T^{20}, for the triple-alpha process the dependence is T^{30}! This makes energy production in the core very centrally concentrated. The helium-burning core soon becomes a helium-burning shell surrounding a carbon and oxygen core (often written 'CO'), outside which the

Figure 5.4 Helium-burning evolutionary tracks, following on from the hydrogen-burning phase shown in Figure 4.5. The red lines show the hydrogen-burning phase, and the blue lines show the helium-burning phase. In stars with $M \leq 2.5\,M_\odot$, helium is ignited in a degenerate core, and the star moves back down from the top of its red-giant track to the red-giant clump halfway up the track. In stars with $2.5\,M_\odot < M \leq 9\,M_\odot$, the core is not degenerate when helium ignites, and these stars move blueward (i.e. to the left) in the H–R diagram, but move redward again (i.e. to the right) once core-helium burning is replaced by shell-helium burning. This path is called a *blue loop*. In stars with $M > 9\,M_\odot$, helium ignites when the star is still close to the main sequence and does not interrupt its redward evolution.

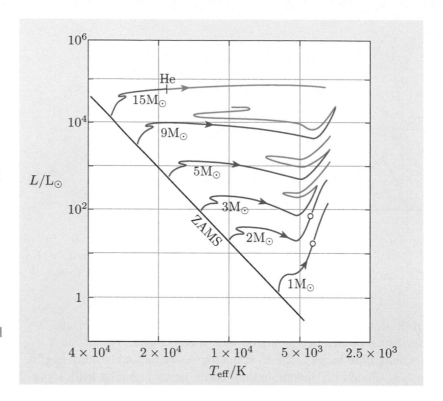

hydrogen-burning shell still burns. The hydrogen-burning shell now burns closer towards the surface, which causes the hydrogen envelope to expand and cool once more. See Figures 5.5 and 5.6.

Low-mass stars ($M \leq 3.5\,M_\odot$) increase in luminosity again as convection again becomes important in energy transport, and make a slow return up the giant branch, the return leg being slightly more luminous than for lower-mass first-ascent red giants, thus delineating a separate **asymptotic giant branch** (AGB), as shown in Figure 5.5a. The AGB is so named because it approaches the earlier red-giant branch, but does not overlay it.

Intermediate-mass stars ($3.5\,M_\odot < M \leq 8\,M_\odot$) reverse their blue loops. Broadly speaking, stars move back towards their helium-ignition points in the H–R diagram, as shown in Figure 5.5b. The envelope expansion causes the envelope to become more convective, just as it did when the star moved from core-hydrogen burning on the main sequence to shell-hydrogen burning on the giant branch.

To distinguish the two phases of giant-branch evolution, the first being from the main-sequence turnoff to the helium ignition point, and the second being the AGB, the former is sometimes referred to as the **first-ascent giant branch**. In AGB stars with $M > (3.5\text{--}4)\,M_\odot$ the convection reaches deep enough to effect another dredge-up episode, called the second dredge-up in recognition of its similarity to the event that occurs on the first ascent of the giant branch. Energy generation during the first portion of the AGB – the early-AGB or **E-AGB** – is dominated by the helium-burning shell. (The hydrogen-burning shell may have extinguished temporarily.) Later in this phase, the hydrogen shell dominates, and the fresh production of helium causes the helium-burning shell periodically to burst into life again. Thermal instabilities in the helium shell drive thermal

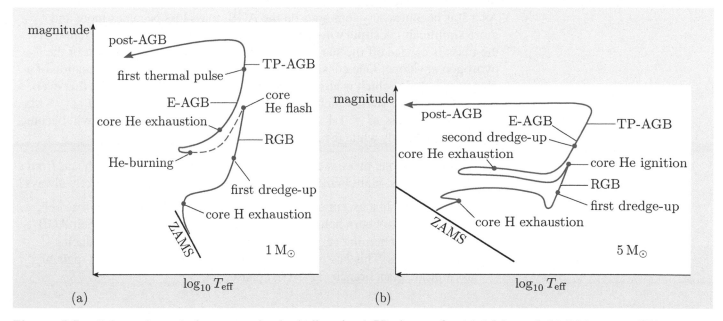

Figure 5.5 Schematic evolutionary tracks, including the AGB phases, for (a) $1\,M_\odot$ and (b) $5\,M_\odot$ stars. The extent of the blueward evolution a star during helium burning depends on its metallicity and the extent of mass loss. The AGB phases are shown divided into the early AGB (E-AGB) and the later thermally pulsing AGB (TP-AGB). The superwind develops as the star reaches the highest luminosities, sending the star into the post-AGB phase.

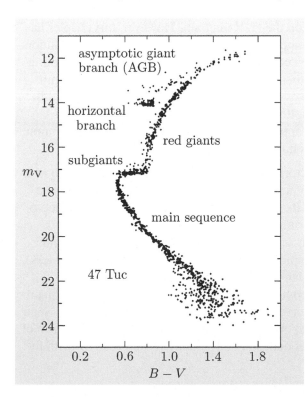

Figure 5.6 Observational H–R diagram of the old globular cluster 47 Tuc, showing a clear asymptotic giant branch (AGB) occupied by shell-helium burning stars making their way from the horizontal branch toward the top of the giant branch.

pulses; this later phase is called the thermally pulsing **TP-AGB**. Several tens of pulses, each lasting only $\sim 10^2$ years, may occur for a star at intervals of $\sim 10^4$ years, and it is during these pulses that conditions appear suitable for *s-process nucleosynthesis*. This process, which you will study in Chapter 6, is responsible for the production of many of the elements heavier than iron.

As a star becomes very luminous on the AGB, mass loss becomes more and more significant. A **superwind** may develop in which a substantial fraction of the mass is carried off the star. In some cases this corresponds to most of the hydrogen envelope! One consequence of extreme mass loss is the formation of a planetary nebula, which is also a topic of Chapter 6. Another effect is that even intermediate-mass stars, which had main-sequence masses $3\,M_\odot \leq M_{ms} \leq 8\,M_\odot$, end up with a mass $M \leq 1.4\,M_\odot$. The significance of this final mass will become clear once we study white dwarfs in Chapter 6.

● What underlying process distinguishes an asymptotic-giant-branch star from a first-ascent red-giant branch star? Does this distinction hold perfectly always?

○ Stars on their first ascent of the giant branch burn only hydrogen in a shell, while AGB stars burn helium in a shell as well. For stars *late* in their AGB phase, helium burning periodically diminishes and re-ignites, producing thermal pulses. This late phase of evolution is called the thermally-pulsing asymptotic giant branch (TP-AGB) phase.

5.9   Other factors influencing stellar evolution

It is clear from the preceding chapters that the mass of a star has a major influence on its evolution. As a first-order approximation, one might imagine that the mass of a star is solely responsible for the evolutionary path it takes throughout its life, and the consequent track it traces on the H–R diagram. However, this chapter has reminded you that the chemical composition of a star also influences its evolution. In Section 5.7, you saw that whilst core-helium burning, shell-hydrogen burning stars with a metallicity similar to that of the Sun will lie in the red-giant clump, stars in this same phase of their lives but with a lower metallicity content will lie further to the left (i.e. bluer and hotter) on the HR diagram in the horizontal branch. Hence, as a second-order approximation one might say that the evolution of a star depends on its mass *and* on its metallicity.

However, there are two further factors which can also influence the evolution of a star. The first is the rotation rate of a star. All stars rotate (the Sun does so once every 25 days), and some stars rotate faster than others. Rapid rotation can act to efficiently mix the internal layers of a star, which in turn alters how it evolves during its life. Recent work has shown that if massive stars (more than several tens of solar masses) rotate very rapidly (with periods of a couple of days or less), then they do not evolve into red giants as their core hydrogen is exhausted. Rather, their internal structure becomes chemically homogenous and they remain blue and close to their main-sequence radius even as hydrogen is exhausted. Such rapidly rotating stars may be found in close binary systems, where the presence of a second star in a close orbit forces both stars into tidally locked rotation with the orbital period.

This in turn leads us to consider the second additional factor which can influence the evolution of a star, namely the presence of a binary companion. Many stars are formed in binary systems – a fact we shall return to at the end of Chapter 8 – and the more massive the star is, the greater the chance of it being formed in a binary. As well as the influence such a companion may have on a star's rotation, as mentioned above, stellar binarity also opens the possibility of **mass transfer** from one star to the other. Firstly, many stars (particularly the more massive ones) may have a substantial stellar wind. A binary companion can thus sweep-up and accrete some of this mass lost from its partner, and increase its own mass as a result. This has the potential to influence the evolution of the accreting star. However, this method is not generally very efficient, as only a small fraction of the wind is likely to be captured.

There are additional ways of transferring mass though. During the evolution of an isolated star, there are three epochs during which its radius increases significantly: the core-hydrogen burning phase (on the main sequence) when a relatively slow and small increase in radius occurs; the rapid expansion towards the first giant branch before ignition of core-helium burning; and the expansion along the asymptotic giant branch after core-helium burning ends. In each of these three phases, a star in a binary system may expand sufficiently that its outer layers have a greater gravitational attraction to the binary companion than to the original star. In such circumstances, mass transfer will occur from the expanding donor star to the companion accreting star and this will likely affect the evolution of one or both stars in the binary system. Binary stars are not limited to a single episode of mass transfer, and indeed mass may be transferred to and fro as first one star, then the other, undergoes a period of radius expansion.

The influence of mass transfer on stellar evolution may be illustrated by the so-called **Algol paradox**. Algol is a well known eclipsing binary star in which the less-massive star of the two has already evolved off the main sequence and is seen as a subgiant, whilst the more-massive star is still on the main-sequence. The two stars must have been born at the same time, so have the same age, and the observation therefore apparently contradicts a basic result of stellar evolution that more-massive stars will evolve more rapidly. The paradox is resolved by the realization that mass transfer has occurred from the initially more-massive star (now the less-massive subgiant) onto the initially less-massive star (now the more-massive main-sequence star) at some stage in the past.

In extreme cases, one star of a binary may evolve to such a state that its extended envelope engulfs its companion star, resulting in two stellar cores co-orbiting inside one stellar envelope. Owing to the enormous drag forces that result, the two cores will tend to rapidly spiral together and once the common envelope is eventually expelled (rather like a planetary nebula – see the next chapter), may emerge as either a very compact binary system, or (in extreme cases) as a single compact object formed from the merger of the two stellar cores.

Summary of Chapter 5

1. The Big Bang produced $^1_1$H, $^2_1$H, $^3_2$He, $^4_2$He and $^7_3$Li only. Nucleosynthesis of heavier elements was prevented by (i) the Coulomb barriers of elements with higher atomic numbers, (ii) the lack of stable isotopes of mass numbers 5 and 8, and (iii) the decreasing density of matter as the Universe expanded. The production of all other elements required the formation of stars.

2. The gravitationally contracting helium cores of stars with $M \geq 0.5\,\mathrm{M_\odot}$ become hot enough to ignite helium burning, which requires $T \approx (1\text{–}2) \times 10^8$ K. As there are no stable isotopes of mass number 5 or 8,

helium burning proceeds via a temporary $^8_4$Be nucleus. A tiny fraction of the $^8_4$Be nuclei fuse with another helium nucleus to form an excited state of $^{12}_6$C, a tiny fraction of which then decays to the $^{12}_6$C ground state. The production rate of ground-state $^{12}_6$C is the equilibrium number density of the excited state divided by the decay lifetime: $dn_{12}/dt = n_{12*}/\tau^*$).

3. The thermodynamic properties of a gas are described by its temperature T, pressure P, and chemical potential μ'. The chemical potential is a measure of the internal energy of the particles; $\mu' = mc^2 - kT \log_e(g_s n_{\mathrm{QNR}}/n)$ (for non-relativistic particles). The first term is the mass-energy equivalent $E = mc^2$ of each particle. The second term comprises a thermal energy, the number of polarizations g_s for each particle, the number of gas particles per unit volume, and $n_{\mathrm{QNR}} = (2\pi mkT/h^2)^{3/2}$, the (non-relativistic) quantum concentration which reflects the number of available quantum states.

4. The sum of chemical potentials on one side of an equilibrium reaction equals the sum of chemical potentials on the other side.

5. While hydrogen burning converts 0.66% of proton mass into energy, the triple-alpha process converts only 0.065% of the helium mass, a factor of ten lower. Helium burning therefore liberates *at most one-tenth* of the energy that hydrogen burning does. This, and the fact that they are brighter during helium burning, explains why stars' helium-burning lifetimes are less than one-tenth of their main-sequence lifetimes.

6. Helium burning also produces oxygen from the newly available $^{12}_6$C via the reaction $^4_2$He $+ ^{12}_6$C $\longrightarrow ^{16}$O $+ \gamma$. The nuclear S-factor and hence the rate of this reaction is one of the most important uncertainties in stellar astrophysics.

7. The number of quantum states available to particles is often very large, but nevertheless finite. In the very dense interiors of some astronomical objects, the finite number of available quantum states is insufficient for the huge number of particles squeezed into the confined volume.

8. Degeneracy can be described in three equivalent ways:

   - when the separation of particles is less than the de Broglie wavelength for their momentum, $\lambda_{\mathrm{dB}} = h/p \approx h/(3mkT)^{1/2}$.

   - when the number of particles per unit volume, n, is higher than the number of available quantum states n_Q at their energy, where n_Q is the quantum concentration.

   - when the temperature T of the particles is much less than the value $h^2 n^{2/3}/(2\pi mk)$. Sometimes a degenerate gas is described as cold, but this can be misleading.

9. Electrons in low-mass main-sequence stars are non-relativistic ($kT \ll m_e c^2$) and classical ($n_e \ll n_{\mathrm{QNR}}$), but $n_e/n_{\mathrm{QNR}} \propto R^{-3/2}$, so as the core contracts, the gas approaches degeneracy.

10. Electrons, along with protons and neutrons, are fermions. The Pauli exclusion principle dictates that no more than one identical fermion can occupy a given quantum state. As spin ($\pm 1/2$) is the only way to distinguish one fermion of a given type from another, at most two can occupy a given

quantum state. In a cold electron gas, the energy of the most energetic degenerate electron is called the Fermi energy, E_F, and its momentum p_F is called the Fermi momentum.

11. Whereas the equation of state for a classical ideal gas depends on the density *and* temperature, e.g. $P = nkT$, for a degenerate gas it depends only on the number density: $P_{NR} = K_{NR}n^{5/3}$ and $P_{UR} = K_{UR}n^{4/3}$ where K_{NR} and K_{UR} are constants for non-relativistic and ultra-relativistic conditions. This decoupling of pressure and temperature in a degenerate gas disables thermostatic regulation. A classical gas responds to an increased temperature by increasing pressure and hence expanding and cooling slightly. In a degenerate gas a thermonuclear runaway can develop.

12. Stars with $M \leq 2.5\,M_\odot$ become electron-degenerate in their cores once core-hydrogen burning ends, and a hydrogen-burning shell develops. Degenerate electrons then provide the pressure support. When helium-burning ignites, the pressure of degenerate gas does not respond to the increased temperature. A thermonuclear runaway called the helium flash occurs. Most of the released energy goes into expanding and heating the core, which goes into lifting the electron degeneracy, rather than into raising the luminosity of the star. Pressure and temperature recouple through the equation of state of a classical gas.

13. The helium flash terminates the red-giant branch evolution of stars with $M \leq 2.5\,M_\odot$. The core-helium burning low-mass star becomes a red-giant clump star or horizontal-branch star depending on whether it has a solar or substantially sub-solar metal abundance and the extent of mass loss.

14. For $M > 2.5\,M_\odot$, the thermal runaway/helium flash is avoided. Helium ignition leads to an expansion of the core and hydrogen-burning shell, the temperature at the base of the envelope decreases, and the envelope contracts, driving up the effective surface temperature. That is, $M > 2.5\,M_\odot$ stars initiate blue loops in the H–R diagram.

15. Energy generation ε for the triple-alpha process varies as T^{30}. A helium-burning shell develops around the new carbon core, beyond which the hydrogen-burning shell burns gradually closer towards the surface, causing the hydrogen envelope to expand and cool. The stars move back towards their helium-ignition points in the H–R diagram as asymptotic giant branch (AGB) stars.

16. On the AGB, low-mass stars ($M \leq 3.5\,M_\odot$) increase in luminosity as convection again becomes important in energy transport; intermediate-mass stars ($3.5\,M_\odot < M \leq 8\,M_\odot$) reverse their blue loops and deepening convection causes an increase in luminosity and the second dredge-up. Thermal instabilities in the helium shell drive pulses during the thermally pulsing-AGB (TP-AGB) phase. Several tens of pulses, each lasting only 10^2 years, may occur at intervals of $\approx 10^4$ years.

17. Although the evolution of a star depends primarily on its mass, and to a lesser extent on its metallicity, a star's rotation rate and its binary character can also strongly affect its evolution. Rapidly-rotating massive stars may become so chemically homogeneous as a result of rotational mixing that they avoid becoming red giants. Stars in binary systems may undergo periods of mass transfer, which affect the subsequent evolution of each component.

Chapter 6　Late stages of stellar evolution

Introduction

So far, we have followed the evolution of stars through their hydrogen- and helium-burning stages. In this chapter we investigate the ejection of a planetary nebula, the white-dwarf remnants, and nucleosynthesis stages late in the evolution of more-massive stars.

6.1　Planetary nebulae

The mass lost from stars at the end of their AGB phase is not immediately visible. In fact, initially the star becomes obscured by the formation of dust grains in the cool ejecta. However, the underlying helium-rich star (with a degenerate carbon and oxygen core) gets smaller and hotter, and a fast, radiation-driven wind develops that induces additional mass loss in a bipolar outflow. Such an object may be described as a *proto*planetary nebula. Two protoplanetary nebulae are shown in the top two images of Figure 6.1, where the breakout of the fast bipolar wind is visible along with the concentric shells of material previously ejected by the star.

Figure 6.1　(*Top row*) Protoplanetary nebulae (≤ 1000 years old). Concentric shells have been episodically (i.e. sporadically) ejected from the AGB star – these are especially obvious in IRAS 17150–3224 – obscuring the central star with newly formed dust. The star is not yet hot enough to ionize the nebula. The fast, radiation-driven wind is just beginning to break out. (*Bottom row*) Planetary nebulae (2500–7000 years old). The central star is now visible and sufficiently hot ($T_{\mathrm{eff}} \geq 10\,000$ K) to ionize the nebula. The central star of NGC 6818 has a surface temperature $T_{\mathrm{eff}} \approx 50\,000$ K. These two nebulae have diameters of $\sim 20\,000$ AU (300 times the diameter of the Solar System) and $55\,000$ AU (1000 times the diameter of the Solar System) respectively.

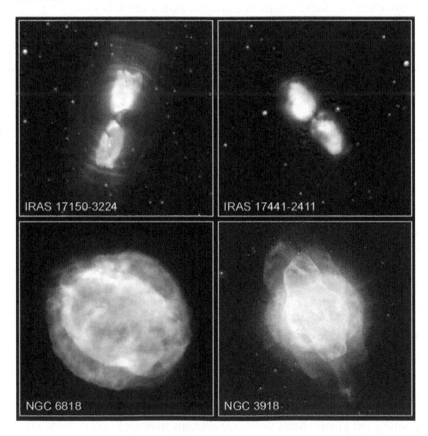

IRAS 17150-3224

IRAS 17441-2411

NGC 6818

NGC 3918

The central star continues to get smaller and hotter, while the nebula expands at typically 30–60 km s^{-1}. Once the central star's effective surface temperature reaches $T_{\text{eff}} \approx 10\,000$ K, it ionizes the ejected envelope, which is then seen as a planetary nebula.

High-resolution images of the slowly ejected material reveal concentric shells, showing that although the slower ejection is isotropic, it is not smooth, which is to say the ejection must be episodic. For example, see especially IRAS 17150–3224 in Figure 6.1, and Figure 6.2a and b. Planetary nebulae may also be compressed by the 'snow-plough' effect of hot winds from the central star which helps to create shells within the nebula structure.

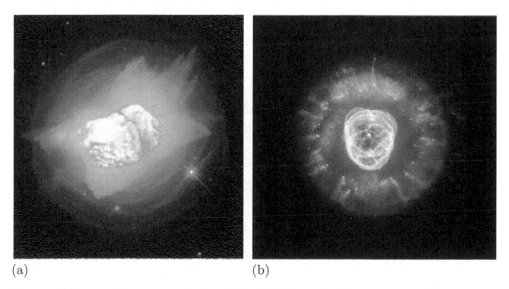

(a)   (b)

Figure 6.2 (a) Planetary nebula NGC 7027. The effective surface temperature of the central star is $T_{\text{eff}} \approx 90\,000$ K, and the nebula diameter is $\sim 20\,000$ AU. The outer part of the nebula shows the concentric shells of material episodically ejected from the AGB star. The centre of the image is dominated by the fast ejection of deeper layers of the star. (b) Planetary nebula, NGC 2392 (The Eskimo nebula), $\sim 10\,000$ years old and with a central star having effective surface temperature $T_{\text{eff}} \approx 40\,000$ K. The comet-like features around the outer part of the nebula are possibly where fast-moving wind particles are overtaking slower-moving matter that was ejected earlier.

The hydrogen-burning shell (if any hydrogen remains) is extinguished when it gets too close to the surface and hence too cool, and the same happens eventually to the helium-burning shell. What is left is a hot CO core ($X_O \approx 80\%$) surrounded by hot helium, but all nuclear burning has been extinguished, and there is no prospect of thermonuclear reactions being re-ignited. Observations show that although the nebula continues to expand, once the central star reaches $T_{\text{eff}} \approx 100\,000$ K it ceases to increase in temperature, drops in luminosity, and begins a long slide to lower luminosity and temperature at almost constant radius (Figure 6.3 overleaf). The star has entered the realm of the cooling, white dwarfs, which you will study in the next section.

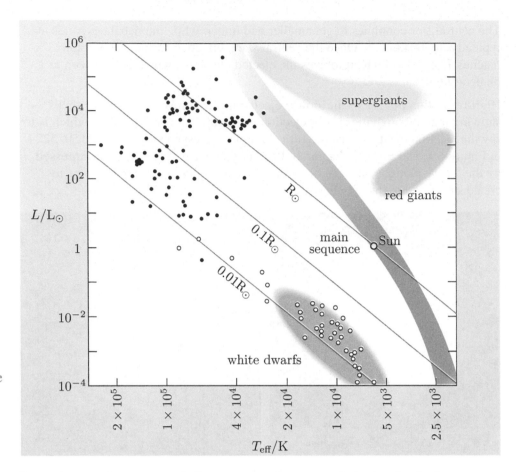

Figure 6.3   H–R diagram showing the locations of the central stars of planetary nebulae (filled circles) and white dwarfs (unfilled circles). Lines corresponding to stars having radii $R = R_\odot$, $0.1\,R_\odot$ and $0.01\,R_\odot$ are shown.

From the angular extent and distance of a planetary nebula, its physical diameter can be derived. As the expansion velocity of the nebula gas (typically 30–60 km s^{-1}) can be measured by its Doppler shift, the duration of the expansion and hence of the post-AGB phase can be calculated. By extrapolating the observed expansion of the nebula backwards, it is shown that the whole evolution of a planetary nebula lasts only about 20 000 years.

Although a standard picture of planetary nebulae has been presented here, it should be noted that the morphology of these objects is complex, and a topic of ongoing study. In particular, binary-star physics may be an important factor in shaping the evolution and structure of these objects.

Ejection of the stellar envelope becomes significant during the AGB phase , but the nebula is not seen until the central star has a surface temperature of around 10 000 K. This is because the hydrogen in the nebula emits light only once its atoms become ionized and then the electrons recombine with their nuclei. Hydrogen atoms do not become ionized until they are irradiated with ultraviolet photons of sufficient energy to remove the electron from the atom, which takes 13.6 eV. The hotter a black body is, the shorter the wavelength and the higher the energy of the photons it radiates. It is not until the effective surface temperature of the central star reaches 10 000 K that UV photons with sufficient energy to ionize the hydrogen are radiated.

6.2 White dwarfs

This topic draws on several areas you have already studied, including stellar structure and electron degeneracy.

6.2.1 The Chandrasekhar mass

Because we will be considering electron degeneracy, we begin by deriving a new way of expressing the electron density. From your study of reaction rates, you will be familiar with the general equation for the number density n_A of some type of particle A in terms of its mass fraction X_A, the particle mass m_A, and the gas density ρ:

$$n_A = \rho X_A / m_A. \qquad \text{(Eqn 3.10)}$$

We can write this for electrons just as easily as for nucleons: $n_e = \rho X_e / m_e$. However, it is more convenient to write

$$X_e \equiv Y_e m_e / m_H \qquad (6.1)$$

where Y_e is the **number of electrons per nucleon**. This leads to the expression

$$n_e = \rho Y_e / m_H. \qquad (6.2)$$

● What is the value of Y_e in (a) pure ionized hydrogen, (b) pure ionized helium?

○ (a) Hydrogen has one electron and one nucleon (its proton), so $Y_c = 1$. (b) Helium has two electrons and four nucleons (two protons and two neutrons), so $Y_e = 0.5$ for helium.

You can readily convince yourself that $Y_e = 0.5$ for the dominant isotopes of ^4_2He, $^{12}_6\text{C}$, and $^{16}_8\text{O}$. For white dwarfs the amount of hydrogen is negligible – it has all been burnt to helium or on to carbon and/or oxygen – so for white dwarfs, $Y_e \approx 0.5$. The simplest type of white dwarf, resulting from the collapse of a star of very low mass ($M \leq 0.5\,\text{M}_\odot$) in which no helium burning has occurred, would be composed entirely of helium and is referred to as a **helium white dwarf**.

For an electron-degenerate gas, you know that the pressure is independent of temperature, and is given by a constant multiplied by some power of the electron density. The constant and the power both depend on whether the degenerate gas is non-relativistic or ultra-relativistic:

- for non-relativistic gas, $P_{NR} = K_{NR} n_e^{5/3}$ (Equation 5.11) where the constant is $K_{NR} = (h^2/5m_e)(3/8\pi)^{2/3}$.

- while for ultra-relativistic gas, $P_{UR} = K_{UR} n_e^{4/3}$ (Equation 5.13) where the constant is $K_{UR} = (hc/4)(3/8\pi)^{1/3}$.

Using Equation 6.2, we can write

- for non-relativistic gas, $P_{NR} = K_{NR}(\rho Y_e / m_H)^{5/3}$

- while for ultra-relativistic gas, $P_{UR} = K_{UR}(\rho Y_e / m_H)^{4/3}$.

As noted in Chapter 2, a particular model (by Donald Clayton) for the internal structure of a star gives the central pressure required to support a star as:

$$P_c \approx (\pi/36)^{1/3} G M^{2/3} \rho_c^{4/3}. \qquad \text{(Eqn 2.18)}$$

By saying that degenerate electrons *provide* the required internal pressure to support the star, i.e. by equating this to the degenerate electron pressure, it is possible to write these equations purely in terms of the mass and central density, which therefore allows us to write one variable in terms of the other.

Exercise 6.1   (a) By equating the core pressure in a star to the degenerate pressure of non-relativistic electrons, derive an expression for the core density in terms of its mass M and the number of electrons per nucleon, Y_e. Leave physical constants unevaluated.

(b) Using the relation between electron number density n_e and gas density ρ_c in the core of the star, express the core electron density as a function of stellar mass. ■

The exercise above shows that in the non-relativistic case, the density in the core of a star may be written as

$$\rho_c = \left(\frac{16\pi^3}{81}\right)\left(\frac{5m_e}{h^2}\right)^3 G^3 \frac{m_H^5}{Y_e^5} M^2 \tag{6.3}$$

or, in terms of the degenerate electron number density, as

$$n_e = \left(\frac{16\pi^3}{81}\right)\left(\frac{5m_e}{h^2}\right)^3 G^3 \frac{m_H^4}{Y_e^4} M^2. \tag{6.4}$$

But, you might ask, is this material really non-relativistic? The Fermi energy E_F of degenerate electrons is given by $E_F \approx p_F^2/2m_e$ (in the purely non-relativistic case) where p_F is the Fermi momentum given by $p_F = (3n_e/8\pi)^{1/3}h$ (Equation 5.8). Therefore, the Fermi energy of the degenerate, non-relativistic electrons in the centre of a white dwarf is

$$E_F \approx \left[\left(\frac{3}{8\pi}n_e\right)^{1/3}h\right]^2 \frac{1}{2m_e}$$

$$\approx \left[\frac{3}{8\pi}\left(\frac{16\pi^3}{81}\right)\left(\frac{5m_e}{h^2}\right)^3 G^3\frac{m_H^4}{Y_e^4}M^2\right]^{2/3}\frac{h^2}{2m_e}$$

$$\approx \frac{25}{2}\left[\frac{48\pi^2}{648}\frac{m_e^{3/2}G^3m_H^4 M^2}{h^3 Y_e^4}\right]^{2/3}$$

$$\approx \frac{25}{2}\left(\frac{2\pi^2}{27}\right)^{2/3}\left(\frac{G}{h}\right)^2\left(\frac{m_H}{Y_e}\right)^{8/3}M^{4/3}m_e. \tag{6.5}$$

Exercise 6.2   Evaluate Equation 6.5 for a low-mass white dwarf with $M = 0.4\,M_\odot$ and $Y_e = 0.5$. Express your answer as a fraction of the electron rest-mass energy $m_e c^2$. ■

Hence, for low-mass white dwarfs, the Fermi energy is $E_F \approx 0.21 m_e c^2$. This is already more than 20% of the rest-mass energy, which makes it doubtful that the non-relativistic treatment is reliable for a white dwarf with $M \geq 0.4\,M_\odot$. In more-massive white dwarfs, the Fermi energy is higher (since $E_F \propto M^{4/3}$) and the non-relativistic treatment will be progressively worse. If we consider a white dwarf with $M = 1.3\,M_\odot$, even the purely non-relativistic treatment indicates $E_F \approx m_e c^2$, indicating that the ultra-relativistic treatment is required.

Exercise 6.3 (a) By equating the core pressure in a star to the degenerate pressure of ultra-relativistic electrons, derive an expression for the mass of a star supported by ultra-relativistic electrons, in terms of the other physical quantities.

(b) Assuming $Y_e = 0.5$, evaluate the stellar mass in kg and $M_\odot$. ∎

The discussion earlier showed that as the mass of a white dwarf is increased, the matter becomes more relativistic.

> The mass for the star when the ultra-relativistic limit is reached should be viewed as the maximum stellar mass that can be supported by degenerate electron pressure. This extremely important limiting mass is called the **Chandrasekhar mass**. The most realistic computations estimate the Chandrasekhar mass as $M_{Ch} \sim 1.4\,M_\odot$.

If the mass is increased further, the pressure *needed* to support the star, given by $P_c \approx (\pi/36)^{1/3} G M^{2/3} \rho_c^{4/3}$ also increases, but the ultra-relativistic degenerate electrons will not be able to increase their pressure to support it. Something has to give ... . That something is the material supporting the star. There *is* a stable configuration of material at higher mass, but it is not degenerate electrons. Rather, the electrons and protons of higher-mass objects are forced to combine as neutrons, and the stable object above the maximum white-dwarf mass is called a *neutron star*. We study such objects in the next chapter.

● White dwarfs with masses in the range $(0.1-0.8)M_{Ch}$ have core densities of order (10^8-10^{10}) kg m^{-3}. Convert this density range into kg cm^{-3}. Give examples of everyday objects with a volume of 1 cm^3 and examples of everyday objects with masses at each end of the range that you have calculated. Hence, what would be comparable densities for white-dwarf matter?

○ 1 cm $= 10^{-2}$ m, so 1 cm$^3 = 10^{-6}$ m^3. Therefore the core density range for these white dwarfs is $(100-10\,000)$ kg cm^{-3}. One cubic centimetre is about the size of a sugar cube or a finger tip. Objects with masses of 100 kg include adult people (typically 50–100 kg), and objects with masses 10 000 kg include lorries. So, a density of 100 kg cm^{-3} would be like a large adult compressed to the size of a sugar cube, while a density of 10 000 kg cm^{-3} would be like a lorry compressed to that size.

6.2.2 The white-dwarf mass–radius relation

The mean density of a star is $\langle \rho \rangle = \text{mass/volume} = 3M/4\pi R^3$. It can be related to the central density if the density profile of the star is known. **Polytropic** stellar models are ones in which the pressure at some radius is proportional to the density at that radius to some power, $P(r) \propto \rho(r)^\gamma$. For a star described by a polytropic model with $P(r) \propto \rho(r)^{5/3}$ (a reasonable approximation), the mean density turns out to be one-sixth of the core density: $\langle \rho \rangle = \rho_c/6$. It is therefore possible to write a relation between the radius, mass, and core density of a star.

Where the core density is dictated by non-relativistic degenerate electrons, for which

$$\rho_c = \left(\frac{16\pi^3}{81}\right)\left(\frac{5m_e}{h^2}\right)^3 G^3 \frac{m_H^5}{Y_e^5} M^2 \qquad \text{(Eqn 6.3)}$$

we can eliminate the density, and find an expression for the radius of the object as a function of mass.

Exercise 6.4   (a) Rearrange the definition of the mean density $\langle\rho\rangle = 3M/4\pi R^3$ to get an expression for the radius in terms of the mass and mean density.

(b) Use the result that $\langle\rho\rangle = \rho_c/6$ for a polytropic stellar model with $P \propto \rho^{5/3}$, to re-express the radius in terms of mass and core density.

(c) Substitute the expression for the core density of non-relativistic degenerate electrons into the result of (b), to derive an expression for the radius (R_{WD}) of a white dwarf as a function of mass. Give your final answer in units of $M_\odot$ and $R_\odot$.

(d) Re-express the result of (c) in units of $M_\odot$ and *Earth* radii.   ∎

The result of Exercise 6.4 is remarkable for two reasons:

$$R_{WD} = \frac{R_\odot}{74} \times \left(\frac{M}{M_\odot}\right)^{-1/3}. \qquad (6.6)$$

First, it shows that the radius of a white dwarf with the same mass as the Sun is almost 100 times smaller. White dwarfs are indeed very compact, comparable in radius to the Earth!

Second, note the mass-dependence of the white-dwarf radius: $R_{WD} \propto M^{-1/3}$.

Compare this to the result for main-sequence stars which you derived earlier. There you found that for main-sequence stars burning hydrogen by the p–p chain, $R_{MS} \propto M^{2/5}$, while for CNO-cycling stars the relation was $R_{MS} \propto M^{4/5}$ (see Table 4.1). Note the important result that the sign of the exponent is negative for white dwarfs!

> Main-sequence stars have radii that *increase* with increasing mass, but white dwarfs have radii that *decrease* with increasing mass!

The derivation of the white-dwarf mass–radius relationship above was based on the assumption that the electrons are non-relativistic, which we showed is incorrect as M approaches $1.3\,M_\odot$. Furthermore, a star is gravitationally unsupportable at the Chandrasekhar mass, $1.4\,M_\odot$, and approaches infinite density (zero radius). There is another formula which gives a more accurate picture of the *radius* of white dwarfs, but it is a fitting formula designed to give you the right result without any explanation of why. This formula, due to Michael Nauenberg in 1972, is

$$R_{WD} = 7.83 \times 10^6 \text{ m} \times \left(\left(\frac{1.44\,M_\odot}{M}\right)^{2/3} - \left(\frac{M}{1.44\,M_\odot}\right)^{2/3}\right)^{1/2}. \qquad (6.7)$$

The formula we derived, while unable to correctly describe white dwarfs near the Chandrasekhar mass, does encapsulate real *physical ideas* which are valid for low-mass stars. There is a place in science for both formulae!

6.2.3 Different types of white dwarf

By the time stars of main-sequence mass $M_{ms} = 8\,M_\odot$ approach carbon ignition late in their life, mass loss has stripped them of $\approx 80\%$ of their mass, with only $\approx 1.4\,M_\odot$ remaining. The similarity of this mass to the Chandrasekhar mass is no mere coincidence. Recall that electron-degenerate white dwarfs approach infinite density at this mass. Only stellar cores that continue to collapse attain the temperatures necessary to ignite carbon. Intermediate-mass stars, with $3\,M_\odot < M_{ms} \leq 8\,M_\odot$ lose so much mass as giants that only $M < 1.4\,M_\odot$ remains to form the white dwarf. These CO cores are supported against further collapse by degenerate electrons, and hence do not attain the temperatures required to initiate carbon burning. They leave carbon–oxygen **CO white dwarfs** as remnants, with masses typically in the range $0.5\,M_\odot < M_{CO,WD} \leq 1.2\,M_\odot$. (These white dwarfs may later become supernovae if they can acquire additional matter from a close companion star.)

There is a narrow range of massive stars, $8\,M_\odot < M_{ms} \leq 11\,M_\odot$, whose remaining cores are massive enough to collapse and ignite carbon, but which ultimately fall below the Chandrasekhar mass due to continued mass loss. When carbon burning terminates they will be supported by degenerate electrons again and will contract no further, leaving oxygen–neon–magnesium **ONeMg white dwarfs**, with masses typically in the range $1.2\,M_\odot < M_{ONeMg,WD} \leq 1.4\,M_\odot$.

Stars initially having $M_{ms} > 11\,M_\odot$ retain enough mass during mass loss that their cores continue to exceed the Chandrasekhar mass, and at the end of each burning phase their cores contract and heat up more. These stars complete all advanced burning phases including silicon burning (see the next section).

- ● What distinguishes helium white dwarfs, CO white dwarfs, and ONeMg white dwarfs?

- ○ A helium white dwarf is the end-point of the evolution of a star of very low mass ($M_{ms} \leq 0.5\,M_\odot$) in which helium burning (by the helium flash) has not occurred. As such, a helium white dwarf will have a mass of typically $0.4\,M_\odot$ or less. A CO white dwarf is the end-point of low-mass and intermediate-mass stars ($0.5\,M_\odot < M_{ms} \leq 8\,M_\odot$) in which helium burning occurred, producing carbon and oxygen in the core, and will have a mass in the range $\sim (0.5\text{–}1.2)\,M_\odot$. ONeMg white dwarfs result from the lightest high-mass stars ($8\,M_\odot < M_{ms} \leq 11\,M_\odot$) in which carbon burning has occurred, producing Ne and Mg, but which have not burnt neon or oxygen. They will have masses typically in the range $\sim (1.2\text{–}1.4)\,M_\odot$.

- ● Why are we unlikely to find isolated (i.e. non-binary) helium white dwarfs in the Galaxy?

- ○ Helium white dwarfs form from main-sequence stars whose mass is less than $0.5\,M_\odot$. Such a star will have a main-sequence lifetime of order 100 billion years, or around ten times the age of the Universe. Hence such a star would not have had time to evolve into a white dwarf yet. Any helium white dwarfs that are seen must have formed as part of a binary star system which has undergone a phase of mass transfer that alters the mass and composition of one or both of the components.

6.2.4   Fading (and cooling) of white dwarfs

Recall from Chapter 1 that it was possible to draw lines of constant radius on the H–R diagram. These are lines sloping from top-left to lower-right. As a white dwarf has a constant mass as it radiates away its energy, it is clear that it will gradually slide down one of these constant radius lines. In the previous section you saw that the central stars of planetary nebulae were to be found at high luminosities and high temperatures, in the top-left of the H–R diagram. The next stage of evolution of these stars is therefore a gradual descent along a line of constant radius in the H–R diagram.

- Substitute the radius R from the mass–radius relation of white dwarfs into the relation $L = 4\pi R^2 \sigma T_{\mathrm{eff}}^4$ to indicate how bright a $1\,\mathrm{M_\odot}$, solar-temperature white dwarf is.

○ We have

$$L_{\mathrm{WD}} = 4\pi R_{\mathrm{WD}}^2 \sigma T_{\mathrm{WD,eff}}^4 \approx 4\pi \left(\frac{\mathrm{R_\odot}}{74}\right)^2 \left(\frac{M_{\mathrm{WD}}}{\mathrm{M_\odot}}\right)^{-2/3} \sigma T_{\mathrm{WD,eff}}^4 .$$

For $M_{\mathrm{WD}} = 1\,\mathrm{M_\odot}$ and $T_{\mathrm{WD,eff}} = \mathrm{T_{\odot,eff}}$ this becomes

$$L_{\mathrm{WD}} \approx 4\pi \left(\frac{\mathrm{R_\odot}}{74}\right)^2 \sigma \mathrm{T_{\odot,eff}^4} .$$

Since $\mathrm{L_\odot} \approx 4\pi \mathrm{R_\odot^2} \sigma \mathrm{T_{\odot,eff}^4}$, we can write $L_{\mathrm{WD}} \approx (1/74)^2\,\mathrm{L_\odot}$, i.e. a $1\,\mathrm{M_\odot}$, solar-temperature white dwarf is approximately 5500 times fainter than the Sun.

Figure 6.4 shows how faint the white dwarfs in the globular cluster M4 are compared to the main-sequence stars and red giants.

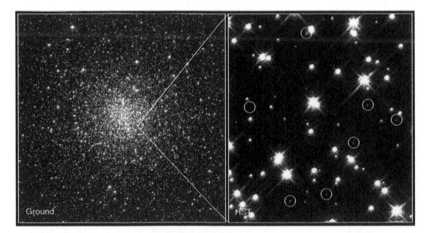

Figure 6.4   (*Left*) The globular cluster M4, which is $\approx 13 \times 10^9$ yr old and contains more than $100\,000$ stars, was the target of a Hubble Space Telescope search for white dwarfs. Ancient red giants are predominant in this view from a ground-based telescope. The field is 47 light-years across. The box (right of centre) shows the small area that the Hubble Telescope probed. (*Right*) This image from the Hubble Space Telescope shows a small portion of the cluster only 0.63 light-years across, and reveals seven white dwarfs (inside circles) among the cluster's much brighter population of yellow Sun-like stars and cooler red dwarfs. The cluster is expected to contain about $40\,000$ white dwarfs in total.

As there are no nuclear energy sources active in a white dwarf, its luminosity is due solely to the slow leakage of thermal energy into space, as radiation. That is, $L = -dE/dt$, where the minus sign indicates that the luminosity is positive when the star is losing energy (dE/dt is negative). This equation can be developed into an expression for the time-dependence of the luminosity, which shows that $L \propto t^{-7/5}$. (The proof is beyond the scope of this book.) It takes $\sim 10^9$ yr for white dwarfs to fade to $L \sim 10^{-3}\,L_\odot$. As white dwarfs fade, they also cool.

● Consider the time-evolution of the white dwarf luminosity, $L \propto t^{-7/5}$. What does this indicate about their rate of evolution in the H–R diagram?

○ White dwarfs decrease in brightness rapidly at the start, but more slowly later on. Therefore they pile up towards the bottom of the H–R diagram. By seeing how bright the faintest white dwarfs are, it is possible to infer the age of the stellar population to which they belong.

● Why do we say that the luminosity of white dwarfs is provided by thermal leakage rather than slow gravitational contraction?

○ White dwarfs are degenerate, so their temperature and pressure are decoupled. As they cool, the pressure is unchanged so the radius is unchanged. This is why they cool and fade in the H–R diagram along a line of constant radius. If the radius is constant, there is *no* release of gravitational potential energy.

6.3 Advanced nuclear burning

In Chapter 5, we examined helium burning, which converts helium into carbon and/or oxygen. The degree of oxygen production depends on the reaction $^{12}_{6}\text{C} + ^{4}_{2}\text{He} \longrightarrow ^{16}_{8}\text{O} + \gamma$, which is more significant at higher temperatures, i.e. in more-massive stars. We followed low- and intermediate-mass helium-burning stars through the planetary nebula phase, and in this chapter have examined their final evolution to become white dwarfs. However, we have yet to deal with massive stars whose energy generation does not finish with helium burning. More advanced stages of nucleosynthesis in massive objects will be discussed here.

6.3.1 Carbon burning

When helium burning ends, the core of a star will consist mainly of carbon and oxygen nuclei. In stars of mass greater than $\sim 8\,M_\odot$ whose cores reach a temperature $T_c > 5 \times 10^8$ K and density $\rho_c > 3 \times 10^9$ kg m^{-3}, carbon burning can begin via reactions such as the following:

$$
\begin{aligned}
^{12}_{6}\text{C} + {}^{12}_{6}\text{C} &\longrightarrow {}^{20}_{10}\text{Ne} + {}^{4}_{2}\text{He} \\
^{12}_{6}\text{C} + {}^{12}_{6}\text{C} &\longrightarrow {}^{23}_{11}\text{Na} + \text{p} \qquad (6.8)\\
^{12}_{6}\text{C} + {}^{12}_{6}\text{C} &\longrightarrow {}^{23}_{12}\text{Mg} + \text{n.}
\end{aligned}
$$

which produce isotopes of neon, sodium and magnesium. The timescale for this phase of nucleosynthesis is only of order five hundred years.

In the following subsections, the temperatures for the onset of various stages of nuclear burning cannot be regarded as absolute, because reaction rates increase gradually with temperature and also depend on the density of the material in which fusion occurs.

6.3.2   Neon burning

In stars of mass greater than $\sim 10\,\mathrm{M_\odot}$ whose cores reach a temperature $T_c > 10^9$ K a **photodissociation** reaction can occur in which neon nuclei are broken down by high-energy gamma-ray photons, as follows:

$$^{20}_{10}\mathrm{Ne} \;+\; \gamma \;\longrightarrow\; ^{16}_{8}\mathrm{O} \;+\; ^{4}_{2}\mathrm{He}. \tag{6.9}$$

This allows the newly produced helium nuclei to react with other existing neon nuclei in the following neon-burning reaction:

$$^{20}_{10}\mathrm{Ne} \;+\; ^{4}_{2}\mathrm{He} \;\longrightarrow\; ^{24}_{12}\mathrm{Mg} \;+\; \gamma. \tag{6.10}$$

The timescale for this phase of nucleosynthesis is only of order a year.

6.3.3   Oxygen burning

After neon burning finishes, a star's core will consist mainly of oxygen and magnesium nuclei. In stars of around $11\,\mathrm{M_\odot}$ or more, whose cores reach a temperature $T_c > 2 \times 10^9$ K, oxygen burning can begin via the reaction:

$$^{16}_{8}\mathrm{O} \;+\; ^{16}_{8}\mathrm{O} \;\longrightarrow\; ^{28}_{14}\mathrm{Si} \;+\; ^{4}_{2}\mathrm{He}. \tag{6.11}$$

The oxygen burning phase of nucleosynthesis will typically last only a few months.

6.3.4   Silicon burning

The final phase of standard nucleosynthesis in stars of $11\,\mathrm{M_\odot}$ or more is silicon burning, which occurs if the stars's core temperature reaches $T_c > 3 \times 10^9$ K. Initially there will be a series of photodissociation reactions whereby high-energy gamma-ray photons break apart the silicon-28 nuclei into helium-4 nuclei and a mixture of free protons and neutrons. These can subsequently recombine to build more-massive nuclei. As an example, one such chain of reactions is as follows, beginning with photodissociation by a gamma-ray photon whose energy exceeds about 10 MeV:

$$
\begin{aligned}
^{28}_{14}\mathrm{Si} \;+\; \gamma \;&\longrightarrow\; ^{24}_{12}\mathrm{Mg} \;+\; ^{4}_{2}\mathrm{He} \qquad (6.12)\\
^{28}_{14}\mathrm{Si} \;+\; ^{4}_{2}\mathrm{He} \;&\longleftrightarrow\; ^{32}_{16}\mathrm{S} \;+\; \gamma\\
^{32}_{16}\mathrm{S} \;+\; ^{4}_{2}\mathrm{He} \;&\longleftrightarrow\; ^{36}_{18}\mathrm{Ar} \;+\; \gamma\\
^{36}_{18}\mathrm{Ar} \;+\; ^{4}_{2}\mathrm{He} \;&\longleftrightarrow\; ^{40}_{20}\mathrm{Ca} \;+\; \gamma\\
^{40}_{20}\mathrm{Ca} \;+\; ^{4}_{2}\mathrm{He} \;&\longleftrightarrow\; ^{44}_{22}\mathit{Ti} \;+\; \gamma\\
^{44}_{22}\mathit{Ti} \;+\; ^{4}_{2}\mathrm{He} \;&\longleftrightarrow\; ^{48}_{24}\mathit{Cr} \;+\; \gamma\\
^{48}_{24}\mathit{Cr} \;+\; ^{4}_{2}\mathrm{He} \;&\longleftrightarrow\; ^{52}_{26}\mathit{Fe} \;+\; \gamma\\
^{52}_{26}\mathit{Fe} \;+\; ^{4}_{2}\mathrm{He} \;&\longleftrightarrow\; ^{56}_{28}\mathit{Ni} \;+\; \gamma.
\end{aligned}
$$

Each of the reactions that produce sulfur, argon, calcium, titanium, chromium, iron and nickel exists in equilibrium (hence the double-headed arrows), so the concentrations in each case may be found by equating the chemical potentials on either side of each reaction. Note also that some of the isotopes in the example chain above (e.g. $^{44}_{22}$Ti, $^{48}_{24}$Cr, $^{52}_{26}$Fe and $^{56}_{28}$Ni shown in italics) are unstable to β-decay or electron capture with half-lives of order a few hours to a few years. So in some cases these nuclei will decay to other stable products (such as $^{44}_{20}$Ca, $^{48}_{22}$Ti, $^{52}_{24}$Cr and $^{56}_{26}$Fe) before capturing another helium nucleus. In this way many different isotopes of the elements from titanium ($Z = 22$) to zinc ($Z = 30$) are produced.

● What simple ratio exists between the number of protons and the number of neutrons in the major nuclei produced by silicon burning?

○ Since $^{28}_{14}$Si has the same number of protons ($Z = 14$) and neutrons ($N = A - Z = 28 - 14 = 14$), and α-particles have the same number of neutrons as protons ($Z = 2$ and $N = A - Z = 4 - 2 = 2$), then so do the major products of the silicon-burning.

Because silicon burning involves a series of photodissociation reactions in which the nuclei are initially broken down, the process is sometimes referred to as silicon melting and can be envisaged as 'melting' the silicon nuclei in a sea of helium nuclei. The timescale for the entire process is typically of order 1 day!

6.3.5 The end-points of carbon, neon, oxygen and silicon burning

As noted above, the final stage of burning, silicon burning, lasts only about a day. It begins with the photodissociation (by thermal photons) of some fraction of the $^{28}_{14}$Si produced previously by oxygen burning. The α-particles ($^4_2$He nuclei) released in these disintegrations can then combine with remaining $^{28}_{14}$Si nuclei. This happens in conditions close to thermodynamic equilibrium, so the differences in binding energy determine the equilibrium ratios of the nuclei involved in the reactions. The equilibrium reactions are governed by a **Boltzmann factor** with the relative nuclear abundances given by

$$\frac{n_A\, n_\alpha}{n_{(A+\alpha)}} \propto \exp\left(-\frac{\Delta Q}{kT}\right) \tag{6.13}$$

where n_A is the number density of the reacting nucleus with mass number A, n_α is the number density of α-particles, $n_{(A+\alpha)}$ is the number density of the product nucleus and ΔQ is the energy change in the α-capture. Where ΔQ is positive (i.e. energy is released) the exponential is < 1, and hence equilibrium favours formation of the heavier ($A + \alpha$) nucleus. Where ΔQ is negative (i.e. energy is *absorbed* in the capture) the exponential is > 1 and equilibrium favours the separate A and α-particles to the ($A + \alpha$) nucleus. As the binding energy per nucleon peaks around $A = 56$ (see Figure 6.5 overleaf), ΔQ is positive for captures at $A < 56$ and negative for captures at $A > 56$. The equilibrium therefore drives both heavier and lighter nuclei towards $A = 56$. Essentially, the nuclear abundances are rearranged, always favouring the production of more stable nuclei, and this accounts for the high abundance of nuclei near $A = 56$, known as the *iron peak*.

The process produces primarily $^{56}_{28}$Ni, which subsequently β^+-decays to $^{56}_{27}$Co, which in turn undergoes β^+-decay to $^{56}_{26}$Fe. Because the process is governed by thermodynamic equilibrium, the nuclei are said to be formed in a process of **nuclear statistical equilibrium** or **NSE**; unofficially, you might like to think of it as natural selection for elements. This process produces elements from Si (the starting point) at atomic number $Z = 14$ up to the so-called **iron-peak nuclei** in the group between $Z = 22$ and 30: titanium (Ti), vanadium (V), chromium (Cr), manganese (Mn), iron (Fe), cobalt (Co), nickel (Ni), copper (Cu), and zinc (Zn).

Figure 6.5 Nuclear binding energy per nucleon. Note that for $A \geq 15$, this changes only slowly with atomic mass, but nevertheless has a peak around $A = 56$. This accounts for the build-up of elements near iron – the iron peak – during silicon burning. Note also that binding energy is a *negative* quantity.

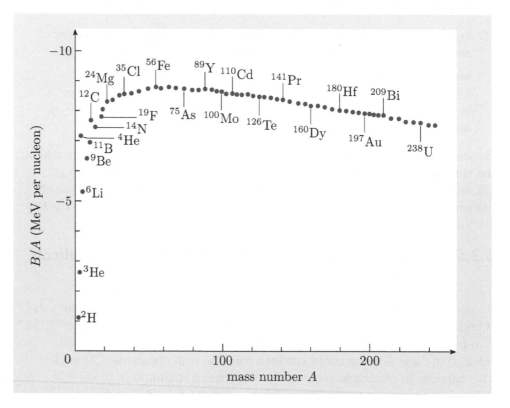

By the end of the various burning phases, e.g. carbon burning, neon burning, oxygen burning, and silicon burning, the star has built up a series of shells of different composition (see Figure 6.6). The outermost shell is unprocessed hydrogen and helium, while the next shell contains helium from hydrogen burning. Progressively deeper shells are composed of carbon and oxygen from helium burning, neon, sodium and magnesium from carbon burning, oxygen and magnesium from neon burning, silicon from oxygen burning, and elements from silicon to the iron-peak elements from silicon burning in NSE. The stratification indicates that the core attained higher peak temperatures than layers further out, and the fact that at higher temperatures, fusion of nuclei with higher atomic numbers is possible. This layering is sometimes referred to as an **onion-skin structure**.

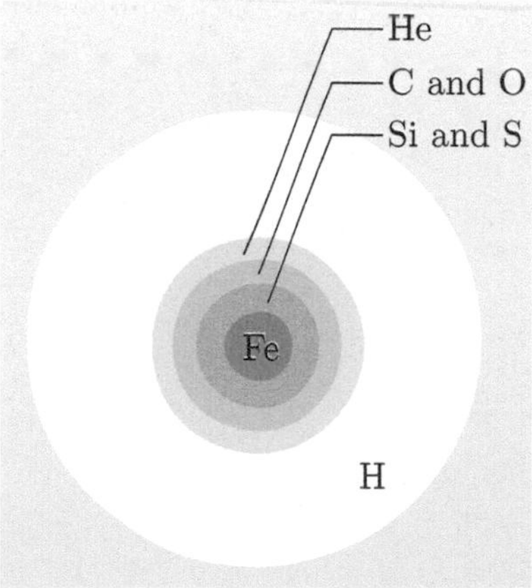

● The next fusion process after carbon burning in massive stars is neon burning, which produces oxygen and magnesium, followed by oxygen burning to produce silicon. Why aren't silicon white dwarfs produced?

○ The cores of stars which are massive enough to burn neon and then oxygen exceed the Chandrasekhar mass, so they are too massive to leave white dwarf remnants. We will examine the fate of such stars in the next chapter.

Figure 6.6 Schematic diagram (not to scale) of onion-skin structure of a massive star ($M \geq 10\,M_\odot$) that has synthesized elements up to the iron peak in its core. Not all detail described in the text is shown.

6.4 Neutron-capture nucleosynthesis

Our discussion of fusion reactions concentrated on charged nuclei, but the heaviest elements cannot be produced that way. First, they would have to overcome huge Coulomb barriers which tend to keep positively charged nuclei apart. Second, silicon burning occurs in thermodynamic equilibrium, producing primarily nuclei with the highest binding energy per nucleon, i.e. around the iron peak. This produces elements up to $Z \approx 30$ (zinc), but with ≈ 100 elements in existence, many cannot be produced this way; some other process(es) must exist.

There is a series of reactions which avoids the Coulomb barrier completely, called **neutron-capture reactions**, where a single neutron is captured by a nucleus. Because the neutron is electrically neutral, there is no Coulomb barrier at all. This makes neutron capture a particularly important mechanism for the production of nuclei with high atomic number, especially as the Gamow energy that characterizes the Coulomb barrier increases as Z^2. Neutron-capture reactions are in fact the main mechanism for the nucleosynthesis of elements heavier than the iron peak. Neutron capture can also be important for some lighter elements.

6.4.1 The chart of nuclides

Nucleosynthesis pathways are often best visualized in the **chart of nuclides**. This plots each isotope separately, with atomic number, Z, on the vertical axis and neutron number, N, on the horizontal axis. It shows unstable as well as stable isotopes. A portion of the chart of nuclides is shown in Figure 6.8 overleaf, where colour distinguishes stable and unstable isotopes. Each increment along the vertical axis corresponds to a new element, e.g. hydrogen at $Z = 1$, helium at $Z = 2$, lithium at $Z = 3$, while each increment along the horizontal axis corresponds to a more neutron-rich isotope of the same element. Lines of constant atomic mass number, A, where $A = Z + N$, lie at 45° between upper-left and lower-right. Stable nuclei occupy a zone stretching from low N and Z to high N and Z, called the **valley of stability**. The valley is clearly seen in an alternative view of the chart, Figure 6.7, where the total energy per nucleon is plotted on the vertical axis. Many heavy elements have several stable isotopes spanning up to ≈ 10 units in N.

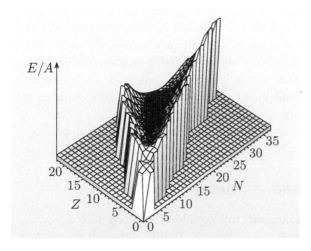

Figure 6.7 A three-dimensional view of the total energy per nucleon as a function of atomic number Z and neutron number N, for elements up to calcium. The valley of stability is clearly illustrated.

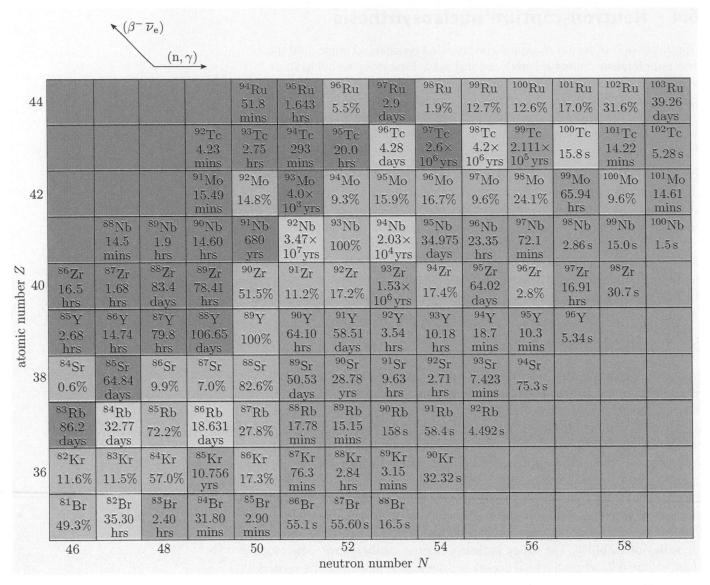

Figure 6.8   A small region of the chart of nuclides. Neutron number N increases from left to right, and atomic number Z increases from bottom to top. Stable nuclei are coloured orange. Isotopes are labelled according to their atomic mass number A, half-life (for unstable nuclei), and the percentage of the element accounted for by the isotope in the Solar System (for stable isotopes). The reactions pathways of neutron capture and β^--decay are indicated. The coloured squares represent the major decay modes for the unstable nuclei. Blue represents β^--decay, dark pink is electron capture, and light pink is either β^--decay or electron capture. To avoid clutter, values of Z are not shown on any isotope.

6.4.2   Neutron capture and β^--decay

The capture of a free neutron by a nucleus, followed by the emission of a γ-ray, leads to a neutron-rich isotope of that same element. That is, although the neutron number N and the atomic mass A both increase by one, the atomic number Z, which is the number of protons in the nucleus, is unchanged. The symbol for an element X with these particle numbers is written ${}^{A}_{Z}X_{N}$, so the neutron-capture reaction could be written

$$
{}^{A}_{Z}X_{N} + \text{n} \longrightarrow {}^{A+1}_{Z}X_{N+1} + \gamma. \tag{6.14}
$$

Viewed in the chart of nuclides (Figure 6.8), it's just a jump of one cell to the right.

Neutron-rich isotopes are generally unstable. Even if they are not unstable after the addition of just *one* neutron, they will be after the capture of *several* more. The radioactive decay mode of neutron-rich unstable nuclei is by β^{-}-decay. This results in the conversion of one of the neutrons in the nucleus into a proton, with the ejection of a β^{-}-particle (an electron). This process does not change the atomic mass number A, as the number of nucleons is unchanged, but it does increase the atomic number (proton count) by one, and thus produces a new element one step further along in the Periodic Table.

Thus, a single neutron capture on to a nucleus ${}^{A}_{Z}X_{N}$ leads initially to a neutron-rich isotope ${}^{A+1}_{Z}X_{N+1}$, and if this is unstable it β^{-}-decays to give a new element ${}^{A+1}_{Z+1}(X+1)_{N}$, by the reaction

$$
{}^{A+1}_{Z}X_{N+1} \longrightarrow {}^{A+1}_{Z+1}(X+1)_{N} + \text{e}^{-} + \bar{\nu}_{\text{e}}. \tag{6.15}
$$

A β^{-}-decay, which converts a neutron into a proton, decreases N while increasing Z by one unit. In the chart of nuclides, the nucleus moves diagonally to the upper-left at $45°$, one step along a line of constant atomic mass number A.

● If ${}^{87}_{38}\text{Sr}$ captured a neutron, what would be the next stable isotope produced?

○ The neutron capture would convert ${}^{87}_{38}\text{Sr}$ into ${}^{88}_{38}\text{Sr}$. According to Figure 6.8, this is stable.

● If ${}^{88}_{38}\text{Sr}$ captured a neutron, what would be the next stable isotope produced?

○ The neutron capture would convert ${}^{88}_{38}\text{Sr}$ into ${}^{89}_{38}\text{Sr}$, which according to Figure 6.8 is unstable to β^{-}-decay. The ${}^{89}_{38}\text{Sr}$ would therefore β^{-}-decay with a half-life of 50.53 days to ${}^{89}_{39}\text{Y}$, which according to Figure 6.8 is stable.

6.4.3 Closed neutron shells and neutron magic numbers

Switching for a moment from nuclear physics to atomic physics, you may be aware that certain electronic configurations are very stable, and elements with these configurations are inert. Such elements are called the **noble gases**, e.g. He, Ne, Ar and are said to have full *electron* shells at $Z = 2, 10, 18$, etc. Returning now to nuclear physics, an analogous behaviour corresponding to full *neutron* shells (**closed neutron shells**) is found for certain values of the neutron number N, including $N = 50, 82$ and 126. These values are called **neutron magic numbers**. The closed-shell nuclei have very low **neutron-capture cross-sections**, which means that nuclei reaching these isotopes have a very low probability of capturing another neutron and hence a high probability of remaining in that state. Nuclei which *form* with magic neutron numbers also tend to be more stable than those with slightly higher neutron numbers, so if a nucleus with a full neutron shell at $N = N_{\text{m}}$ captures another neutron, the new nucleus at $N = N_{\text{m}} + 1$

generally β^--decays very rapidly, to produce an element with a higher atomic number Z, but again with neutron number N_m. As a result of this, nuclei with closed neutron shells are amongst the most abundant elements above the iron peak, and distinct peaks in the abundances of heavy elements can be ascribed to this aspect of nuclear physics. We will return to this point later in this section.

- Consider the closed-shell neutron numbers given in the previous paragraph. Identify which stable nuclei in Figure 6.8 are closed-neutron-shell nuclei.

○ Nuclides with neutron number $N = 50$ are closed-shell nuclei. The stable ones are $^{86}_{36}$Kr, $^{87}_{37}$Rb, $^{88}_{38}$Sr, $^{89}_{39}$Y, $^{90}_{40}$Zr and $^{92}_{42}$Mo.

6.4.4   Competition between neutron capture and β^--decay

Once a nucleus has captured a neutron, what happens next? Does it β^--decay or does it capture another neutron first? The timescale for β^--decay depends on the isotope in question and is governed by the nuclear physics. The timescale for neutron capture, on the other hand, depends also on the environment, and in particular on the probability that a collision with a neutron occurs. In Chapter 3, you saw that the fusion rate for dissimilar nuclei is given by

$$R_{AB} = n_A n_B \langle \sigma v_r \rangle. \tag{Eqn 3.17}$$

For neutron capture, the nuclear physics enters via the cross-section, σ, while the environment influences the rate via the number density of free neutrons, n_n.

Consider the initial state of the solar material, which is composed typically of 70% hydrogen, 28% helium, and 2% heavier elements (by mass). What is the fraction of free neutrons? Zero; all neutrons are locked up in helium and heavier elements. Furthermore, *free* neutrons are unstable and spontaneously decay $(n \longrightarrow p + e^- + \overline{\nu}_e)$ with a half-life of only ≈ 10 minutes. Consequently, neutron-capture reactions can only be activated once other nuclear reactions liberate neutrons. Clearly, the density of free neutrons can vary greatly depending on the particular environment. That is, the neutron-capture rate, and hence the competition between neutron capture and β^--decay, depends on the stellar environment more than the nuclear physics. It has been common to consider one of two cases which represent the extremes of the possibilities: one in which the neutron-capture rate is much slower than the β^--decay rates, so the β^--decay dominates, and one in which the neutron-capture rate is much faster than the β^--decay rate, so that neutron capture dominates.

Low neutron number densities: the s-process   The *slow* neutron-capture process, or **s-process**, occurs when neutron-capture number densities are sufficiently low that the neutron-capture rate is well below the β^--decay rate. Any unstable nucleus formed β^--decays to a stable nucleus of higher atomic number before another neutron capture occurs. That stable nucleus is then able to capture a neutron, but once an unstable isotope is formed, β^--decay occurs prior to another neutron capture. In this way, the s-process involves the temporary production of unstable nuclei no more than one neutron-capture event from the valley of stability. The trajectory taken by a nucleus undergoing the s-process in the chart of nuclides is called the **s-process pathway**. Most of the connected stable nuclei (zigzag line in Figure 6.9) are at least partially produced by this process. They are produced primarily in TP-AGB stars between the hydrogen-burning and helium-burning shells, and during core-helium burning in more-massive stars ($M \geq 10\,\mathrm{M_{\odot}}$). The timescale for which the s-process is active is typically $\sim 10^4$ yr.

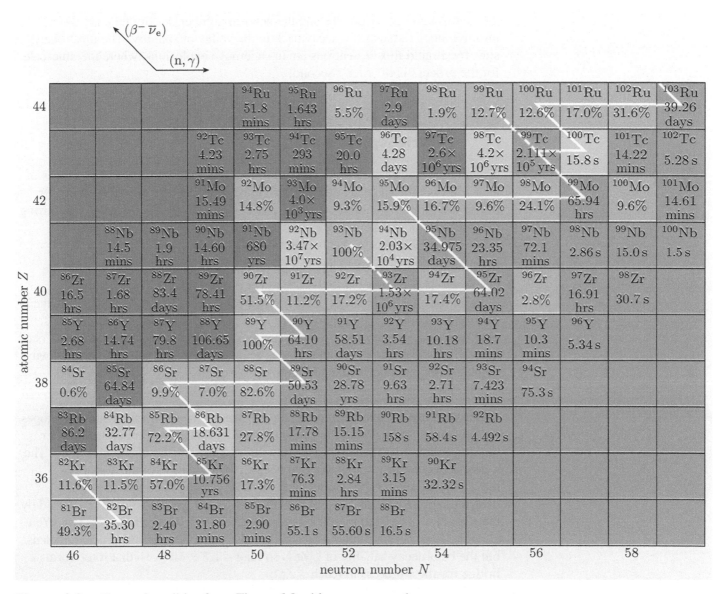

Figure 6.9   Chart of nuclides from Figure 6.8 with s-process pathway highlighted. (The two dashed lines indicate the decays of the long-lived s-process isotopes $^{93}_{40}$Zr and $^{99}_{43}$Tc *after* the s-process terminates.)

Although the s-process pathway remains close to the valley of stability, it is not quite a *single* route. It has a small number of **branching points** where the β^--decay timescales for some unstable nuclei (e.g. $^{79}_{34}$Se, which is *not* on Figure 6.9) depend on whether they are in their ground state or an excited state. Such branching points lead to a splitting of the s-process pathway into two routes that rejoin after typically a few steps.

High neutron number densities: the r-process   The *rapid* neutron-capture process, or **r-process**, is when neutron number densities are sufficiently high that the neutron-capture rate is well above the β^--decay rate. Multiple successive neutron captures occur, temporarily producing very neutron-rich unstable isotopes. Whereas the s-process pathway remains close to the valley of stability, the r-process proceeds far to the right (neutron-rich) side of the valley. Only after the neutron source terminates and the neutron number-density falls are the unstable nuclei able to β^--decay back to the valley of stability. The most likely sites for a rapid flux of neutrons are in supernova explosions, where the timescale for the process is of order 1 second!

6.4.5   Different products of s- and r-process nucleosynthesis

The most massive stable nuclei produced by slow neutron capture are isotopes of lead and bismuth ($Z = 82$ and 83) with mass numbers just over 200. However, due to the high flux of neutrons, the r-process can also produce naturally occurring radioactive isotopes beyond this limit, such as those of thorium and uranium ($Z = 90$ and 92) which have half-lives of billions of years.

Most neutron-capture elements can be produced by both the s-process and the r-process. However, because the s- and r-processes take different pathways through the chart of nuclides, there are some differences in the isotopes which can be produced, and in the proportions of each isotope produced by each process. One feature common to both processes though is that nuclei with *even* numbers of protons (i.e. even atomic numbers) are produced in greater abundance than their neighbouring isotopes with odd numbers of protons.

Isolated neutron-rich isotopes of the r-process   There are some stable isotopes on the neutron-rich side of the valley of stability that are separated from lower neutron-number stable isotopes of the same elements by an *unstable* isotope. The s-process is unable to jump the gap, because by definition of the s-process the β^--decay rate of the intermediate unstable isotope is greater than the neutron-capture rate. Such isolated neutron-rich isotopes can only be produced by the r-process. The examples in Figure 6.8 are $^{86}_{36}$Kr, $^{87}_{37}$Rb, $^{96}_{40}$Zr, and $^{100}_{42}$Mo. (You could be forgiven for thinking that $^{94}_{40}$Zr is also an isolated pure-r-process nucleus, but the beta-decay half-life of $^{93}_{40}$Zr is so long – 1.5×10^6 yr – that it too forms a bridge for the s-process to reach $^{94}_{40}$Zr.)

Shielded isotopes of the s-process   As noted previously, the β^--decay path is along a $45°$ line in the chart of nuclides, from lower-right to upper-left. In the r-process, β^--decays begin from very neutron-rich unstable nuclei. Once the β^--decay reaches a *stable* nucleus with atomic number Z and neutron number N, no further β^--decay of that nucleus occur. Any nuclei on the s-process path which are shielded in this way from r-process β^--decays must originate purely from the s-process and *not also* the r-process. The examples in Figure 6.8 are $^{86}_{38}$Sr (which is shielded from the r-process by $^{86}_{36}$Kr), $^{87}_{38}$Sr (which is shielded by $^{87}_{37}$Rb), $^{94}_{42}$Mo (shielded by $^{94}_{40}$Zr), $^{96}_{42}$Mo (shielded by $^{96}_{40}$Zr), and $^{100}_{44}$Ru (shielded by $^{100}_{42}$Mo).

Abundance peaks due to closed neutron shells As noted above, nuclei having closed neutron shells ($N = 50, 82, 126$) are more stable than more neutron-rich species and have lower neutron-capture cross-sections.

As the s-process pathway is close to the valley of stability, the s-process produces many stable nuclei with those neutron numbers. This produces peaks in the abundances of the s-process nuclei around $N = 50, 82$ and 126, which correspond to $A \approx 90, 138$ and 208; see Figure 6.10. These correspond to elements such as zirconium & molybdenum ($A \sim 90$); barium & cerium ($A \sim 138$); lead & bismuth ($A \sim 208$).

The r-process, on the other hand, follows a path far to the neutron-rich side of the valley of stability. Nuclei with closed neutron shells are still produced abundantly, but in the r-process these are unstable and β-decay at constant atomic mass number A. For a given closed-shell neutron number N, the unstable r-process nuclei have much lower atomic number Z and atomic mass number A than s-process nuclei. Consequently, the abundance peaks of the r-process are found at lower A than the peaks of the s-process. The r-process peaks at $N = 50, 82$ and 126 correspond to $A \approx 85, 130$ and 195 (see Figure 6.10). These correspond to elements such as krypton & strontium ($A \sim 85$); tellurium & xenon, ($A \sim 130$); osmium, platinum, gold & mercury ($A \sim 195$).

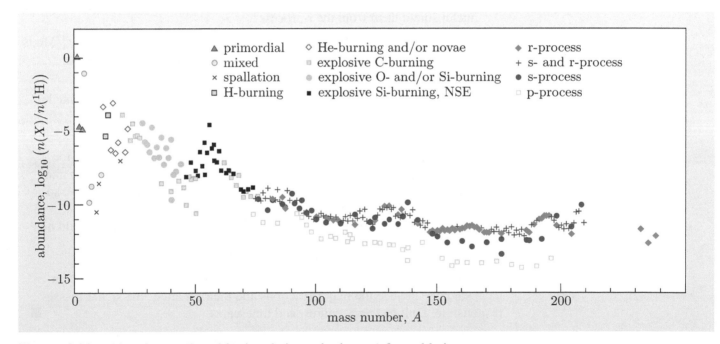

Figure 6.10 Abundances (logarithmic relative to hydrogen) for stable isotopes. These are grouped according to their main nucleosynthetic origin. Of particular interest in this section are the neutron-capture elements above the iron peak, i.e. with $A \geq 75$. These are grouped into primarily r-process, primarily s-process, mixed s- and r-process, and the p-process. The s-process gives rise to abundance peaks at $A \approx 90, 138$ and 208, whereas the r-process leads to broader, more rounded peaks at $A \approx 85, 130$ and 195.

As the s-process remains close to the valley of stability, only a small *range* of atomic numbers Z is associated with any particular closed neutron shell. The r-process, on the other hand, brings a larger range of atomic numbered nuclei to a particular closed shell N-value. The result is that the closed neutron shell abundance peaks for the r-process are not only at lower A than for the s-process, but are also *wider* than the s-process peaks.

The p-process exceptions　Several stable nuclei in Figure 6.9 lie on the neutron-poor (= proton-rich) side of the valley of stability where they are shielded from the r-process but do not lie on the s-process path either. The five nuclei are $^{84}_{38}\text{Sr}$, $^{92}_{42}\text{Mo}$, $^{94}_{42}\text{Mo}$, $^{96}_{44}\text{Ru}$ and $^{98}_{44}\text{Ru}$. $^{94}_{42}\text{Mo}$ is off the s-process path because the β^--decay half-lives of $^{93}_{40}\text{Zr}$ and $^{94}_{41}\text{Nb}$ are so long compared to typical neutron capture intervals that those isotopes appear stable to the s-process.

These five are called **p-process** nuclei, where the 'p' signifies a proton-rich nucleus compared to the valley of stability. p-process isotopes are very rare, but the fact that they exist at all indicates that the s- and r-processes are not the only mechanisms for the production of heavy nuclei. You *might* think that the name 'p-process' *implies* proton capture, but in fact as many as three different processes have been considered for the synthesis of these trace p-isotopes; the details are beyond the scope of this book.

- Spend a moment considering these p-process nuclei in Figure 6.9. Which nuclei shield them from the r-process?

○ $^{84}_{38}\text{Sr}$ is shielded from the r-process by $^{84}_{36}\text{Kr}$, $^{92}_{42}\text{Mo}$ is shielded by $^{92}_{40}\text{Zr}$, $^{94}_{42}\text{Mo}$ is shielded by $^{94}_{40}\text{Zr}$, $^{96}_{44}\text{Ru}$ is shielded by $^{96}_{40}\text{Zr}$ (*not* by $^{96}_{42}\text{Mo}$), and $^{98}_{44}\text{Ru}$ is shielded by $^{98}_{42}\text{Mo}$.

- Neutron-capture reactions are responsible for the production of most nuclei heavier than zinc. Why don't normal fusion reactions involving nuclei, protons, and/or α-particles produce these?

○ There are two main reasons. First, the Coulomb barriers of nuclei are too large to permit charged-particle fusion of elements more massive than zinc. Second, the binding energy per nucleon reaches a maximum near $^{56}_{26}\text{Fe}$, so production of heavier nuclei requires that energy be extracted from the environment rather than contributed to the environment, and they would be very rare in an equilibrium process like silicon burning.

Exercise 6.5　Produce a table summarizing the nucleosynthesis stages you have studied. Include the major reactions and their products, the stellar masses responsible, ignition temperatures, and timescales.　■

Summary of Chapter 6

1. As AGB stars become very luminous, a superwind removes most of the hydrogen envelope. A *proto*planetary nebula develops as the mass-losing star becomes obscured by dust. The underlying helium-rich star gets smaller and hotter, developing a fast, radiation-driven wind that causes mass loss in a bipolar outflow.

2. Once the central star's surface temperature reaches $T_{\text{eff}} \approx 10\,000$ K, it ionizes the ejected envelope, which is then seen as a planetary nebula

expanding at 30–60 km s^{-1}. The star's hydrogen burning and helium-burning shells (if any hydrogen remains) are extinguished when they get too close to the surface and hence too cool, leaving a hot, degenerate CO core surrounded by hot helium, but with all nuclear burning extinguished. Once the central star reaches $T_{\text{eff}} \approx 100\,000$ K, it begins to fade and cool at almost constant radius as a white dwarf. The whole evolution of a planetary nebula lasts only about 20 000 years.

3. A helium white dwarf is the end-point of the evolution of a star of very low mass ($M_{\text{ms}} \leq 0.5\,M_{\odot}$) in which helium burning (by the helium flash) has not occurred. As such, a helium white dwarf will have a mass of typically $0.4\,M_{\odot}$ or less. A CO white dwarf is the end-point of low- and intermediate-mass stars ($0.5\,M_{\odot} < M_{\text{ms}} \leq 8\,M_{\odot}$) in which helium burning occurred, producing carbon and oxygen in the core, and will have a mass in the range $\sim (0.5\text{–}1.2)\,M_{\odot}$. ONeMg white dwarfs result from the lightest high-mass stars ($8\,M_{\odot} < M_{\text{ms}} \leq 11\,M_{\odot}$) in which carbon burning has occurred, producing Ne and Mg, but which have not burnt neon or oxygen. They will have masses typically in the range $\sim (1.2\text{–}1.4)\,M_{\odot}$.

4. The number density n_{A} of some type of particle A can be written in terms of its mass fraction X_{A}, mass m_{A} and the gas density ρ, as $n_{\text{A}} = \rho X_{\text{A}}/m_{\text{A}}$. Hence $n_{\text{e}} = \rho Y_{\text{e}}/m_{\text{H}}$, where Y_{e} is the number of electrons per nucleon defined by $X_{\text{e}} \equiv Y_{\text{e}} m_{\text{e}}/m_{\text{H}}$. For pure hydrogen, $Y_{\text{e}} = 1$, whereas for helium, carbon and oxygen (and hence for white dwarfs), $Y_{\text{e}} = 0.5$.

5. By saying that degenerate electrons provide the pressure support of a star, it is possible to express the Fermi energy, E_{F}, in terms of the mass.

   For *non-relativistic* degenerate electrons, $E_{\text{F}} \propto M^{4/3}$, so degenerate electrons become more relativistic in more-massive white dwarfs.

   For *ultra-relativistic* degenerate electrons, the core density becomes infinite as the mass increases to the Chandrasekhar mass, $\approx 1.4\,M_{\odot}$, the maximum value which can be supported by degenerate electron pressure.

6. By ascribing the core density of a star to non-relativistic degenerate electrons, a relation between the radius and mass of a white dwarf can be derived: $R_{\text{WD}} \approx (R_{\odot}/74)(M/M_{\odot})^{-1/3}$, almost 100 times smaller than the Sun and comparable to the radius of the Earth. While more-massive main-sequence stars have *larger* radii, more-massive white dwarfs have *smaller* radii! As the radius depends only on the mass, a white dwarf cools and fades along a line of constant radius in the H–R diagram.

7. Combining the mass–radius relation $R_{\text{WD}} \approx (R_{\odot}/74)(M/M_{\odot})^{-1/3}$ with the relation $L = 4\pi R^2 \sigma T_{\text{eff}}^4$ indicates that a white dwarf with the same mass and effective surface temperature as the Sun would be ≈ 5500 times fainter.

8. White dwarfs can tap neither nuclear nor gravitational energy sources, so are reservoirs of thermal energy which is lost at a rate $L = -dE/dt$. The time-dependence of the temperature and luminosity indicate that white dwarfs cool very slowly, reaching $L \sim 10^{-3}\,L_{\odot}$ after $\sim 10^9$ yr.

9. High-mass stars ($M_{\text{ms}} > 8\,M_{\odot}$) become hot (and dense) enough in their cores to fuse carbon and heavier nuclei. The sequence of nucleosynthesis is:

carbon burning (at temperature $> 5 \times 10^8$ K) on timescale ~ 500 yr:

$$^{12}_{6}\text{C} + {}^{12}_{6}\text{C} \longrightarrow {}^{20}_{10}\text{Ne} + {}^{4}_{2}\text{He}$$

$$^{12}_{6}\text{C} + {}^{12}_{6}\text{C} \longrightarrow {}^{23}_{11}\text{Na} + \text{p} \qquad \text{(Eqn 6.8)}$$

$$^{12}_{6}\text{C} + {}^{12}_{6}\text{C} \longrightarrow {}^{23}_{12}\text{Mg} + \text{n}$$

neon burning (at temperature $> 10^9$ K) on timescale ~ 1 yr:

$$^{20}_{10}\text{Ne} + \gamma \longrightarrow {}^{16}_{8}\text{O} + {}^{4}_{2}\text{He} \qquad \text{(Eqn 6.9)}$$

and then

$$^{20}_{10}\text{Ne} + {}^{4}_{2}\text{He} \longrightarrow {}^{24}_{12}\text{Mg} + \gamma \qquad \text{(Eqn 6.10)}$$

oxygen burning (at temperature $> 2 \times 10^9$ K) on timescale $\sim$ a few months:

$$^{16}_{8}\text{O} + {}^{16}_{8}\text{O} \longrightarrow {}^{28}_{14}\text{Si} + {}^{4}_{2}\text{He}. \qquad \text{(Eqn 6.11)}$$

silicon burning (at temperature $> 3 \times 10^9$ K) on timescale ~ 1 day:

$$^{28}_{14}\text{Si} + \gamma \longrightarrow {}^{24}_{12}\text{Mg} + {}^{4}_{2}\text{He} \qquad \text{(Eqn 6.12)}$$

elements between silicon and the iron group form in an equilibrium (NSE) between photodissociation and α-, p- and n-capture reactions.

10. The silicon burning equilibrium reactions are governed by a Boltzmann factor. When energy is *released* it favours formation of the heavier ($A + \alpha$) nucleus. Where energy is *absorbed* it favours separate A and α-particles. The binding energy per nucleon peaks around $A = 56$. The equilibrium therefore drives the nuclei towards $A = 56$, which accounts for the high abundance of nuclei near the iron peak.

11. By the end of silicon burning, the star has built up a series of shells of different composition.

12. The nucleosynthesis of elements above the iron peak faces two problems: very high Coulomb barriers and a decreasing binding energy per nucleon. Neutron capture avoids the Coulomb barrier and is the main mechanism for the nucleosynthesis of elements heavier than the iron-peak group.

13. The chart of nuclides records atomic number, Z, on the vertical axis and neutron number, N, on the horizontal axis. Lines of constant atomic mass number, A, lie at $45°$ between upper-left and lower-right. Stable nuclei occupy a zone stretching from low N and Z to high N and Z, called the valley of stability.

14. A neutron-capture reaction leads to a neutron-rich isotope of an element;

$$^{A}_{Z}\text{X}_N + \text{n} \longrightarrow {}^{A+1}_{Z}\text{X}_{N+1} + \gamma. \qquad \text{(Eqn 6.14)}$$

Unstable neutron-rich isotopes β^--decay. This increases the atomic number by one while decreasing the neutron number N, and thus produces a new element

$$^{A+1}_{Z}\text{X}_{N+1} \longrightarrow {}^{A+1}_{Z+1}(\text{X}+1)_N + \text{e}^- + \bar{\nu}_{\text{e}}. \qquad \text{(Eqn 6.15)}$$

β^--decay moves diagonally to the upper-left at $45°$, along a line of constant atomic mass number A, in the chart of nuclides.

15. Nuclei with closed neutron shells (magic numbers $N = 50, 82$ and 126) are more stable than those with slightly higher neutron numbers and have low neutron-capture cross-sections, so they are more abundant.

16. The β^--decay timescale depends solely on the isotope, whereas the neutron-capture timescale also depends on the environment (the number-density of free neutrons). Two extremes are the s- (slow) process in which the neutron-capture rate is much lower than the β^--decay rates, so the β^--decay dominates, and the r- (rapid) process in which the neutron-capture rate is much faster than the β-decay rate, so that neutron capture dominates. The s-process pathway thus remains close to the valley of stability. The r-process temporarily produces very neutron-rich unstable isotopes far to the right (neutron-rich) side of the valley of stability. Only after the neutron number-density falls are the unstable nuclei able to β^--decay back to the valley of stability.

17. Some slightly neutron-rich stable isotopes are separated from the rest of the valley of stability by short-lived unstable nuclei. The s-process cannot jump the gap and they can only be produced by the r-process.

18. Any nucleus on the s-process path that is shielded from r-process β^--decays by another stable nucleus must originate purely from the s-process.

19. Some very rare neutron-poor (proton-rich) isotopes cannot be made by the s- or r-processes; they are p-process nuclei.

20. The abundance peaks of the r-process are found at lower A than the peaks of the s-process. The r-process peaks for $N = 50, 82$ and 126 appear at $A \approx 85$, 130 and 195, while the s-process peaks are at $A \approx 90, 138$ and 208. The closed neutron shell abundance peaks for the r-process are also *wider* than the s-process peaks.

Chapter 7 Supernovae, neutron stars & black holes

Introduction

This chapter considers the end-points in the lives of massive stars. However, as noted at the beginning of the book, this process also marks the beginning of the life cycles of new stars, which form from the raw materials expelled in supernovae explosions.

7.1 Supernovae

In Chapter 6 we found that a star whose final mass is below the Chandrasekhar mass is supported against collapse by the pressure of degenerate electrons. It ends its days as a white dwarf which slowly cools and fades. In this section we see what happens to a star whose final mass *exceeds* this limit and in the subsequent sections of this chapter we discover what remnant it may leave behind.

A star with a main-sequence mass greater than about $11\,M_\odot$ will complete all the stages of nuclear burning that we have discussed in previous chapters. Silicon burning will result in a core composed mainly of iron-56, surrounded by concentric shells of silicon, oxygen, neon, carbon, helium and hydrogen. No energy can be released by the thermonuclear fusion of iron, so the core collapses and the degenerate electrons within it become more and more relativistic. When the mass of the core exceeds the Chandrasekhar mass (about $1.4\,M_\odot$), the degenerate electrons are no longer able to support the core, and a catastrophic collapse follows. The core will essentially collapse on a free-fall timescale given by

$$t_{ff} = \left(\frac{3\pi}{32G\rho}\right)^{1/2}.$$

(Eqn 2.5)

● Calculate the free-fall timescale for a stellar core with a density of $10^{14}\ \mathrm{kg\,m^{-3}}$.

○ The free-fall timescale is

$$\left(\frac{3\times\pi}{32\times 6.673\times 10^{-11}\,\mathrm{N\,m^2\,kg^{-2}}\times 10^{14}\,\mathrm{kg\,m^{-3}}}\right)^{1/2} \sim 0.007\,\mathrm{s}.$$

The initiation of **exothermic** (energy-liberating) fusion reactions provides pressure *support* for stars during their long-lasting burning phases. However, the initiation of **endothermic** (energy-absorbing) reactions draws kinetic energy out of the surrounding medium and hence reduces the pressure support. There are two processes which can absorb energy in the collapsing core. These are photodisintegration of nuclei by high-energy gamma-rays and electron-capture processes. In the first, the energy is used to unbind the nuclei, whilst in the second, energy is liberated as the kinetic energy of neutrinos which stream out of the star largely unhindered. We now consider these two processes in turn.

7.1.1 Nuclear photodisintegration

There are many ways in which high-energy gamma-rays can disintegrate nuclei of iron-56, but as an illustration of the physics involved, and the amount of energy that may be absorbed by the process, we consider the situation where a nucleus of iron-56 is broken down into helium nuclei:

$$\gamma \; + \; {}^{56}_{26}\text{Fe} \; \longleftrightarrow \; 13\,{}^{4}_{2}\text{He} \; + \; 4\text{n}.$$

The amount of energy absorbed by this process is given by $\Delta Q = (13m_4 + 4m_1 - m_{56})c^2 = 124.4$ MeV. Since this is an equilibrium reaction, we may calculate the proportion of iron-56 nuclei that are disintegrated by equating chemical potentials: $\mu'_{56} = 13\mu'_4 + 4\mu'_1$, as shown in the following example.

Worked Example 7.1

Derive an equation for the proportion of iron-56 nuclei that are dissociated into thirteen helium-4 nuclei and four neutrons, in terms of ΔQ, the non-relativistic quantum concentrations and the number of polarizations for each particle.

Solution

We begin by balancing the chemical potentials:

$$\mu'_{56} = 13\mu'_4 + 4\mu'_1$$

and since $\mu' = mc^2 - kT \log_e(g_s\, n_{\text{QNR}}/n)$ we have

$$m_{56}c^2 - kT \log_e\left(\frac{g_{56}\, n_{\text{Q}56}}{n_{56}}\right) = 13m_4c^2 - 13kT \log_e\left(\frac{g_4\, n_{\text{Q}4}}{n_4}\right) + 4m_1c^2$$

$$- 4kT \log_e\left(\frac{g_1\, n_{\text{Q}1}}{n_1}\right),$$

where m_{56}, m_4 and m_1 are the masses of an iron-56 nucleus, a helium-4 nucleus and a neutron respectively; $n_{\text{Q}56}$, $n_{\text{Q}4}$ and $n_{\text{Q}1}$ are the non-relativistic quantum concentrations of an iron-56 nucleus, a helium-4 nucleus and a neutron respectively; g_{56}, g_4 and g_1 are the number of polarizations of iron-56 nuclei, helium-4 nuclei and neutrons respectively; and n_{56}, n_4 and n_1 are the number densities of iron-56 nuclei, helium-4 nuclei and neutrons respectively. This may be rearranged as

$$13m_4c^2 + 4m_1c^2 - m_{56}c^2 = kT \log_e\left[\left(\frac{g_4\, n_{\text{Q}4}}{n_4}\right)^{13}\left(\frac{g_1\, n_{\text{Q}1}}{n_1}\right)^4\left(\frac{g_{56}\, n_{\text{Q}56}}{n_{56}}\right)^{-1}\right].$$

The left-hand side is simply ΔQ, so taking the exponential of both sides and rearranging slightly, we have:

$$\exp\left(\frac{\Delta Q}{kT}\right) = \frac{(g_4\, n_{\text{Q}4}/n_4)^{13}\, (g_1\, n_{\text{Q}1}/n_1)^4}{g_{56}\, n_{\text{Q}56}/n_{56}}.$$

Since we are interested in the proportion of iron-56 nuclei that are dissociated, we take this fraction onto the left-hand side to get:

$$\frac{n_4^{13}\, n_1^4}{n_{56}} = \frac{g_4^{13}\, g_1^4}{g_{56}}\, \frac{n_{\text{Q}4}^{13}\, n_{\text{Q}1}^4}{n_{\text{Q}56}}\, \exp\left(-\frac{\Delta Q}{kT}\right).$$

This equation can be solved at a given temperature, and yields the result that about three-quarters of the iron nuclei are dissociated in this way when the temperature and density are about 10^{10} K and 10^{12} kg m^{-3} respectively.

Worked Example 7.2

If each nucleus of iron can absorb 124.4 MeV of energy by photodisintegration in the process $\gamma + {}^{56}_{26}\text{Fe} \longrightarrow 13\,{}^{4}_{2}\text{He} + 4\text{n}$, and three-quarters of the core of mass $1.4\,\text{M}_{\odot}$ is dissociated in this way, calculate the total energy absorbed by this process.

Solution

The mass M of a sample of material is given by the total number of particles N multiplied by their individual masses m, i.e. $M = Nm$, so the number of particles is $N = M/m$. For a total core mass of $M = (3/4) \times 1.4\,\text{M}_{\odot}$, the number of iron-56 nuclei is $N = (3/4) \times 1.4\,\text{M}_{\odot}/56u = (1.05 \times 1.99 \times 10^{30}\,\text{kg})/(56 \times 1.661 \times 10^{-27}\,\text{kg}) = 2.25 \times 10^{55}$ nuclei.

Each nucleus absorbs 124.4 MeV $= 124.4 \times 10^{6}\,\text{eV} \times 1.602 \times 10^{-19}\,\text{J eV}^{-1}$ $= 1.993 \times 10^{-11}\,\text{J}$.

So the core absorbs $2.25 \times 10^{55} \times 1.993 \times 10^{-11}\,\text{J} = 4.5 \times 10^{44}\,\text{J}$ by the photodisintegration of iron-56 nuclei.

At still higher temperatures, helium nuclei will also undergo photodisintegration via:

$$\gamma + {}^{4}_{2}\text{He} \longleftrightarrow 2\text{p} + 2\text{n}$$

and ΔQ for this process is 28.3 MeV.

Exercise 7.1   Following the pattern of Worked Example 7.1, write down an expression for the proportion of helium-4 nuclei that are dissociated: $n_{\text{p}}^{2}\, n_{\text{n}}^{2}/n_{4}$ where n_{n} is the number density of neutrons, n_{p} is the number density of protons and n_{4} is the number density of helium-4 nuclei.   ∎

At the temperature and density involved in the stellar core, namely about 10^{10} K and 10^{12} kg m^{-3} respectively, roughly half the helium-4 nuclei are dissociated.

Exercise 7.2   Following the pattern of Worked Example 7.2, if each nucleus of helium-4 can absorb 28.3 MeV of energy by photodisintegration in the process $\gamma + {}^{4}_{2}\text{He} \longrightarrow 2\text{p} + 2\text{n}$, and half of the core of mass $1.4\,\text{M}_{\odot}$ is dissociated in this way, calculate the total energy absorbed by this process.   ∎

The total amount of energy absorbed by nuclear photodisintegration is therefore $4.5 \times 10^{44}\,\text{J}$ from dissociation of iron-56 plus $9.5 \times 10^{44}\,\text{J}$ from dissociation of helium-4, or $1.4 \times 10^{45}\,\text{J}$ in total.

7.1.2   Electron capture

The second mechanism by which energy is absorbed in the iron core of a star is that of electron capture, allowing energy to be carried away by neutrinos. The conversion of protons (in nuclei) to neutrons by electron capture is possible if the gas is sufficiently dense for degenerate electrons to have an energy above the 1.3 MeV mass–energy excess of neutrons compared to protons ($m_{\text{p}}c^{2} = 938.3$ MeV, while $m_{\text{n}}c^{2} = 939.6$ MeV). Actually, the energy excess required for electron capture by bound nuclear protons, rather than free protons, is usually somewhat higher and depends on the nucleus, but is nevertheless usually a few MeV. The general reaction may be written as:

$$p + e^- \longrightarrow n + \nu_e$$

and is sometimes referred to as **neutronization**.

At densities above about 10^{12} kg m^{-3}, iron-56 nuclei will capture electrons in the reaction:

$$e^- + {}^{56}_{26}\text{Fe} \longrightarrow {}^{56}_{25}\text{Mn} + \nu_e$$

and when the density exceeds 10^{14} kg m^{-3}, a further electron capture proceeds rapidly:

$$e^- + {}^{56}_{25}\text{Mn} \longrightarrow {}^{56}_{24}\text{Cr} + \nu_e.$$

Yet further electron-capture reactions then occur, converting more and more protons into neutrons.

As electrons are rapidly used up by these reactions, the pressure support provided by degenerate electrons quickly disappears, and the core collapses rapidly, as noted earlier. The neutrinos produced by the electron-capture reactions carry away most of the energy. Although the passage of neutrinos is hindered by the high density in the core which raises their interaction probability, the neutrinos are still all able to escape within a few seconds.

● If each neutrino produced in the reactions above carries away 10 MeV of energy, how much energy (in joules) is removed by neutrinos if the whole core (of $1.4\,M_\odot$) undergoes neutronization?

○ The number of protons contained in a stellar core of mass of $1.4\,M_\odot$ is $\frac{1}{2} \times 1.4\,M_\odot/u = (\frac{1}{2} \times 1.4 \times 1.99 \times 10^{30}$ kg$) / (1.661 \times 10^{-27}$ kg$)$ $\sim 8.4 \times 10^{56}$ protons. Assuming charge neutrality, there will also be $\sim 8.4 \times 10^{56}$ electrons. If each electron undergoes neutronization, this will produce $\sim 8.4 \times 10^{56}$ neutrinos. If each neutrino carries away 10 MeV of energy, the amount of energy removed by each neutrino is $10 \times 10^6 \times 1.602 \times 10^{-19}$ J $= 1.6 \times 10^{-12}$ J. So the total amount of energy removed is $1.6 \times 10^{-12} \times 8.4 \times 10^{56} = 1.3 \times 10^{45}$ J.

So, neutronization, like nuclear photodisintegration, can remove of order 10^{45} J of energy from the collapsing stellar core within a few seconds!

Over its lifetime, the star has been supported against collapse by the thermal energy derived from converting hydrogen to iron, via intermediate species. Some of this energy is radiated away by the surface of the star over its long lifetime. Then, at the end of its life, it seeks to reverse that entire process by converting the iron in the core back into hydrogen. Things are not looking good for the star, which is destined to collapse from a lack of pressure support. If this wasn't bad enough, neutrinos are also being liberated by neutronization.

● The combined processes of photodisintegration and neutronization remove around 3×10^{45} J from the star in a few seconds. How long would a $12.5\,M_\odot$ star take to emit this much energy, radiating at its main-sequence luminosity?

○ Since $L \propto M^3$, the luminosity of a $12.5\,M_\odot$ star on the main sequence is $L \sim 12.5^3\,L_\odot \sim 2000\,L_\odot$ or $\sim 2000 \times 4 \times 10^{26}\,J\,s^{-1} \sim 8 \times 10^{29}\,J\,s^{-1}$.

So radiating at this rate, it would take $3 \times 10^{45}\,J\,/\,8 \times 10^{29}\,J\,s^{-1} \sim 4 \times 10^{15}$ seconds to release the same amount of energy. This is equivalent to about 100 million years, which is comparable to the main-sequence lifetime of such a star.

7.1.3   Supernovae explosions

Photodisintegration and neutronization reactions trigger a rapid collapse of a stellar core composed largely of iron. The collapse will continue until the core reaches a density that is comparable to that of an atomic nucleus.

● Given that a nucleus of mass number A has a radius roughly $R \sim 1.2 \times 10^{-15}\,m \times A^{1/3}$, what is its density?

○ The density is $\rho = M/V = (Au)/(4\pi R^3/3) = (3Au/4\pi \times (1.2 \times 10^{-15}\,m)^3 \times A)$. Hence $\rho \sim (3 \times 1.661 \times 10^{-27}\,kg)/(4\pi \times (1.2 \times 10^{-15}\,m)^3) \sim 2.3 \times 10^{17}\,kg\,m^{-3}$.

When the core approaches nuclear densities, nuclear forces resist further compression and the collapsing core will rebound. This sends a shockwave through the infalling material which produces an expulsion of much of the stellar envelope. This is a **supernova** explosion.

Observationally, supernovae are detected by the sudden brightening of a previously invisible star. The optical light output is typically about 10^{42} J over the course of a year or so. However, it can also be observed that a further 10^{44} J of energy is carried away as kinetic energy of the exploding debris, at velocities of tens of thousands of kilometres per second. These are certainly vast amounts of energy, but it turns out they are not the whole story for the energy release during a supernova.

The equation for the gravitational potential energy (E_{GR}) released in the collapse of a stellar core of mass M, from an initial radius R_1 to a final radius R_2, is $E_{GR} = GM^2/R_2 - GM^2/R_1$. In the case of a supernova, the core will collapse from a radius of around 1000 km to 10 km. Since $R_2 \ll R_1$, we have $E_{GR} \sim GM^2/R_2$. So the energy released by the collapse of a $1.4\,M_\odot$ stellar core is $E_{GR} \sim (6.673 \times 10^{-11}\,N\,m^2\,kg^{-2} \times (1.4 \times 1.99 \times 10^{30}\,kg)^2)/10^4\,m \sim 5 \times 10^{46}$ J.

This is two orders of magnitude greater than the observed energy carried away by the expanding debris, and it is at least 10 times the energy required to photodisintegrate the iron core, or 10 times the energy that can be carried away by neutronization. So where does all the gravitational energy released by a supernova actually go?

It seems that there is an intermediate stage between the iron core and the production of a neutron star remnant. This takes the form of a hot, dense plasma of neutrons, protons, electrons, neutrinos and photons. At a temperature of 10^{11} K and a density of $10^{14}\,kg\,m^{-3}$, this plasma is opaque to electromagnetic radiation, but not to neutrinos. It is believed that neutrino–antineutrino pairs are produced in the plasma, and each type of particle (electron neutrino, electron antineutrino, muon neutrino, muon antineutrino, tauon neutrino, tauon antineutrino) carries away typically one-sixth of the gravitational binding energy of the collapsing core.

● What are the main ways in which liberated energy is carried away from a supernova? Quantify them.

○ Photons provide the *least* effective means of removing the energy, only $\sim 10^{42}$ J over the first year or so after the explosion. The expansion of the ejected material carries 100 times as much in the form of kinetic energy, $\sim 10^{44}$ J. However, the binding energy of the neutron star is $\sim 5 \times 10^{46}$ J, so much more energy must be removed than the photons and kinetic energy carry. The majority is carried away by neutrinos.

We end this discussion of supernova explosions by noting that the mechanism by which the outer layers of a core-collapse supernova are ejected is uncertain, and computer models of the process currently fail to reproduce what is believed to happen. It is clear that there is the *energy* to do it, but the means by which that energy and momentum is *imparted* to the outer layers is unclear. When the collapsing envelope of the star reaches the collapsed neutron core which resists further compression, a rebounding shock-wave is set up that helps reverse the infall of the envelope. This begins to convert the collapse into an explosion, but current models fall short of completing the process. Current research in this area is investigating the possibility that interactions between neutrinos and the stellar envelope above the newly formed neutron star deposit some of the neutrino energy into the envelope. (Usually we think of neutrinos as not interacting strongly with matter, but the flux of neutrinos is so large above the newly formed neutron star, and the matter so dense, that sufficient interactions may nevertheless take place.) This rapid heating of the stellar matter by the intense neutrino flux may then drive the expansion that ejects the envelope into space. This same event is also widely regarded as the probable site of r-process neutron-capture reactions responsible for synthesizing many of the elements above the iron peak.

7.1.4 Types of supernovae

Observationally, astronomers distinguish two broad categories of supernovae, referred to as type I and type II, depending whether hydrogen is absent (type I) or present (type II) in their visible spectra. The core-collapse supernovae described above are supernovae type II, if the progenitor still retained its hydrogen envelope at the moment it exploded, or the rarer type Ib or Ic if its hydrogen envelope had been lost due to a strong wind or some other stripping mechanism. The other main type, supernovae type Ia, are believed to arise in binary stars in which a CO white dwarf accretes material from its companion, pushing the white dwarf over the Chandrasekhar mass. The resulting thermonuclear explosion occurring under degenerate conditions causes a catastrophic explosion of the white dwarf which most likely destroys it and its companion star completely. A different kind of supernova, called a pair-instability supernova (PISN), can result from the final evolutionary stages of a very massive star ($M > 140 \, M_\odot$). In these objects, electron–positron pairs form spontaneously from thermal gamma-rays in the core, and undermine the pressure support of the star. It collapses suddenly, triggering explosive burning of oxygen and silicon which disrupts the star.

The main distinctions between type Ia and type II supernovae are listed in Table 7.1.

Table 7.1   Main types of supernovae

Supernovae type Ia (SN Ia)	Supernovae type II (SN II)
hydrogen *not* in visible spectrum	hydrogen in visible spectrum
associated with old stars	associated with young stars
standard lightcurve; used as distance indicator	less predictable lightcurve
probably leaves no compact remnant, only a fireball	leaves neutron star or black hole remnant
due to accretion on to white dwarf, which then exceeds Chandrasekhar mass $(1.4\,\mathrm{M_\odot})$	due to core collapse of massive star $(M_{ms} > 11\,\mathrm{M_\odot})$

● Why might core-collapse supernovae be associated with young stars, while SN Ia are associated with old stars?

○ Core-collapse supernovae involve massive stars with $M_{ms} > 11\,\mathrm{M_\odot}$, whose main-sequence lifetimes are typically $t_{ms} \approx 10(M/\mathrm{M_\odot})^{-2.5} \times 10^9$ yr $\leq 30 \times 10^6$ yr (see Chapter 4). SN Ia, on the other hand, are associated with lower-mass stars that become white dwarfs, probably with initial masses $M_{ms} \leq 2\,\mathrm{M_\odot}$, and therefore having main-sequence lifetimes $t_{ms} \sim 10(M/\mathrm{M_\odot})^{-2.5} \times 10^9$ yr $\geq 2 \times 10^9$ yr.

In February 1987, the nearest supernovae for centuries was observed in our neighbouring galaxy, the Large Magellanic Cloud (see Figure 7.1). This became

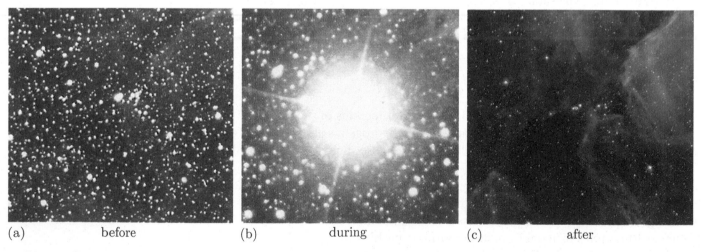

(a)   before   (b)   during   (c)   after

Figure 7.1   Before, during, and after (10 years on) images of the best studied supernova, SN1987A. This was the first supernova whose progenitor (parent star) was identifiable from archival photographs, and the first supernova from which neutrinos were detected. Although it occurred in a neighbouring galaxy, the Large Magellanic Cloud, it was visible to the unaided eye at maximum brightness. Because of its proximity, circumstellar material ejected before the supernova explosion can be seen as a bright equatorial ring surrounding the SN. Around 1999, the slowly expanding equatorial ring was struck by the fastest ejecta from the supernova explosion, causing the ring to vary in brightness. Also around 1999, the envelope of the expanding supernova became large enough to resolve. (a) and (b) are to the same scale; (c) is a $1.5\times$ enlargement showing only the central region.

the best studied supernovae to date, and through Earth-based detections of (a mere) 20 neutrinos from the supernova, confirmed many of the theories about how supernovae behave.

● The SNR Cas A (see Figure 7.2) had for a long time shown no evidence for a compact stellar remnant, but new images taken in X-rays seem to reveal one. If this is correct, was Cas A a type Ia or type II supernova?

○ Type II. Type Ia completely disrupt the stars, leaving a brilliant 'fireball' but no compact remnant.

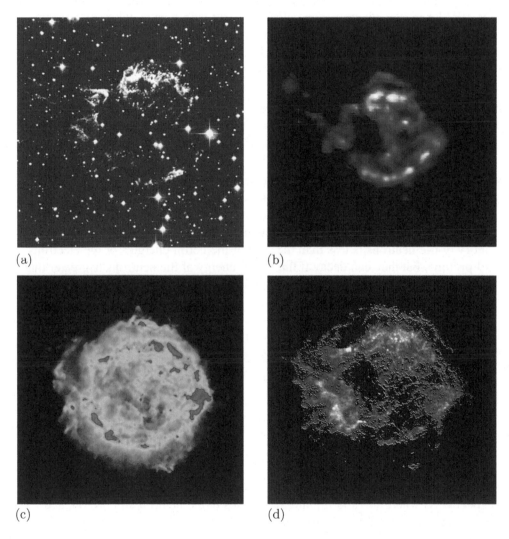

(a) (b)

(c) (d)

Figure 7.2 The appearance of a supernova remnant (SNR) depends greatly on the waveband observed. Cassiopeia A (known as Cas A), the remnant of a ~ 300 year-old supernova in the Galaxy, is inconspicuous at optical wavelengths (a), where many stars in the field radiate with similarly brightness, but in infrared (b), radio (c), and X-ray wavelengths (d) it is the only source visible. The X-ray image, which is the most recent, shows a probable stellar remnant for the first time, and also shows that the gaseous remnant consists of two shock fronts, an inner one caused by the collision between the supernova ejecta with the circumstellar shell, which is heated to $\sim 10^7$ K, and an outer shock caused possibly by a sonic boom running ahead of the expanding ejecta.

7.2 Neutron stars

Core-collapse supernovae leave behind one of two **remnants**: either a **neutron star** or a *black hole*. These two classes of object will now be examined in turn.

A newly formed neutron star will have a temperature of $(10^{11}-10^{12})$ K but it will quickly cool, by emission of neutrinos, to around 10^9 K on a timescale of a day, and to around 10^8 K within a hundred years. It will also have a mass of order $\sim M_\odot$ contained within a sphere only ~ 10 km in radius, hence it will have a density of a few $\times 10^{17}$ kg m^{-3}. These extreme temperatures and densities lead to some extraordinary consequences for the neutron star's properties.

7.2.1   Neutron star composition

For normal matter, the most stable configuration, with the most negative binding energy per nucleon, is nuclei around iron-56. When the density approaches 10^{14} kg m^{-3}, the most stable nuclei are actually far more neutron-rich, namely those such as nickel-78 and iron-76. At still higher densities, above 4×10^{14} kg m^{-3}, a phenomenon called **neutron drip** occurs, whereby neutrons 'leak out' of nuclei and an equilibrium mixture of nuclei, neutrons and electrons exists. When the density exceeds that of normal nuclear matter (at about 2×10^{17} kg m^{-3}), the nuclei begin to merge and a dense gas of protons, neutrons and electrons is produced. Under these conditions there are complicated and uncertain interactions between nuclei, so the equation of state for material with these properties is not known.

Despite this, a reasonable idea of the composition of a neutron star can be gained by considering a crude model in which the neutron star is treated as an ideal gas of degenerate electrons, protons and neutrons. The normal β^--decay of free neutrons is blocked in this situation because of the Pauli exclusion principle: there are no available quantum states left for the protons and electrons to occupy, so the neutrons are not able to decay. In particular, neutrons cannot decay if the Fermi energy of the neutrons is less than the sum of the Fermi energies of the electrons and protons, but they can decay if the Fermi energy of the neutrons is greater than that of the protons plus electrons. An equilibrium will therefore exist when:

$$E_F(n) = E_F(p) + E_F(e).$$

This is equivalent to a relationship between the chemical potentials of the neutrons, protons and electrons.

Now, the neutrons and protons can be considered to be non-relativistic, hence their Fermi energies are:

$$E_F(n) = m_n c^2 + \frac{p_F^2(n)}{2m_n}$$

$$E_F(p) = m_p c^2 + \frac{p_F^2(p)}{2m_p}$$

respectively, where m_n and m_p are the masses of the neutron and proton, and $p_F(n)$ and $p_F(p)$ are their Fermi momenta.

Conversely, the electrons are much less massive then the protons and neutrons, so they can be considered to be ultra-relativistic. Their Fermi energy is:

$$E_F(e) = p_F(e)c$$

where $p_F(e)$ is the electron's Fermi momentum.

Now, since the Fermi momentum is $p_F = [3n/8\pi]^{1/3} h$, we can write the equilibrium condition as:

$$m_n c^2 + \left(\frac{3n_n}{8\pi}\right)^{2/3} \frac{h^2}{2m_n} = m_p c^2 + \left(\frac{3n_p}{8\pi}\right)^{2/3} \frac{h^2}{2m_p} + \left(\frac{3n_e}{8\pi}\right)^{1/3} hc.$$

Since the gas will be neutral, $n_p = n_e$. Furthermore, the mass energy difference between the neutron and proton is $m_n c^2 - m_p c^2 = 1.3$ MeV, so we can rearrange this as:

$$\left(\frac{3n_p}{8\pi}\right)^{1/3} hc + \left(\frac{3n_p}{8\pi}\right)^{2/3} \frac{h^2}{2m_p} - \left(\frac{3n_n}{8\pi}\right)^{2/3} \frac{h^2}{2m_n} = 1.3 \, \text{MeV}. \quad (7.1)$$

This equation can be solved numerically to find the ratio of neutrons to protons at a given density. For instance, at a density of 2.5×10^{17} kg m^{-3}, the ratio is $n_n/n_p \sim 200$.

Worked Example 7.3

For a density of 2.5×10^{17} kg m^{-3}, verify that Equation 7.1 predicts a neutron-to-proton ratio of about 200 to 1.

Solution

The number density and mass density are related by $n = \rho X/m$, where X is the mass fraction. Since the neutron star is dominated by neutrons, $X_n \sim 1$ and we have $n_n \sim 2.5 \times 10^{17}$ kg m^{-3} / 1.675×10^{-27} kg $\sim 1.49 \times 10^{44}$ m^{-3}.

Hence the stated ratio implies $n_p \approx 7.45 \times 10^{41}$ m^{-3}.

So, the first term on the left-hand side of Equation 7.1 becomes:

$$\left(\frac{3 \times 7.45 \times 10^{41} \, \text{m}^{-3}}{8\pi}\right)^{1/3} \times (6.626 \times 10^{-34} \, \text{J s}) \times (2.998 \times 10^8 \, \text{m s}^{-1})$$

$$= 8.87 \times 10^{-12} \, \text{J}.$$

Similarly, the second term on the left-hand side of Equation 7.1 becomes:

$$\left(\frac{3 \times 7.45 \times 10^{41} \, \text{m}^{-3}}{8\pi}\right)^{2/3} \frac{(6.626 \times 10^{-34} \, \text{J s})^2}{2 \times 1.673 \times 10^{-27} \, \text{kg}} = 0.26 \times 10^{-12} \, \text{J}$$

and the third term on the left-hand side of Equation 7.1 becomes:

$$\left(\frac{3 \times 1.49 \times 10^{44} \, \text{m}^{-3}}{8\pi}\right)^{2/3} \frac{(6.626 \times 10^{-34} \, \text{J s})^2}{2 \times 1.675 \times 10^{-27} \, \text{kg}} = 8.93 \times 10^{-12} \, \text{J}.$$

So, the entire left-hand side of Equation 7.1 is $(8.87 + 0.26 - 8.93) \times 10^{-12}$ J $= 0.20 \times 10^{-12}$ J or $(0.20 \times 10^{-12}/1.602 \times 10^{-19})$ eV which is 1.25 MeV. This is close enough to the 1.3 MeV energy difference between the neutron and proton, given the approximate ratio of 200 : 1 which we used as the starting point for the calculation.

So, as long as there is one proton and one electron for every 200 neutrons, the neutrons are prevented from decaying, and the neutron star remains supported, essentially by the pressure of degenerate neutrons. We see, then, that neutron stars are aptly named, and why they are not, for example, called nucleon stars.

7.2.2   The radius of a neutron star

When deriving equations to describe neutron stars we should be wary of oversimplifying the assumptions about conditions inside a neutron star. In reality, Newtonian gravity should be replaced by Einstein's general-relativistic treatment, because for a neutron star of mass M and radius R_{NS}, a neutron's gravitational potential energy, GMm_n/R_{NS}, is comparable to its rest-mass energy, $m_n c^2$.

● Calculate the ratio of a neutron's gravitational potential energy to its rest-mass energy, inside a neutron star of mass 1.4 $M_\odot$ and radius 10 km.

○ The ratio is $GMm_n/R_{NS}m_n c^2 = GM/R_{NS}c^2$. This is equal to $(6.673 \times 10^{-11} \text{ N m}^2 \text{ kg}^{-2} \times 1.4 \times 1.99 \times 10^{30} \text{ kg})/10^4 \text{ m} \times (2.998 \times 10^8 \text{ m s}^{-1})^2 = 0.2$.

The momenta of neutrons also approach the relativistic limits, requiring special relativity. Nevertheless, the non-relativistic approximation can lead to useful insights into the size of the star, and an understanding of the dominant physics. In particular, it is interesting to compare calculations of the radius of a white dwarf, supported by degenerate electrons, and a neutron star, supported by degenerate neutrons. The following question takes you through such a calculation. (You may find it helpful to revise your study of white dwarfs as you complete this exercise.)

Exercise 7.3   Set up two columns on a sheet of paper, one for a white dwarf and one for a neutron star. Then in each column:

(a) Write the equation for the pressure of non-relativistic degenerate material as a function of particle *number* density.

(b) Rewrite the equation to give the pressure in terms of the *mass* density for degenerate matter at the core of the star.

(*Hint*: In a neutron star, neutrons are ~200 times more abundant than protons or electrons, so treat the neutron star as composed solely of neutrons.)

(c) Rearrange the expression for the core pressure of a Clayton stellar model $P_c \approx (\pi/36)^{1/3} GM^{2/3}\rho_c^{4/3}$, to give an expression for the core density. (This equation is the same for both objects.)

(d) As the core pressure required to support the object is provided by the degenerate particles, use your result from (b) to substitute for the core pressure in (c), and obtain a new expression for the core density in terms of the other variables. (Actually, instead of finding an expression for ρ_c, it will be more convenient to find an expression for $\rho_c^{-1/3}$.)

(*Hint*: As the density appears explicitly in the pressure term, you have to rearrange the equation to get all of the density terms on the left-hand side. You may find it easier first to obtain an expression for $\rho_c^{-1/4}$, then take the (4/3)-power to find an expression for $\rho_c^{-1/3}$.)

(e) Rearrange the definition of the mean density $\langle \rho \rangle = 3M/4\pi R^3$ to get an expression for the radius in terms of the mass and mean density, and then use the result that a polytropic star with $P \propto \rho^{5/3}$ has a core density six times the mean density, to re-express the radius in terms of mass and *core* density. (This is the same for white dwarfs and neutron stars.)

(f) Substitute in to this equation your result from (d) to find an expression for the radius of the star. ∎

The final result for the previous exercise shows that

$$R_{\mathrm{WD}} = \left(\frac{729}{32\pi^4}\right)^{1/3} \frac{1}{G}\left(\frac{h^2}{5m_{\mathrm{e}}}\right)\left(\frac{Y_{\mathrm{e}}}{m_{\mathrm{H}}}\right)^{5/3} M^{-1/3} \tag{7.2}$$

$$R_{\mathrm{NS}} = \left(\frac{729}{32\pi^4}\right)^{1/3} \frac{1}{G}\left(\frac{h^2}{5m_{\mathrm{n}}}\right)\left(\frac{1}{m_{\mathrm{n}}}\right)^{5/3} M^{-1/3}. \tag{7.3}$$

Taking the ratio of the second to the first, noting that $m_{\mathrm{H}} \approx m_{\mathrm{n}}$, gives

$$\frac{R_{\mathrm{NS}}}{R_{\mathrm{WD}}} = \left(\frac{m_{\mathrm{e}}}{m_{\mathrm{n}}}\right) \times \left(\frac{1}{Y_{\mathrm{e}}^{5/3}}\right) \approx \frac{1}{580}. \tag{7.4}$$

There are two differences which affect the two types of star, one minor and one major.

Minor factor: Y_{e} appears in the white-dwarf radius but not in the neutron-star radius. Recall that Y_{e} is the number of degenerate particles (electrons) per nucleon in a white dwarf. This is relevant in the white dwarf because the degenerate particles do not dominate the composition of the material. Typically $Y_{\mathrm{e}} \approx 0.5$, so this affects the mass ratio by only a factor of ≈ 3.

Major factor: The radius of a degenerate star is scaled according to the mass of the particle whose degenerate pressure supports the star. (The mass of the particle also determines its de Broglie wavelength, so you could also say that the de Broglie wavelength determines the radius.) As the pressure in a neutron star is provided by neutrons that are almost 2000 times more massive than the electron, neutron stars are almost a factor of 2000 smaller in radius. In an earlier exercise, you showed that the radius of a white dwarf is $R_{\mathrm{WD}} = 1.5 R_{\mathrm{Earth}}(M/M_{\odot})^{-1/3}$. From the above ratio, we can therefore write

$$\begin{aligned}
R_{\mathrm{NS}} &= R_{\mathrm{WD}} \left(\frac{m_{\mathrm{e}}}{m_{\mathrm{n}}}\right) \times \left(\frac{1}{Y_{\mathrm{e}}^{5/3}}\right) \\
&= 1.5 \left(\frac{m_{\mathrm{e}}}{m_{\mathrm{n}}}\right) \times \left(\frac{1}{Y_{\mathrm{e}}^{5/3}}\right) R_{\mathrm{Earth}} \left(\frac{M}{M_{\odot}}\right)^{-1/3} \\
&= 0.0026\, R_{\mathrm{Earth}} \left(\frac{M}{M_{\odot}}\right)^{-1/3}.
\end{aligned}$$

That is, whereas white dwarfs are comparable in size to the Earth, neutron stars have radii of only ≈ 5–10 km!

7.2.3 The maximum mass of a neutron star

It is more difficult to specify the upper-mass limit for neutron stars than for white dwarfs for a couple of reasons.

First, interactions between neutrons are significant. They repel one another at separations $< 1.4 \times 10^{-15}$ m, which makes them less compressible, but also their energies are so high that they produce hyperons (baryons containing one or more strange quarks, as well as up and down quarks) and pions (mesons made from up and down quarks) which reduce the pressure and makes them more compressible. The net effect is to allow more-massive neutron stars than one might expect based on simpler assumptions.

Secondly, the gravitational fields are so strong that Einstein's theory of gravity, general relativity, must be used instead of Newton's theory of gravity, and under general relativity, the gravity also depends on pressure. Whereas in a normal star internal pressure resists gravity, under general relativity it *strengthens* the gravity, which reduces the maximum stable mass.

The competing mechanisms of these two effects mean that it is very difficult indeed to calculate the upper-mass limit for neutron stars, and indeed no consensus exists on the matter. Various calculations and theories give maximum masses between $M_\odot$ and $3\,M_\odot$.

A *very* crude approximation for the upper-mass limit for a neutron star may be gained by following a similar analysis to that which yielded the Chandrasekhar mass limit for white dwarfs. Exercise 6.3 showed that by equating the core pressure in a star to the degenerate pressure of ultra-relativistic electrons, we were able to derive an upper-mass limit for white dwarfs. The resulting equation was

$$M = \frac{1}{\pi}\left(\frac{27}{128}\right)^{1/2}\left(\frac{hc}{G}\right)^{3/2}\left(\frac{Y_e}{m_H}\right)^2.$$

For neutron stars, we simply replace Y_e/m_H by $1/m_n$ and so

$$M = \frac{1}{\pi}\left(\frac{27}{128}\right)^{1/2}\left(\frac{6.626\times10^{-34}\text{ J s}\times2.998\times10^8\text{ m s}^{-1}}{6.673\times10^{-11}\text{ N m}^2\text{ kg}^{-2}}\right)^{3/2}\left(\frac{1}{1.675\times10^{-27}\text{ kg}}\right)^2$$
$$= 8.48\times10^{30}\text{ kg}.$$

So, a very approximate upper-mass limit for neutron stars is $M \le 4.3\,M_\odot$.

Observationally, neutron-star masses may be measured in various binary star systems. In binary pulsars (see the next section), neutron-star masses may be measured very accurately, and many of their measured masses cluster around $\sim 1.35\,M_\odot$. However, there are indications in some systems for neutron-star masses as high as $2\,M_\odot$.

7.3 Pulsars

Neutrons had been discovered by the English physicist Sir James Chadwick in 1932, and as early as 1934 the existence of neutron stars had been suggested by Lev Landau (in the Soviet Union) and by Fritz Zwicky and Walter Baade (in the USA). Zwicky and Baade coined the term 'supernova' and suggested that these events marked the conversion of a normal star into a neutron star at the end of its life. Theoretical models for neutron stars had even been worked out as early as 1939 by Robert Oppenheimer and George Volkoff. However it was not until thirty years later that the first observational evidence for neutron stars was found.

7.3.1 The discovery of pulsars

In 1967, radio astronomers at Cambridge were carrying out a search for quasars (very distant radio-emitting active galaxies). Because of their compactness, such sources emit highly coherent radiation, and this radiation is then strongly diffracted in the streams of ionized particles emitted by the Sun (the solar wind). As a result of this, the intensity of such radiation at the Earth's surface exhibits rapid fluctuations, known as interplanetary scintillations. Because the solar wind is a fast-changing and random process, the radio astronomers had to build an instrument capable of responding to rapid and random changes in radio intensity and capable of recording these signals for later subsequent analysis.

While she was analysing some of these records, Jocelyn Bell, a graduate student supervised by Antony Hewish, discovered the presence of some mysterious signals that were astonishingly regular and quite unlike the normal random fluctuations. The regular pulses repeated every 1.337 seconds and each pulse was only about 20 ms long.

The first reaction of all those concerned was to dismiss the signals as interference of human origin. But their regular appearance and disappearance in the records kept pace with the sidereal time (time relative to distant stars) and not with the calendar time, as would have been the case if the signals had anything to do with ordinary human activities. Hence, it was argued, these signals must be coming from the Universe. If so, could they be signals from some other intelligent creatures? This exciting possibility unfortunately had to be discarded when it was established that the source of these signals was fixed with respect to distant stars and did not show any signs of planetary motion. (Presumably any intelligent life would inhabit a planet in orbit around a star.) The only remaining possibility was that the signals were produced by some new, hitherto unknown, kind of celestial object. Moreover, this new object would have to be very small – much smaller than any ordinary star – because of the extremely short duration of each pulse.

● Why does a short pulse length imply a compact source of radiation?

○ Imagine a spherical source of radius R emitting an infinitesimally short flash of radio waves simultaneously from all points of its surface. Because of their finite speed c, the waves from different parts of the source would not reach the Earth simultaneously. If the observed duration Δt of individual pulses were entirely due to this effect, then the source could not possibly have a radius larger than $R = c\,\Delta t = (3 \times 10^8 \text{ m s}^{-1}) \times (20 \times 10^{-3} \text{ s}) = 6 \times 10^6 \text{ m} = 6000$ km. This is at least two orders of magnitude less than the size of a typical star.

More realistic assumptions about the mechanism generating the pulses lead to even lower limits on the radius.

The acceptance of such new objects became considerably easier when several other sources were discovered, all with very short periods of pulsation (less than a couple of seconds). The discovery of these objects, now known as **pulsars**, was to become a touchstone for many new ideas in astronomy in subsequent years. In particular, it was soon shown that pulsars are rotating neutron stars.

In retrospect, the name 'pulsar' may well have been somewhat unfortunate because it may have seemed to imply that pulsars are objects exhibiting real contractions and expansions of their volume, as if they were continually 'breathing' in and out with infallible regularity, sending out a burst of radio waves with each 'pulse'. Although this was at one stage one of several alternative models for the mechanism of pulsars, it has now been rejected. The reason is that the period of such oscillations in volume would be determined by the size, and by the mean density, of the object. From the known or well-estimated densities and sizes of stars of different types, we can conclude that if normal stars and white dwarfs were to pulse then they would have periods much longer than those observed, and that oscillations in the volume of neutron stars would lead to very much shorter periods. Thus some other explanation must be found.

Short of any totally new phenomenon hitherto unforeseen by scientists, there remained one hypothetical explanation which was compatible with our knowledge of the laws of nature: pulsars could be rotating neutron stars.

7.3.2   The rotation period of pulsars

The simple light-travel-time argument in the previous bulleted question does not entirely rule out white dwarfs as a possible cause of the pulsar phenomenon. However, white dwarfs can be ruled out if we consider how fast a body is able to rotate without breaking up. If an object rotates quickly enough, matter can be thrown off its surface by the centrifugal force. The maximum angular frequency (or minimum rotation period) may therefore be found by equating the gravitational attraction acting at the surface of the object to the centrifugal force acting there. This gives:

$$\frac{GM}{R^2} = R\omega_{\mathrm{max}}^2.$$

So the minimum rotation period to avoid break-up is:

$$P_{\mathrm{min}} = \frac{2\pi}{\omega_{\mathrm{max}}} = 2\pi \left(\frac{R^3}{GM}\right)^{1/2}. \tag{7.5}$$

Exercise 7.4   (a) What is the maximum radius for a $1\,M_\odot$ object rotating with a period of 1 s?

(b) What is the minimum rotation period for a $1\,M_\odot$ neutron star with a radius of 10 km?   ■

The suggestion that pulsars could be rotating neutron stars was made in 1968 by Thomas Gold. He suggested that the radio pulses are produced in the envelope of ions and electrons that surrounds the neutron star and rotates with it. In the outer regions of this envelope, the speed of individual electrons will be approaching the speed of light. As they are breaking away at such speeds from the magnetic field of the star, they can emit radio waves by the mechanism of synchrotron radiation. Although many working details of such a model are still a matter of discussion, one important consequence of this model was considered by Gold: if a rotating neutron star must continually support all the losses of energy due to the emission of fast particles and electromagnetic waves, its rate of rotation must be slowing down as it loses mass. Thus, if the period of radio pulses is determined by the period of rotation, it too must be changing – the pulsars should be slowing down.

Strong support for this theory was provided in November 1968 by the discovery of a pulsar in the central region of the Crab nebula. The Crab nebula is a supernova remnant (see Figure 7.3) which is the result of a supernova explosion that was observed by Chinese astronomers in 1054. The rotation period of the Crab pulsar is about 33 milliseconds, but more importantly, it was found that this pulsar was indeed slowing down. In 24 hours the interval between 2 successive pulses was longer by some 36 nanoseconds (the period would, at this rate, double in about a million days, or roughly every 2500 years).

Figure 7.3 An optical image of the Crab nebula, the gaseous remnant of a supernova in 1054 (i.e. $\approx$ 1000 years ago). At the heart of the gaseous remnant lies a pulsar with a 33 ms period, being the neutron-star remnant of the collapsed core of the massive parent star (progenitor).

● If the period of the Crab pulsar is 33 ms and it increases in period by 36 ns in 24 hours, what are its angular frequency ω, and its rate of change of angular frequency $d\omega/dt$ (often written as $\dot{\omega}$)?

○ Angular frequency and period are related by $\omega = 2\pi/P$, so $\omega = 2\pi/(33 \times 10^{-3} \text{ s}) = 190 \text{ s}^{-1}$.

By the chain rule, the rate of change of angular frequency is $d\omega/dt = d\omega/dP \times dP/dt$. Now, $d\omega/dP = -2\pi/P^2$, so $\dot{\omega} = (-2\pi/P^2) \times \dot{P}$.

An increase in period by 36 ns in 24 hours corresponds to a rate of change of period of $\dot{P} = (36 \times 10^{-9} \text{ s})/(24 \times 3600 \text{ s}) = 4.2 \times 10^{-13}$.

So the rate of change of angular frequency is $d\omega/dt \equiv \dot{\omega}$ $= (-2\pi/(33 \times 10^{-3} \text{ s})^2) \times 4.2 \times 10^{-13} = -2.4 \times 10^{-9} \text{ s}^{-2}$.

Notice that if the pulsar is slowing down its period derivative $\dot{P}$ is positive and dimensionless (or could be given units of s s^{-1}), but the derivative of its angular frequency $\dot{\omega}$ is negative and has units of s^{-2}. The period derivative $\dot{P}$ and the angular deceleration $\dot{\omega}$ are sometime's referred to as the pulsar's *spin-down rate*.

The supernova that produced the Crab nebula happened around 1000 years ago, and conditions within the nebula are no longer favourable for nuclear reactions to occur. So the question arises, how does it shine? Suspicion naturally falls on the pulsar, which is clearly not conserving energy, and this leads to two further questions: (i) what is the mechanism by which the pulsar loses energy and (ii) is this energy loss rate enough to power the nebula? We address these questions in the following subsections.

7.3.3   The energy loss from a rotating pulsar

To test whether the slowing down of the pulsar in the Crab nebula is powering the observed nebula we begin by considering the moment of inertia of a neutron star. For a uniform sphere, the moment of inertia is $I = 2MR^2/5$. The rotational energy of such as sphere is then $E_{\text{rot}} = \frac{1}{2}I\omega^2$.

Worked Example 7.4

The angular frequency of the Crab pulsar is $\omega = 190$ s^{-1} and the rate of change of angular frequency $\dot{\omega} = -2.4 \times 10^{-9}$ s^{-2}.

(a) Calculate the moment of inertia of the Crab pulsar assuming it to have a mass of 1.35 $M_\odot$ and a radius of 10 km.

(b) Calculate the rotational energy of the Crab pulsar.

(c) Differentiate the equation for rotational energy to find an expression for the rate of change of rotational energy with time as a function of the rate of change of angular frequency with time (assuming the moment of inertia is constant).

(d) Calculate the current rate of loss of energy from the Crab pulsar.

Solution

(a) The moment of inertia is $I = 2MR^2/5 =$
$2 \times 1.35 \times 1.99 \times 10^{30}$ kg $\times (10^4$ m$)^2/5 \sim 10^{38}$ kg m^2.

(b) The rotational energy is $E_{\text{rot}} = \frac{1}{2}I\omega^2 = \frac{1}{2} \times 10^{38}$ kg m$^2 \times (190$ s$^{-1})^2$
$= 1.8 \times 10^{42}$ J.

(c) Differentiating the expression for rotational energy with respect to time gives $\dot{E}_{\text{rot}} = I\omega\dot{\omega}$ (where we have used the 'dot' notation to signify a derivative with respect to time).

(d) So in this case

$$\dot{E}_{\text{rot}} = (10^{38} \text{ kg m}^2) \times (190 \text{ s}^{-1}) \times (-2.4 \times 10^{-9} \text{ s}^{-2}) = -4.6 \times 10^{31} \text{ J s}^{-1}.$$

The observed luminosity of the Crab nebula is about 5×10^{31} W, so this ties in extremely well with the observed rate of energy loss from the Crab pulsar. The loss of rotational energy from the Crab pulsar powers the Crab nebula.

7.3.4   The magnetic field strength of a pulsar

The magnetic axis of the star is parallel to the direction of the star's magnetic dipole moment.

The rotational energy of pulsars is believed to be lost as a result of **magnetic dipole radiation**. If the magnetic axis of the neutron star is offset from its spin axis by an angle θ (much as the Earth's magnetic axis is offset from its spin axis), then it will emit electromagnetic radiation at a rate

$$\dot{E}_{\text{rad}} = \frac{2}{3c^3}\frac{\mu_0}{4\pi}m^2\omega^4\sin^2\theta \tag{7.6}$$

where m is the magnitude of the neutron star's magnetic dipole moment and μ_0 is the permeability of free space.

- Assuming that all the rotational energy lost by the Crab pulsar is turned into magnetic dipole radiation, calculate the magnetic dipole moment m of the Crab pulsar.

○ In this case $\dot{E}_{\text{rad}} = -\dot{E}_{\text{rot}}$. So rearranging Equation 7.6, we have

$$m\sin\theta = \left[-\frac{\dot{E}_{\text{rot}}}{\omega^4}\frac{3c^3}{2}\frac{4\pi}{\mu_0}\right]^{1/2}.$$

So in the case of the Crab pulsar

$$m\sin\theta = \left[\frac{4.6\times10^{31}\text{ W}}{(190\text{ s}^{-1})^4}\times\frac{3\times(2.998\times10^8\text{ m s}^{-1})^3}{2}\times\frac{4\pi}{4\pi\times10^{-7}\text{ T m A}^{-1}}\right]^{1/2}$$

$$\sim 4\times10^{27}\text{ J}^{1/2}\text{ m T}^{-1/2}\text{ A}^{1/2}$$

$$\sim 4\times10^{27}\text{ (T A m}^2)^{1/2}\text{ m T}^{-1/2}\text{ A}^{1/2}$$

$$\sim 4\times10^{27}\text{ A m}^2.$$

Since $\sin\theta$ must be ≤ 1, we can say $m \geq 4\times10^{27}\text{ A m}^2$.

The magnitude B of the surface magnetic field of a neutron star is related to the magnitude of its magnetic dipole moment m and radius R by

$$B = \frac{\mu_0 m}{4\pi R^3}. \tag{7.7}$$

- Calculate the surface magnetic field strength of the Crab pulsar, assuming it to have a radius of 10 km.

○ We have

$$B \geq \frac{4\pi\times10^{-7}\text{ T m A}^{-1}\times4\times10^{27}\text{ A m}^2}{4\pi\times(10^4\text{ m})^3}$$

$$B \geq 4\times10^8\text{ T}.$$

Hence the surface magnetic field strength of the Crab pulsar is in excess of 400 megateslas.

Exercise 7.5 Combine the equation for the rotational energy of a uniform sphere with Equations 7.6 and 7.7 to obtain an expression for the magnetic field strength of a neutron star in terms of its rotation period, rate of change of rotation period, mass, radius and other physical constants. ∎

We can now evaluate the constants in the expression derived in the previous exercise:

$$B = \left(\frac{\mu_0}{4\pi}\right)^{1/2}\frac{1}{R^2}\left(\frac{2M}{5}\right)^{1/2}\left(\frac{3c^3}{2}\right)^{1/2}\frac{(\dot{P}P)^{1/2}}{2\pi}\frac{1}{\sin\theta}.$$

We can assume that $R \sim 10$ km and $M \sim 1.35$ M$_\odot$, and we know that $\sin\theta \leq 1$. Therefore a lower limit for a neutron star's magnetic field strength is

$$B \geq \left(\frac{4\pi \times 10^{-7} \text{ T m A}^{-1}}{4\pi}\right)^{1/2} \frac{1}{10^8 \text{ m}^2} \left(\frac{2 \times 1.35 \times 1.99 \times 10^{30} \text{ kg}}{5}\right)^{1/2}$$

$$\times \left(\frac{3 \times (2.998 \times 10^8 \text{ m s}^{-1})^3}{2}\right)^{1/2} \frac{(\dot{P}P)^{1/2}}{2\pi}.$$

Simplifying this we obtain

$$B/\text{tesla} \geq 3.3 \times 10^{15}(P\dot{P}/\text{seconds})^{1/2}. \tag{7.8}$$

● Confirm that Equation 7.8 gives the same result for the magnetic field strength of the Crab pulsar as you derived earlier from calculating its moment of inertia, rotational energy, rate of change of rotational energy and magnetic dipole moment. (The observed properties of the Crab pulsar are $P = 33 \times 10^{-3}$ s and $\dot{P} = 4.2 \times 10^{-13}$.)

○ $B/\text{tesla} \geq 3.3 \times 10^{15} \times (33 \times 10^{-3} \times 4.2 \times 10^{-13})^{1/2}$. So the magnetic field strength is $B \geq 3.9 \times 10^8$ T or about 400 megateslas as found earlier.

7.3.5   The ages of pulsars

There is good evidence that pulsars do indeed lose rotational energy via magnetic dipole radiation, and this is provided by a calculation of their ages. If we equate the rate of loss of rotational energy to the power of magnetic dipole radiation, we have

$$\dot{E} = I\omega\dot{\omega} = \frac{2}{3c^3}\frac{\mu_0}{4\pi}m^2\omega^4 \sin^2\theta.$$

So the angular frequency of the pulsar obeys the following differential equation

$$\dot{\omega} = -C\omega^3 \tag{7.9}$$

where C is a constant. For the Crab pulsar for instance, given that $\omega = 190$ s^{-1} and $\dot{\omega} = -2.4 \times 10^{-9}$ s^{-2}, clearly $C = -\dot{\omega}/\omega^3 = 3.5 \times 10^{-16}$ s.

Now, in order to solve Equation 7.9, we rewrite it as:

$$\frac{d\omega}{dt} = -C\omega^3$$

then invert it to give:

$$\frac{dt}{d\omega} = -\frac{1}{C\omega^3}.$$

We then integrate this with respect to ω from $\omega = \omega_0$ (the initial angular speed) to $\omega = \omega_1$ (the angular speed at some later time). These limits correspond to the times $t = 0$ and $t = t_1$ (i.e. the elapsed time) respectively.

$$\int_{\omega=\omega_0}^{\omega=\omega_1} \frac{dt}{d\omega}\, d\omega = -\frac{1}{C}\int_{\omega=\omega_0}^{\omega=\omega_1} \frac{1}{\omega^3}\, d\omega$$

$$\int_{t=0}^{t=t_1} dt = -\frac{1}{C}\left[-\frac{1}{2\omega^2}\right]_{\omega=\omega_0}^{\omega=\omega_1}$$

$$t_1 = \frac{1}{2C}\left[\frac{1}{\omega_1^2} - \frac{1}{\omega_0^2}\right].$$

Reverting to the symbol ω to indicate the current angular speed and letting τ be the age of the pulsar, we may write the relationship as follows, and substitute in for C using the original differential equation:

$$
\begin{aligned}
\tau &= \frac{1}{2C}\left[\frac{1}{\omega^2} - \frac{1}{\omega_0^2}\right] \\
&= -\frac{\omega^3}{2\dot{\omega}}\left[\frac{1}{\omega^2} - \frac{1}{\omega_0^2}\right] \\
&= -\frac{1}{2}\frac{\omega}{\dot{\omega}}\left[1 - \frac{\omega^2}{\omega_0^2}\right]
\end{aligned}
\tag{7.10}
$$

where ω_0 is the angular frequency of the pulsar at its birth. If we assume that the pulsar has undergone a continuous angular deceleration as it has aged, then $\omega < \omega_0$. An upper limit for the age of a pulsar can be found by assuming the initial angular frequency was very high and setting $\omega^2/\omega_0^2 \sim 0$. So the age of a pulsar is related to its angular frequency and rate of change of angular frequency (or to its period and rate of change of period) by

$$
\tau < -\frac{1}{2}\left(\frac{\omega}{\dot{\omega}}\right) \equiv \frac{1}{2}\left(\frac{P}{\dot{P}}\right).
\tag{7.11}
$$

● What is an upper limit for the age of the Crab pulsar, assuming a constant rate of change of angular frequency?

○ An upper limit to the age is $\tau < -\omega/2\dot{\omega}$ hence $\tau < -(190\ \text{s}^{-1})/(2 \times -2.4 \times 10^{-9}\ \text{s}^{-2})$ or $\tau < 4 \times 10^{10}\ \text{s}$. So the Crab pulsar is less than about 1250 years old.

This age limit ties in well with the known age of the Crab pulsar, which is about 950 years.

Exercise 7.6   Assuming the age of the Crab pulsar is 950 years, what must its initial angular frequency have been?   ■

7.3.6   Types of pulsar

A pulsar may be characterized observationally by two numbers: its rotation period (P) and the rate of change of its rotation period ($\dot{P}$). Remarkably these two numbers enable limits to be placed on both the pulsar's magnetic field strength and its age, via Equations 7.8 and 7.11.

● Why can we only put limits on a pulsar's magnetic field strength and age from measurements of its period and period derivative?

○ The equation for the pulsar's magnetic field strength depends on $1/\sin\theta$ where θ is the (usually unknown) angle between the pulsar's magnetic field axis and spin axis. Since $\sin\theta \le 1$, so $1/\sin\theta \ge 1$ and we only have a lower limit on B.

The equation for the pulsar's age depends on $[1 - \omega^2/\omega_0^2]$ where ω_0 is the (usually unknown) initial angular frequency of the pulsar. Since the pulsar continually slows down, so $\omega_0 \ge \omega$, and the expression $[1 - \omega^2/\omega_0^2]$ is therefore ≤ 1. Hence we only have an upper limit on τ.

Despite these limitations, it is instructive to plot the positions of pulsars on a diagram of $\log \dot{P}$ versus $\log P$, as shown in Figure 7.4. Lines of constant magnetic field strength and constant age have been drawn on this graph, according to Equations 7.8 and 7.11. Notice that the pulsars still associated with supernova remnants (including the Crab) are *young* pulsars. This is because supernova remnants fade and disperse on a timescale of one hundred thousand years, so they are not in evidence for older pulsars.

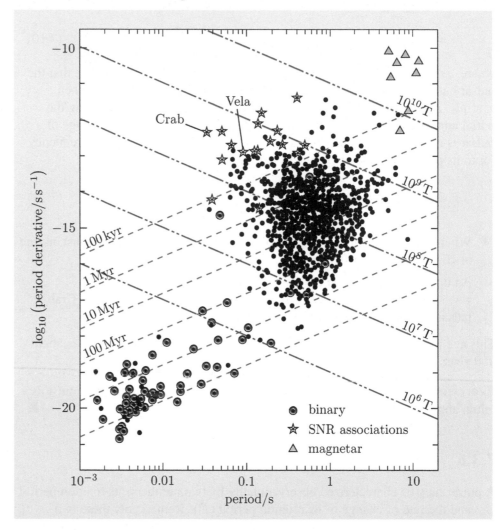

Figure 7.4   The periods and period derivatives of known pulsars. Lines of constant magnetic field strength and constant age have been drawn, according to Equations 7.8 and 7.11. These indicate lower limits to the magnetic field strength and upper limits to the age respectively. Pulsars in binary systems (mostly recycled, millisecond pulsars) are indicated by circles, and magnetars (pulsars with extremely high magnetic field strengths, see next page) are indicated by triangles. Those pulsars still associated with supernova remnants (SNR) are indicated by stars.

● What are the gradients of lines of constant pulsar magnetic field strength and constant pulsar age on a graph of $\log \dot{P}$ versus $\log P$? What are the spacings between these lines of constant field strength and age?

○ Taking logs of Equation 7.8, we have $\log(B/\text{tesla}) = \log(3.3 \times 10^{15}) + \frac{1}{2}\log(P/\text{seconds}) + \frac{1}{2}\log\dot{P}$. This may be rearranged as: $\log\dot{P} = -\log(P/\text{seconds}) + 2\log(B/\text{tesla}) - 2\log(3.3 \times 10^{15})$. So on a graph of $\log\dot{P}$ versus $\log P$, lines of constant B will have a gradient of -1. For every factor of 10 increase (or decrease) in B, the vertical interval will be $+2$ (or -2) logarithmic units.

Taking logs of Equation 7.11, we have $\log\tau = \log(1/2) + \log P - \log\dot{P}$. This may be rearranged as: $\log\dot{P} = \log P + \log(1/2) - \log\tau$. So on a graph of $\log\dot{P}$ versus $\log P$, lines of constant τ will have a gradient of $+1$. For every factor of 10 increase (or decrease) in τ, the vertical interval will be -1 (or $+1$) logarithmic units.

The 'zoo' of known pulsar types has expanded somewhat since the discovery of the Crab pulsar. Amongst the ~ 2000 radio pulsars now known, some pulsars have now been discovered that spin much faster than the Crab pulsar. The fastest of these objects, which are known as **millisecond pulsars,** has a period of about 1.5 milliseconds (i.e. the neutron star spins on its axis more than 660 times a second). The evolutionary history of millisecond pulsars is, however, slightly different from that of pulsars like the Crab. Millisecond pulsars, of which about 100 are currently known, are thought to have spent some time during their history as one component of a binary star system. Accretion of material from the companion star onto the neutron star has transferred large amounts of angular momentum to the neutron star, so causing it to speed up to the very fast rotation rate we now see. For this reason millisecond pulsars are often known as **recycled pulsars**. In some millisecond pulsars we can still see the companion star; in others the companion has been 'blown away' or 'evaporated' by high-energy emissions from the pulsar. The Crab pulsar is however probably the youngest (i.e. the most recently formed) pulsar that has been discovered, and as such is one of the fastest non-recycled pulsars that is known.

In contrast, there are now about a hundred **X-ray pulsars** known, most of which have periods much longer than the (approximately) 8 second upper limit for radio pulsars. These X-ray pulsars are known as **accretion-powered pulsars** to distinguish them from the rotation-powered pulsars of which all radio pulsars (both recycled and non-recycled) are examples. Accretion-powered pulsars, as their name suggests, currently receive their power via accretion of material from a companion star in a binary system. The known systems have pulse periods in the range 0.07 seconds to 850 seconds and (in general) are spinning ever faster as time passes (i.e. the period gets shorter). This is in contrast to the Crab pulsar and other rotation-powered pulsars which (in general) are spinning ever more slowly as time goes by. It is possible, and perhaps likely, that some accretion-powered X-ray pulsars eventually evolve into millisecond, rotation-powered, radio pulsars.

A final class of pulsars consists of objects known as **magnetars,** of which around 15 are currently known. These are believed to be neutron stars with magnetic field strengths $B \geq 10^{10}$ T. In these objects, it is the decay of the magnetic field itself which powers the high-energy radiation that is observed. Two different phenomena are observed. Some magnetars are detected as X-ray pulsars with spin periods in the range (5–12) s. Unlike conventional X-ray pulsars however, these so-called **anomalous X-ray pulsars** are not powered by accretion and show no evidence for companion stars. Other magnetars are seen to emit enormous, but brief, bursts of low-energy gamma-rays. These bursts last for only a few seconds but recur on irregular timescales of months or years. In this regard they are sometimes recorded as **soft gamma-ray repeaters**. The bursts of emission are believed to be caused by re-arrangements of the neutron star crust which twist magnetic field lines and release energy in the process. A couple of soft gamma-ray repeaters have been seen to emit pulsations during their bursts of emission, which appear to confirm that soft gamma-ray repeaters and anomalous X-ray pulsars are indeed both sub-types of the same object, namely magnetars.

● What are the sources of energy in radio pulsars, X-ray pulsars and magnetars?

○ The sources of energy are rotational energy, accretion energy and magnetic energy respectively.

7.4   Stellar-mass black holes

So far we have studied two stellar remnants: white dwarfs and neutron stars. One other stellar remnant remains to be discussed: black holes. Black holes have many unusual properties, but in this final topic on end-points of stellar evolution, we will consider them only briefly.

If the core of a collapsing star is too massive to be supported by the pressure of degenerate neutrons (i.e. if its mass is greater than about $3\,M_\odot$), then it will continue collapsing. Its gravitational field will increase, and the pressure within the core will increase. However, in general relativity, an increase in pressure leads to an increase in the gravitational field, so the collapse will accelerate still further. Nothing can halt this collapse and it will proceed until nothing, not even light, can escape from the intense gravitational field. The collapsing core becomes infinitely compressed and nothing is left except a gravitational field. This is a region of spacetime known as a **black hole**.

The radius from within which nothing can escape is known as the **Schwarzschild radius** and is defined by

$$R_S = \frac{2GM}{c^2}. \tag{7.12}$$

● Evaluate the Schwarzschild radius for a $5\,M_\odot$ black hole.

○ $R_S = 2GM/c^2 = (2 \times 6.673 \times 10^{-11}\,\mathrm{N\,m^2\,kg^{-2}} \times 5 \times 1.99 \times 10^{30}\,\mathrm{kg})/(2.998 \times 10^8\,\mathrm{m\,s^{-1}})^2 = 1.5 \times 10^4$ m or about 15 km.

Clearly if no electromagnetic radiation can escape from black holes they are likely to be rather difficult to detect. However, there are cases where stellar-mass black holes *can* be detected. The technique relies on observing the behaviour of matter orbiting the black hole, before it is consumed.

If a black hole is formed in a binary star system, then it may evolve to a state whereby the black hole is able to accrete material from its companion. As material is stripped from the companion star, it carries with it a certain amount of angular momentum. This prevents the material from falling directly into the black hole, and instead it will form a flattened, circular structure known as an *accretion disc*, around the black hole. Viscous forces operate within the accretion disc as material spirals in towards the black hole, causing the disc to heat up to temperatures of typically 10^8 K such that it will emit X-rays. These X-rays are observed and provide the signature of a high-energy accretion process, possibly indicating a black hole. The only proof that a black hole is present is to determine the mass of the (unseen) compact object by studying the reflex motion of the companion star as the two orbit their common centre of mass. If the measured mass of the compact object turns out to be in excess of $3\,M_\odot$, then it is likely to be a black hole. There are currently around 20 binary star systems known where there is dynamical evidence for a compact object with a mass indicative of a black hole, and more are being discovered all the time.

One other observational signature of some black holes, or at least their formation, are gamma-ray bursts. These are short, intense bursts of gamma-rays, lasting typically only a few seconds. The burst is thought to arise from the collapse of a massive, rapidly rotating stellar core into a black hole.

● The maximum mass of a neutron star is $(1.0\text{–}3.0)\,M_\odot$; all more massive remnants become black holes. We also know that a star with $M_{\mathrm{ms}} \approx 8\,M_\odot$ burns carbon but leaves a ONeMg *white dwarf* remnant, not a black hole. How do such stars avoid becoming black holes?

○ Don't forget mass loss! Stars lose a large fraction of their mass during their giant phases, so although a star must *begin* life with a *main-sequence mass* in excess of $8\,M_\odot$ if it is to ignite carbon, it is much less massive by the time it actually does so.

The physics and observational properties of black holes are discussed in the companion volumes to this book: *Relativity, Gravitation and Cosmology* by R. J. A. Lambourne and *Extreme Environment Astrophysics* by U. C. Kolb.

Summary of Chapter 7

1. Stars with $M_{\mathrm{ms}} \geq 11\,M_\odot$ achieve silicon burning, which forms an iron core supported by degenerate electrons that become ultra-relativistic. When the Chandrasekhar limit ($\approx 1.4\,M_\odot$) is reached, the electrons can no longer support the star. Nuclear photodisintegration by thermal photons and electron capture by nuclear protons (neutronization) absorb energy so efficiently they send the core into free fall.

2. Photodisintegration of iron into hydrogen proceeds as
   $\gamma + {}^{56}_{26}\mathrm{Fe} \longrightarrow 13{}^{4}_{2}\mathrm{He} + 4\mathrm{n}$, and $\gamma + {}^{4}_{2}\mathrm{He} \longrightarrow 2\mathrm{p} + 2\mathrm{n}$, which could photodissociate about three-quarters of the iron core. Up to $\approx 1.4 \times 10^{45}$ J could be lost in this way.

   Neutronization is the conversion of nuclear protons into neutrons via electron capture: $\mathrm{e}^- + \mathrm{p} \longrightarrow \mathrm{n} + \nu_\mathrm{e}$. The electron capture $\mathrm{e}^- + {}^{56}_{26}\mathrm{Fe} \longrightarrow {}^{56}_{25}\mathrm{Mn} + \nu_\mathrm{e}$ occurs once the electron density n_e is high enough. The neutrinos carry away the energy. Up to $\approx 1.3 \times 10^{45}$ J could be lost this way.

   The total amount of energy lost via photodisintegration and neutronization in a few seconds is comparable to the energy previously liberated via nuclear burning over the star's entire main-sequence lifetime!

3. The collapse of a stellar core is halted when the density reaches that of nuclear matter. The compression and rebounding of the nuclear matter sends

a shock wave through the rest of the star that partially, if not fully, reverses the stellar collapse, leading to the ejection of the outer layers as a supernova. (Heating by neutrinos may also be required to eject the envelope.) The typical luminous energy is $\sim 10^{42}$ J, and the kinetic energy of the ejecta is $\sim 10^{44}$ J. The gravitational binding energy ($\sim 5 \times 10^{46}$ J) is an order of magnitude more than the energy (i) required to photodissociate the iron core or (ii) that is lost by neutronization, and (iii) two orders of magnitude more than the kinetic energy of the ejecta. Thus most of the liberated energy is probably lost in an intense neutrino burst from the surface of the new, hot neutron star, such as that measured for SN1987A.

4. The distinctions between the two main types of supernovae are as follows.

   - SN Ia show no hydrogen in their visible spectra and are associated with old stellar populations. They probably leave no compact remnant, only a fireball. They are most probably due to a CO white dwarf accreting material from a companion such that it exceeds the Chandrasekhar limit.

   - SN II do show hydrogen in their visible spectra and are associated with young stellar populations. They leave a neutron star or black hole behind and are due to core collapse of a massive ($M_{ms} \geq 11\,M_{\odot}$) star.

5. Neutronization converts nuclear protons into neutrons. When the density reaches 4×10^{14} kg m^{-3}, neutrons drip from the nuclei, giving rise to free neutrons, nuclei, and electrons. Once the density exceeds the density of normal nuclear matter, 2.3×10^{17} kg m^{-3}, nuclei merge into a dense gas of electrons, protons and neutrons.

6. Neutron stars survive because the normal β^--decay of free neutrons is blocked by the Pauli exclusion principle – the energy states which the resulting proton and electron would occupy are full. Neutrons greatly outnumber protons: $n_n \sim 200 n_p$.

7. The radii of white dwarfs and neutron stars are set *primarily* by the mass of the degenerate particle providing pressure support. Since the neutron mass is 1840 times the electron mass, the radius of a neutron star is much smaller than a white dwarf of comparable mass. The radii of neutron stars are ≈ 580 times smaller, or ≈ 5–10 km.

8. A maximum mass exists for neutron stars, analogous to that for white dwarfs, corresponding to the neutrons becoming ultra-relativistic. Unfortunately it is more difficult to calculate the neutron-star value because of neutron interactions (which make the maximum mass greater) and the need to use general relativity (in which gravity is strengthened at high energy density and pressure, and makes the maximum mass smaller).

9. Pulsars are recognized as rapidly rotating, magnetized neutron stars. The Crab pulsar is a particularly well-studied example, being the remnant of a star which underwent a supernova explosion in 1054. It has a rotation period of 33 ms and is steadily undergoing angular deceleration.

10. By ascribing the observed angular deceleration rate $\dot{\omega}$ of a pulsar to an energy loss via magnetic dipole radiation (caused by the spin axis and

magnetic axis being misaligned), we obtain:

$$\dot{E} = I\omega\dot{\omega} = \frac{2}{3c^3}\frac{\mu_0}{4\pi}m^2\omega^4\sin^2\theta \qquad \text{(Eqn 7.6)}$$

where the magnetic field strength B of the pulsar is related to the magnetic dipole moment m by

$$B = \frac{\mu_0 m}{4\pi R^3}. \qquad \text{(Eqn 7.7)}$$

This leads to an estimate for the magnetic field strength of pulsars as

$$B/\text{tesla} \geq 3.3 \times 10^{15}(P\dot{P}/\text{seconds})^{1/2}. \qquad \text{(Eqn 7.8)}$$

We can also estimate an upper limit to the age of a pulsar as

$$\tau < -\frac{1}{2}\left(\frac{\omega}{\dot{\omega}}\right) \equiv \frac{1}{2}\left(\frac{P}{\dot{P}}\right) \qquad \text{(Eqn 7.11)}$$

where P and $\dot{P}$ are a pulsar's spin period and period derivative respectively.

11. Around ~ 2000 rotation-powered (radio) pulsars are now known. About 5% of these are in binary star systems, where it is likely that the pulsar has been recycled by accretion from its companion star and is currently observed as a millisecond pulsar. About a hundred accretion-powered (X-ray) pulsars are also known, where accretion from the companion star is currently accelerating the pulsar to a shorter spin period. A handful of magnetars are also observed, powered by the decay of their immense magnetic field.

12. Objects exceeding the mass of a neutron star collapse to a black hole, a region in spacetime with a Schwarzschild radius $R_\text{S} = 2GM/c^2$. Black hole candidates are recognized from the high masses inferred from observations of companion stars and from the high-energy phenomena (e.g. X-ray emission) from material heated as it flows towards a black hole.

Chapter 8   Star formation

Introduction

Having followed stellar evolution to its end-point and seen how supernovae seed the interstellar medium with nuclear-processed material, we now look at the physics that determines which gas clouds collapse to form stars and which do not, and how the protostars formed in this way evolve towards the main sequence. You will see that the energy balance and pressure balance within the cloud hold the key.

The brief overview of star formation given at the beginning of Chapter 1 conveys little of the physics that determines which clouds become stars, how that transformation occurs, how long the process takes, or the characteristics of the resulting star. The aim of this chapter is to provide that understanding. Figure 8.1 shows a known birthplace of stars. The interstellar medium contains different regions that have an enormous variation in temperature and density; even the densest of these though are much more tenuous than the air around us. Nonetheless it is clear that stars form in cold, dense, dark interstellar gas clouds if the gravity within the cloud is sufficient to make the cloud collapse.

Figure 8.1   Dense, cool clouds of hydrogen gas made opaque by dust grains, in M16 – the Eagle nebula. They are silhouetted against the emission nebula behind, which is excited by the UV flux from young, massive, hot stars off the top of the image. The UV radiation is also photo-evaporating the exposed ends of the dense gas clouds, giving rise to fine wisps of boiled-off gas. Within the dense, dark clouds new stars are forming.

Broadly speaking, the collapse of a gas cloud to form stars may be a spontaneous event or it may be triggered by an external influence. As you will see in the next few sections, spontaneous collapse of an interstellar cloud will occur if the gas pressure within the cloud is insufficient to support it against gravity, and this will depend on the temperature, density and composition of the cloud itself. Triggered star formation can occur when an external event occurs to compress an interstellar cloud. This may increase its density sufficiently so that the cloud's internal gas pressure is no longer large enough to support it, and the cloud begins its gravitational collapse. Such external triggers can include nearby supernova explosions, collisions between interstellar clouds, or even collisions between entire galaxies which can induce enormous bursts of star formation throughout the galaxies involved.

Most of what follows concentrates on the formation of a single star. We believe that the physics for most of this is reasonably well understood. However, star formation is still an active research field, for while the contraction of a single star can be described realistically, there are major uncertainties in the formation of groups of stars, and it is believed that stars generally form in clusters rather than alone. Star clusters may range in size from a few tens to almost a million stars. Even for different ages and chemical compositions, the range of masses of stars that form is very similar from one cluster to the next. That is, the distribution of star masses that arises from a collapsing cloud is far from random, and the same pattern, many low-mass stars and few high-mass stars, is found everywhere. However, our understanding of how this **initial mass function** (**IMF**) arises is far from complete.

8.1 The Jeans criterion for gravitational collapse

The total energy of a cloud of gas is the sum of its kinetic energy and gravitational potential energy. The gravitational potential energy between two masses M and m is negative ($E_{GR} = -GMm/r$) and approaches zero as the separation r becomes infinite. If their kinetic energy is positive when the potential energy has become zero (in which case $E_{TOT} > 0$), they are not bound to one another. If, on the other hand, the kinetic energy is less than the magnitude of the potential energy ($E_K < |E_{GR}|$ and consequently $E_{TOT} < 0$) then as the objects move apart by converting kinetic energy into potential energy, the kinetic energy reduces to zero before the potential energy increases to zero; in this case the objects are confined to a negative potential and hence are bound to one another.

The conditions under which a cloud of gas will collapse to form a star were first quantified by the British physicist Sir James Jeans in 1902. The **Jeans criterion** for collapse simply restates that a gas cloud whose total energy is negative is bound and collapses, whereas a cloud whose total energy is positive does not. That is to say, for collapse to occur, the cloud must be bound, i.e. $E_{TOT} < 0$. Since $E_{TOT} = E_K + E_{GR}$, this condition is equivalent to saying $E_K + E_{GR} < 0$, or $E_K < -E_{GR}$. Since the gravitational potential energy is negative by definition, we can rewrite the Jeans criterion for collapse as $E_K < |E_{GR}|$. This simple criterion can be developed into limiting values for the mass and density of potential star-forming regions.

● If the kinetic energy of the gas must be less than some critical value if it is to collapse, what does this suggest about the temperature of star-forming clouds?

○ The kinetic energy limit requires that clouds be relatively cold if they are to be able to form stars. Hot clouds have too much kinetic energy and will not collapse to form stars unless they cool.

The Jeans condition can also be expressed in terms of limits on the mass and density of clouds that are to form stars, as follows.

The gravitational potential energy of a sphere of uniform density is $E_{GR} = -3GM^2/5R$ (Equation 1.3). However, if the density is greater towards the centre, then the numerical factor is greater than $3/5$, so for simplicity we assume that $E_{GR} \sim -GM^2/R$.

The kinetic energy of the particles in the cloud is given by $E_K = \frac{3}{2}NkT$ where N is the total number of particles.

Hence the Jeans criterion may be written as:

$$\frac{3}{2}NkT < \frac{GM^2}{R}.$$

Noting that the mass of the cloud M is given by $N\overline{m}$ where $\overline{m}$ is the average mass per particle, this equation may be rearranged as

$$\frac{3}{2}kT < \frac{GM\overline{m}}{R}.$$

So, if the mass of a cloud exceeds some value for its temperature and radius, called the **Jeans mass**,

$$M > \frac{3kT}{2G\overline{m}}R \equiv M_J \tag{8.1}$$

then the cloud is bound and collapses due to self-gravity. The Jeans mass is therefore specified for a cloud of given radius R, temperature T and composition $\overline{m}$.

The **Jeans length** is simply the radius that encloses the Jeans mass. By rearranging Equation 8.1, this is simply

$$R_J = \frac{2G\overline{m}}{3kT}M_J. \tag{8.2}$$

An equivalent expression is to consider the density of the cloud. For a sphere of uniform density, $\rho = 3M/4\pi R^3$, and this may be written as $\rho = (M/R)^3 \times (3/4\pi M^2)$. But, the equation for the Jeans mass or Jeans radius may be rearranged as $M_J/R = (3kT/2G\overline{m})$, so we can write the **Jeans density** as

$$\rho > \left(\frac{3kT}{2G\overline{m}}\right)^3 \frac{3}{4\pi M^2} \equiv \rho_J. \tag{8.3}$$

The Jeans density is therefore specified for a cloud of given mass M, temperature T and composition $\overline{m}$.

Exercise 8.1   (a) What would be the radius (in parsecs, astronomical units and solar radii) of a spherical cloud of molecular hydrogen (H_2) at a temperature of $20\,K$, with a Jeans mass equal to the mass of the Sun?

(b) What would be the Jeans density of this cloud?

(*Hint*: Assume that the mass of a hydrogen molecule, H_2, is 2 atomic mass units $= 2u$.) ∎

Exercise 8.2 Using Equation 2.5 calculate the free-fall time for a sphere of uniform-density molecular hydrogen gas having the mass of the Sun, starting at a temperature of 20 K and the Jeans density. ■

Exercise 8.3 Compute the Jeans mass (in units of $M_\odot$) for two cases:

(a) Neutral atomic hydrogen gas typical of the cool interstellar medium ($T = 100$ K, and number density $n = 10^6$ atoms per cubic metre).

(b) Molecular hydrogen gas typical of cold, dense molecular clouds ($T = 10$ K, and number density $n = 10^9$ particles per cubic metre).

(*Hint 1*: Start with the equation for the Jeans density, and rearrange that to give an alternative expression for the Jeans mass.)

(*Hint 2*: The mass per unit volume ρ is just the number of particles per unit volume n multiplied by the mass of each particle $\overline{m}$, i.e. $\rho = n\overline{m}$.) ■

The results from Exercise 8.3 show that the different environments lead to very different structures; the warmer, diffuse interstellar medium will only form very massive clusters of stars, whereas much smaller clusters can form from cold, dense molecular clouds.

8.2 Collapse and cooling in the temperature–density diagram

The H–R diagram provides a vital framework in which to follow the evolution of stars. In studying the collapse of self-gravitating gas, a different diagram – the temperature versus density diagram – turns out to be equally helpful in tracing the evolution of contracting gas.

8.2.1 The temperature–density diagram

From Exercise 8.3, you saw that the Jeans mass, rewritten in terms of the temperature and density of a gas cloud, is

$$M_J = \sqrt{\frac{3}{4\pi\rho_J} \left(\frac{3kT}{2G\overline{m}}\right)^3}. \tag{8.4}$$

● For a *given* Jeans mass, what is the proportionality between the temperature and density of gas?

○ Since we are considering a *given* mass, we treat M_J as a *constant*, so the quantity within the square root of Equation 8.4 must also be constant. In that case,

$$\frac{3}{4\pi\rho_J} \left(\frac{3kT}{2G\overline{m}}\right)^3 = \text{constant}, \quad \text{so} \quad \rho_J = \frac{3}{4\pi \times \text{constant}} \left(\frac{3kT}{2G\overline{m}}\right)^3$$

i.e. $\rho_J \propto T^3$.

Remember that the Jeans mass describes objects at the transition between bound and unbound states; more massive objects collapse, whereas less massive objects expand. The quantities T and ρ_J along this boundary are related by $T \propto \rho_J^{1/3}$.

Just as we used the relation between luminosity, effective temperature and radius to draw lines of fixed radius in the Hertzsprung–Russell diagram (Chapter 1), we can draw lines of fixed critical mass in the temperature–density diagram.

Exercise 8.4   (a) Rearrange the expression for the Jeans mass, written in terms of the temperature and density (Equation 8.4), to give the temperature as a function of the mass and density of the gas of a cloud satisfying the Jeans condition.

(b) Take logarithms of both sides of the resulting equation, write an expression for $\log_{10} T$ in terms of one constant term and two variable terms. Describe the shape of the curve for a *given* Jeans mass in a diagram of $\log_{10} T$ versus $\log_{10} \rho$.

(*Hint*: Recall that $\log_{10} AB = \log_{10} A + \log_{10} B$, and $\log_{10} A^k = k \log_{10} A$.)   ■

Figure 8.2 shows a plot of the temperature–density plane on which are plotted the curves for neutral hydrogen gas for three protostellar masses, namely $1 \, M_\odot$, $10 \, M_\odot$, and $100 \, M_\odot$. As derived in Exercise 8.4, the lines of constant Jeans mass have a slope of 1/3 in the log–log plane.

● Referring to the temperature–density plane in Figure 8.2, what does it mean if a $100 \, M_\odot$ cloud lies above the line for a Jeans mass of $100 \, M_\odot$?

○ A cloud lying above the Jeans line for its mass has a temperature that is too high for collapse to occur, and hence its total energy E_{TOT} is positive. Internal pressure overcomes self-gravity and the cloud expands unless it cools.

Conversely to the case considered above, clouds that lie below the Jeans line for their mass tend to collapse because their temperatures are too low to resist gravity. The Jeans line for a given mass can therefore be regarded as a critical boundary. Objects above the line for their mass do not collapse, but objects below the line do. The Jeans line marks the boundary in the temperature–density plane between collapsing and expanding objects for any given mass.

The boundary between collapse and expansion may also be thought of in terms of timescales. We have already met the free-fall time (Equation 2.5). At the other extreme, when internal pressure forces greatly exceed the gravitational force, the cloud expands essentially at the speed of a pressure wave in the gas, i.e. at the velocity of sound c_s, and an expansion time τ_e may be defined as the time taken for sound to travel the radius R of the cloud, $\tau_e = R/c_s$. The critical boundary between collapse and expansion therefore corresponds to the case where the expansion time and free-fall time are equal, i.e. the opposing processes balance one another.

● Figure 8.2 shows that the Jeans line for a $100 \, M_\odot$ cloud lies above the line for a $10 \, M_\odot$ cloud (at a given density). Using words, explain why.

○ The Jeans line marks the boundary between clouds with total energy $E_{\mathrm{TOT}} > 0$ and $E_{\mathrm{TOT}} < 0$, where the total energy is the sum of the gravitational potential energy and the kinetic energy. The gravitational potential energy depends on the *square* of the mass, whereas the kinetic energy is proportional to the number of particles and hence is only *linearly* proportional to the mass. In a more massive cloud, the effect of the gravitational potential energy is greater than that of the kinetic energy, and so the cloud sits in a deeper (more negative) gravitational potential energy well. It can therefore sustain higher temperatures before E_{TOT} becomes positive, so the Jeans line moves to higher temperature at higher mass.

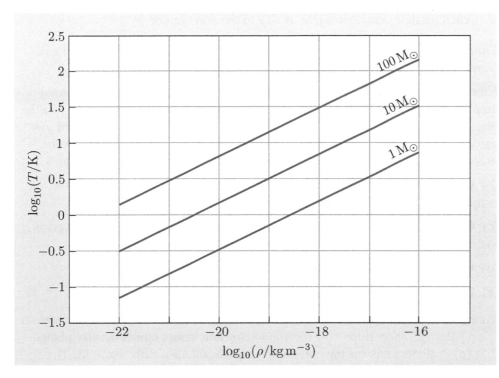

Figure 8.2 The temperature–density plane indicating the Jeans mass lines for neutral hydrogen gas for three protostellar masses, namely $1\,\text{M}_\odot$, $10\,\text{M}_\odot$, and $100\,\text{M}_\odot$. Notice the axes are plotted on logarithmic scales.

For a cloud below the Jeans line for its mass, there is therefore a tendency to collapse. However, cooling of the cloud plays a major role in determining what actually happens. We will consider two cases below: first, the evolution of a gas cloud when there are no exchanges of heat (thermal energy) with its surroundings, and secondly the more usual case where the cloud radiates into its environment.

8.2.2 Adiabatic collapse

An **adiabatic process** is one in which there is no heat exchange by any means (neither radiation, convection nor conduction) between a system and its environment. As you saw in Section 4.7, in such a system, the pressure P, volume V, and density ρ of an ideal gas obey the relations

$$PV^\gamma = \text{constant} \quad \text{and} \quad P \propto \rho^\gamma \qquad \text{(Eqn 4.10)}$$

where γ is the adiabatic index. The adiabatic index takes a value between 1 and 5/3 depending on the number of degrees of freedom s of the particles making up the gas, with

$$\gamma = \frac{1 + (s/2)}{(s/2)}. \qquad \text{(Eqn 4.11)}$$

The concept of degrees of freedom is very valuable because each degree of freedom contributes $kT/2$ to the average energy of each gas particle. Thus the average energy of each gas particle in a monatomic ideal gas is $3kT/2$ and (thanks to rotational and vibrational movement) a diatomic ideal gas can have average energies up to $7kT/2$ per gas particle (molecule). However, at low temperature ($T \leq 100\,\mathrm{K}$) the rotational and vibrational degrees of freedom seem to disappear, leaving just the same number as for a monatomic gas. Consequently, in cold gas clouds, the energy of molecular hydrogen gas (H_2) may be calculated as if each particle contributes just $3kT/2$ to the average energy.

Exercise 8.5   (a) Show that for an adiabatic process affecting an ideal gas, temperature $T \propto \rho^{\gamma-1}$.

Hint: An ideal gas conforms to the relation $PV = NkT$, where N is the total number of particles in the gas, k is Boltzmann's constant, and P, V and T are the pressure, volume and temperature. Note that N/V is the number density of gas particles, n, and multiplying this by the mean mass of a molecule $\overline{m}$ gives the mass density of particles, $\rho = n\overline{m} = N\overline{m}/V$. Thus $P = \rho kT/\overline{m}$ (Equation 1.11).

(b) Comment on the shape of the curve $T \propto \rho^{\gamma-1}$ (called an *adiabat*) in the log temperature versus log density diagram.

(c) Calculate the value of the slope of the adiabat in the log temperature versus log density diagram for an ideal gas with three degrees of freedom.

(d) Compute the range of all possible slopes for adiabats in this diagram.

(e) Calculate the slope for the case $\gamma = 4/3$.   ∎

There are several important results in Exercise 8.5. Part (b) shows that the adiabat of an ideal gas has a slope $\gamma - 1$ in the logarithmic temperature–density plane. Part (e) evaluated this for the particularly significant case with $\gamma = 4/3$. The adiabats for such a gas have a slope $\gamma - 1 = 1/3$, and they will lie *parallel to* the Jeans line. The gas still expands or contracts depending on whether it is hot and tenuous or cold and dense respectively, i.e. whether it lies above or below the Jeans line in the temperature–density diagram. However, because the adiabats run parallel to the Jeans line, they do not intersect it, and hence no stable equilibrium state will be reached. (Recall that the critical Jeans line is where the total energy $E_{\mathrm{TOT}} = 0$, or alternatively where the free-fall time equals the expansion time, $\tau_{\mathrm{ff}} = \tau_{\mathrm{e}}$.)

We now add the adiabats to the logarithmic temperature–density diagram shown in Figure 8.2. The key thing to note is that adiabats for an ideal gas with three degrees of freedom have a different slope (2/3) from the Jeans line (1/3), and therefore must cross it at some point. This is shown in Figure 8.3.

Consider a Jeans line first. Clouds lying below (which is the same as to the right of) the Jeans line *for their mass* are too cool and dense to be stable and so they collapse. Conversely, clouds lying above (i.e. to the left of) the Jeans line for their mass are too hot and tenuous to be stable, and so they expand. The next question is, how will that collapse or expansion proceed? Systems that are collapsing adiabatically (e.g. point A in the figure) move to higher density, which is towards the right. As the adiabat slope $\gamma - 1$ for an ideal gas is positive, the temperature also increases. The adiabat eventually crosses the Jeans line at point E. Therefore, the position of the object in the temperature–density diagram moves from point A towards point E, in the direction indicated by the arrow. Now consider an object lying above the Jeans line, where the tendency is to expand. Such an object, e.g. at point B, expands to lower density, thus moving to the left in the figure, and since the adiabats have positive slope, it also moves to lower temperature. Hence it moves from point B towards point E. In both cases, once the objects obtain temperatures and densities lying on the Jeans line, any tendency to move off the line is met by expansion or contraction to move it back onto the line. That is, the Jeans line indicates a stable equilibrium configuration for *adiabatic* processes, and objects reaching this position cease to collapse (nor will they expand) under adiabatic conditions.

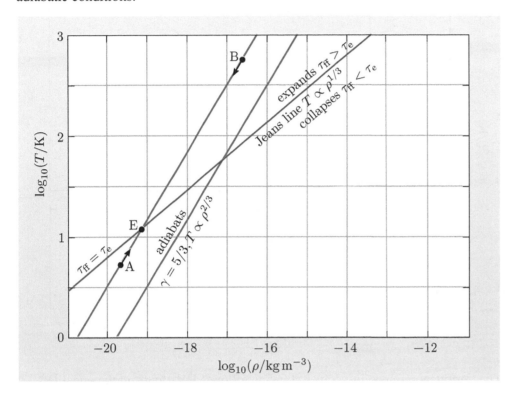

Figure 8.3   Schematic of the logarithmic temperature–density diagram, showing the Jeans line (for $100\,M_\odot$) with $T \propto \rho^{1/3}$, and two adiabats for $\gamma = 5/3$, giving $T \propto \rho^{\gamma-1} = \rho^{2/3}$.

● The Jeans line is stable under *adiabatic* processes, but stars exist. What does this tell us about the collapse of gas clouds to form stars?

○ The fact that stars do collapse from gas clouds indicates that they must do so *non*-adiabatically. That is, they must radiate energy into their environment (or in rare cases, heat up) while they collapse.

As foreshadowed above, the contraction of gas clouds is not adiabatic, and for this reason we must consider cooling processes as well.

8.2.3   Non-adiabatic collapse and the importance of cooling

In reality, gas clouds are subjected to a variety of heating and cooling mechanisms. Heating occurs by **cosmic rays** (mostly completely ionized atoms) and incoming photons. Cooling is achieved via escaping photons, emitted either when atoms and molecules de-excite or as thermal infrared radiation from dust. The latter tends to be more important, because dense, collapsing clouds become opaque to their own radiation except in the infrared.

The temperature–density diagram can be divided into upper and lower regions, where cooling and heating dominate respectively; see Figure 8.4 overleaf. Just as the Jeans line could be viewed as the state at which *dynamical* timescales for expansion and collapse, τ_e and τ_{ff}, balanced, so the division between net heating and net cooling can be viewed as the balance between *thermal* timescales for heating and cooling, τ_h and τ_c. Note that this balance occurs at $T \approx 20$ K.

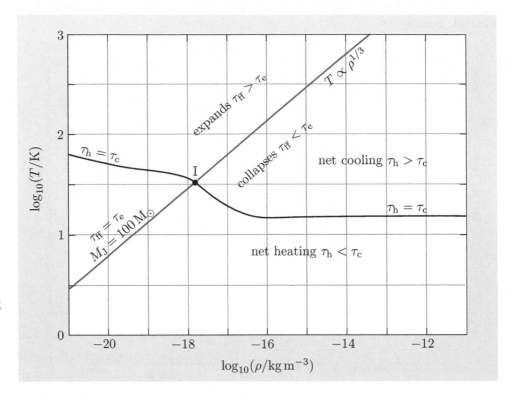

Figure 8.4   Division of the temperature–density diagram into regions in which net cooling (upper zone) and net heating (lower zone) occur. The Jeans line for a $100\,\mathrm{M}_\odot$ cloud is shown for reference.

The importance of cooling comes about as follows. For clouds in the upper left portion of the temperature–density diagram, we know already that $\tau_c < \tau_h$ and $\tau_e < \tau_{ff}$. In addition, *the cooling time is substantially shorter even than the expansion time*, $\tau_c < \tau_e$, so the net trajectory of the cloud in this part of the figure is almost vertically downwards, not along the adiabats of slope $2/3$ (see Figure 8.5). Conversely, for very cold clouds (below the Jeans line and the thermal line), the heating timescale is shorter than the collapse timescale, $\tau_h < \tau_{ff}$, so the clouds move almost vertically upwards in the diagram. Cooling (or heating) comes to a halt once the protostellar cloud reaches the curve where heating and cooling timescales are equal.

There are then two possibilities for a given cloud:

- Clouds having a density *lower* than the intersection of the Jeans line and the curve for balanced thermal timescales, point I in Figure 8.4 and Figure 8.5, reach the curve where heating and cooling timescales are equal when they are *above* the Jeans line, so end up in a state of continual expansion. These clouds are not about to form stars.

- At densities *higher* than the intersection, I, clouds reach the curve for balanced thermal timescales below the Jeans line. They then collapse, move to the right towards higher density, at essentially constant temperature (i.e. *isothermally*) until the gas becomes opaque to its own radiation, i.e. **optically thick**. This inhibits further energy release from the system, and thereafter the process begins to resemble an adiabatic one. Evolution in the temperature–density diagram then follows the adiabats mapped out in the previous subsection.

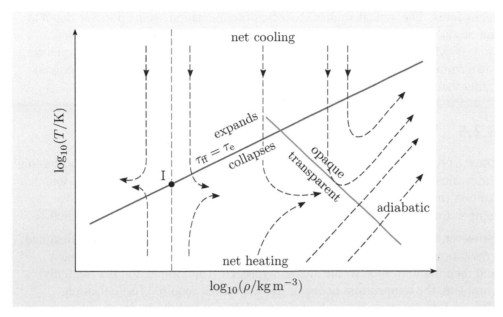

Figure 8.5 Schematic of evolutionary tracks of protostellar clouds in temperature–density diagram. The diagonal line at lower right indicates where the matter becomes opaque to its own radiation.

The evolution of the protostellar clouds can therefore be summarized as follows:

Initially, all objects have thermal timescales (τ_h and τ_c) that are much shorter than their dynamical timescales (τ_{ff} and τ_e), and so they move vertically in the temperature–density diagram without significant expansion or contraction. Once dynamical timescales become comparable to the thermal ones, the evolution changes. Objects of lowest initial density remain too hot to collapse, and ultimately expand. On the other hand, warm objects of higher initial density cool, then collapse at almost constant temperature until they become opaque to their own radiation, after which they collapse adiabatically as they approach stardom.

We study the contraction stage in more detail in the next Section.

8.2.4 Fragmentation

As a protostellar cloud cools (moves vertically downwards in the temperature–density diagram), it crosses Jeans lines for successively lower masses. This means that smaller subregions of the cloud become able to collapse, independently of what the cloud as a whole is doing. This continues even when the object ceases to cool and collapses at uniform temperature – moves to the right in the diagram. More and more subregions become able to collapse in their own right. This process is called **fragmentation**, and describes the pattern whereby a large cloud collapses as a whole initially, but subsequently smaller subregions independently satisfy the Jeans criterion for collapse, and collapse independently of one another, on progressively smaller scales. Fragmentation also marks the transition of the object from being a protostellar cloud to a **protostar**.

Fragmentation can occur throughout the cooling and optically thin contraction phase. It is not until the adiabatic phase of evolution begins, once the cloud has become opaque to its own radiation, that the star ceases crossing lower-mass Jeans lines. Theoretical studies show that fragmentation seems possible down to masses as small as $\sim 0.01\,\mathrm{M}_\odot$, which is below the minimum mass of stars ($\approx 0.08\,\mathrm{M}_\odot$) but is higher than typical planetary masses, suggesting that planets form from the coalescence of smaller bodies (beginning with mere dust grains) rather than from the fragmentation of larger systems.

8.2.5   The non-adiabatic free-fall collapse

Once a protostellar cloud has cooled to $\sim$20 K, whereupon the heating and cooling timescales are similar, and fragmented, contraction of the fragment/protostar follows (assuming the object is below the Jean's line for its mass in the temperature–density diagram), and gravitational potential energy is liberated.

However, the temperature of the object does not rise freely. Instead, the liberated gravitational potential energy causes the dissociation of molecular hydrogen and then the ionization of the atomic hydrogen. Until ionization is essentially complete, the temperature does not rise above that required for ionization. The total energy required to dissociate and then ionize the hydrogen (E_{DI}) is calculated from the dissociation and ionization energies and the number of molecules $M/2m_{\mathrm{H}}$ and (later) atoms M/m_{H} present in the cloud:

$$E_{\mathrm{DI}} = \frac{M}{2m_{\mathrm{H}}}E_{\mathrm{D}} + \frac{M}{m_{\mathrm{H}}}E_{\mathrm{I}} \tag{8.5}$$

where the energy required to dissociate a hydrogen molecule is $E_{\mathrm{D}} = 4.5$ eV and the energy required to ionize a hydrogen atom is $E_{\mathrm{I}} = 13.6$ eV.

Now, we can simply assume that this energy comes from the gravitational energy of the cloud as it collapses from radius R_1 to radius R_2. Hence:

$$\frac{GM^2}{R_2} - \frac{GM^2}{R_1} \approx \frac{M}{2m_{\mathrm{H}}}E_{\mathrm{D}} + \frac{M}{m_{\mathrm{H}}}E_{\mathrm{I}}.$$

Exercise 8.6   Consider a collapsing fragment with a mass $1\,\mathrm{M}_\odot$.

(a) What is the energy required to dissociate the hydrogen molecules and then ionize the hydrogen atoms of which it is composed?

(b) Assuming it starts at a radius of 10^{15} m (i.e. about $1.5 \times 10^6\,\mathrm{R}_\odot$ or 7000 AU), what is its radius by the time the hydrogen is all ionized?   ∎

As Exercise 8.6 showed, a $1\,\mathrm{M}_\odot$ fragment will collapse to a radius of about $150\,\mathrm{R}_\odot$ by the time it has ionized all its hydrogen. This requires a release of $\sim 3 \times 10^{39}$ J (from gravitational potential energy) and the timescale for this process (from Equation 2.5) is of order 100 000 years as calculated in Exercise 8.2.

The temperature of the fragment at this point may be estimated as follows. The kinetic energy of the ions and electrons due to their thermal motion is $E_{\mathrm{K}} = (\frac{3}{2}N_{\mathrm{p}}kT + \frac{3}{2}N_{\mathrm{e}}kT) \approx 3kT(M/m_{\mathrm{H}})$; and the gravitational potential energy is $E_{\mathrm{GR}} \approx -(M/m_{\mathrm{H}})(E_{\mathrm{D}}/2 + E_{\mathrm{I}})$. But from the virial theorem, we know $2E_{\mathrm{K}} + E_{\mathrm{GR}} = 0$ (Equation 2.9), so combining these expressions we have

$$2 \times 3kT\frac{M}{m_{\mathrm{H}}} - \frac{M}{m_{\mathrm{H}}}(E_{\mathrm{D}}/2 + E_{\mathrm{I}}) = 0.$$

The masses all cancel out, leaving

$$kT \approx \frac{1}{12}(E_{\mathrm{D}} + 2E_{\mathrm{I}}).$$

This is easily evaluated as $kT \approx 2.6$ eV.

- If $kT \approx 2.6$ eV, what is the equivalent temperature in kelvin?
- First converting to joules, $kT = 2.6$ eV $\times 1.602 \times 10^{-19}$ J eV^{-1} = 4.17×10^{-19} J. Hence $T = 4.17 \times 10^{-19}$ J$/1.381 \times 10^{-23}$ J K$^{-1} \approx 30\,000$ K.

Since the masses cancelled out, this temperature is *independent* of the mass of the fragment being considered. All fragments will become ionized at around 30 000 K.

Once hydrogen is completely ionized, the gas has become a plasma of protons and (more importantly) electrons, and hence becomes increasingly opaque to its own radiation. Further liberation of gravitational potential energy, which is not required for ionizing the atoms and which can no longer readily escape, results in the heating of the protostar. As the radiation cannot escape easily, the collapse becomes more like an adiabatic one. This increases the temperature and pressure, and hence the star begins its approach to hydrostatic equilibrium. Note also that once the hydrogen is ionized, the adiabatic index achieves the value $\gamma = 5/3$, whereas in the molecular state lower values are obtained. With the material being more opaque, it also comes into **thermodynamic equilibrium**, which is to say that the temperature characterizing the *radiation* is simply the temperature of the gas.

To summarize the findings above, for a $1\,\mathrm{M}_\odot$ protostar, this process takes some 10^5 years, which is the free-fall time from an initial density $\rho \sim 4 \times 10^{-16}$ kg m^{-3}. Over that time, the radius decreases from $\sim 10^{15}$ m ($\sim$7000 AU) to $\sim 10^{11}$ m ($\approx 150\,\mathrm{R}_\odot$). The average internal temperature reaches $\sim 30\,000$ K by the time the protostar attains hydrostatic equilibrium.

The attainment of hydrostatic equilibrium slows the contraction of the protostar, but does not halt it, as energy is being radiated. The phase of evolution where the optically thick protostar is powered by the liberation of gravitational potential energy, is called the Kelvin–Helmholtz contraction. It involves two stages, the Hayashi phase followed by the Henyey phase, which we study in the next Section.

8.3 Kelvin–Helmholtz contraction of the protostar

In this section, we follow protostars as their contraction proceeds, and study the stages they pass through as they approach stardom. There are two phases during this **Kelvin–Helmholtz contraction**, they are known as the **Hayashi phase** and the **Henyey phase**. The overall timescale for the process is the Kelvin–Helmholtz timescale, which you have already calculated in Chapter 2, as the time for a star to collapse to the main sequence assuming its luminosity is provided solely by energy liberated as a result of gravitational collapse.

8.3.1   The Hayashi track

Work by Chushiro Hayashi in the 1960s showed that a star cannot achieve hydrostatic equilibrium if its outer layers are too cool. Two important concepts in understanding Hayashi's work are **opacity** and **convection**.

> If you need reminding about these two concepts, you should re-read the boxes about them in Chapter 1 (page 20) and Chapter 4 (page 92) respectively.

Hayashi showed that there is a boundary on the right-hand side of the H–R diagram cooler than which hydrostatic equilibrium is impossible, and hence stable stars cannot exist. It lies at an effective surface temperature of around $T_{eff} \approx 3000$ to 5000 K (depending on the star's mass, chemical composition and luminosity). Objects to the right of that boundary are out of hydrostatic equilibrium, and collapse rapidly until the surface temperature reaches the value corresponding to stability. So collapsing protostars enter the H–R diagram near that boundary.

Stars at the **Hayashi boundary** are fully convective. The reason, given what you know about opacity and convection, is easy to understand. First, since the opacity of protostellar matter decreases with temperature according to the Kramers opacity law, approximately as $\kappa \propto \rho T^{-3.5}$, *cool objects have high opacity*. Second, high opacity leads to steep radiative temperature gradients, and third, steep radiative temperature gradients lead to convective instability. Putting these three things together, the coolest stars are more likely to be unstable to convection.

A new protostar continues to contract, remaining at the Hayashi convective boundary. As that boundary is almost vertical in the H–R diagram, stars up to a few times the mass of the Sun ($M \leq 3\,M_{\odot}$) collapse with almost constant effective surface temperature. Recall from Chapter 1 that the luminosity, effective surface temperature and radius of a star are related by $L = 4\pi R^2 \sigma T_{eff}^4$ (Equation 1.1). Since the star's temperature changes very little during this stage, Equation 1.1 tells us that the luminosity is proportional to the square of the radius. As its radius R is still decreasing due to contraction, the luminosity decreases significantly. It is therefore expected that protostars enter the H–R diagram around $T_{eff} \approx 4000$ K at high luminosity, and fade as they contract. The path a protostar takes in the H–R diagram during this evolutionary stage is called a **Hayashi track** (Figure 8.6).

As the protostar continues to contract, its *core* gets hotter and its opacity *decreases*, because for a Kramers opacity, $\kappa \propto \rho T^{-3.5}$. As a result of the decreasing opacity, the radiative temperature gradient becomes shallower and the condition for convection eases. Eventually the core becomes non-convective (radiative).

8.3.2   The Henyey contraction to the main sequence

Stars of mass $M \leq 0.5\,M_{\odot}$ reach the main sequence at the bottom of their Hayashi tracks. Higher-mass pre-main-sequence stars evolve to higher effective temperatures (i.e. from right to left across the H–R diagram) before initiating thermonuclear reactions to supply the energy radiated from the surface. As the

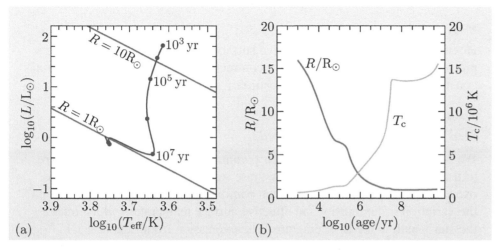

Figure 8.6 (a) H–R diagram with pre-main-sequence (Hayashi) track for the Sun. The red dots indicate elapsed times of 10^3, 10^4, 10^5, 10^6, 10^7, 10^8 and 10^9 years. (b) Evolution of radius and core temperature with time. (Data kindly provided by D. A. VandenBerg.)

star evolves from the base of its nearly vertical Hayashi track (where stars are fully convective) to the main sequence, it is in hydrostatic equilibrium, and its luminosity is fuelled by the slow gravitational collapse. This is called the Henyey phase, in recognition of the work of Louis Henyey on this topic in the 1950s. We now examine that evolution, and seek to calculate the path taken by pre-main-sequence stars in the H–R diagram.

● Given that a pre-main-sequence star is still contracting during the Henyey phase, is it safe to say that it is in hydrostatic *equilibrium*?

○ Yes, it is safe to say this. Although the star must continue to contract to release gravitational potential energy to make up for surface loses due to radiation, the contraction time is much longer than the free-fall time, and therefore the star is in hydrostatic equilibrium during this phase. (We calculate the contraction time explicitly later in this section.)

To produce an evolutionary track for a pre-main-sequence star, we need to calculate how its luminosity and temperature change. It is important to remember that the temperature, pressure, and density change with radius r inside a star. To derive such an evolutionary track, we begin by making the homology assumption. In Chapter 4 we used this assumption to calculate how the luminosity, core temperature and radius of a star on the main sequence depend on its mass and chemical composition. Here, we want to investigate the effect of changing a star's radius (at constant mass and chemical composition) on a star's luminosity and effective surface temperature. Remember, homology implies that as the star gets smaller, its structure changes by *the same factor at all radii*. The consequence is, if the radius contracts by some factor A, causing the core temperature to increase by a factor B, then the temperature *everywhere* in the star increases by a factor B. The density and pressure profiles also scale in this fashion, though by *different* factors. In Worked Example 8.1 we derive these scaling factors for protostars approaching the main sequence.

Worked Example 8.1

How do the luminosity and effective surface temperature of a pre-main-sequence star change as it contracts? (Carry out the analysis for a constant mass and chemical composition.)

Solution

Step 1: *What parameters do we need?*

We begin by identifying what stellar parameters we need to know if we are to track a star's evolution, find how those stellar parameters scale with the overall radius of the star, R, and then combine these scaling relations to find the variation of luminosity and effective surface temperature which trace the star's path in the H–R diagram. The expression from Equation 1.1, $L = 4\pi R^2 \sigma T_{\text{eff}}^4$ gives us one equation with three variables, and hence two unknowns. We need another equation relating the star's luminosity L to its radius R, or its effective surface temperature T_{eff} to its radius R, in order to specify R in terms of L or T_{eff} and eliminate it from the equation, leaving a relation between L and T_{eff}.

One such equation is that provided by an expression for energy transport within the star. If energy transport is dominated by **radiative diffusion**, the luminosity $L(r)$ at some radius r within in the star depends on the temperature $T(r)$, temperature gradient $dT(r)/dr$, density $\rho(r)$ and opacity $\kappa(r)$ at that radius according to:

$$L(r) = -4\pi r^2 \frac{16\sigma}{3} \frac{T^3(r)}{\rho(r)\,\kappa(r)} \frac{dT(r)}{dr}. \qquad \text{(Rearrangement of Eqn 1.14)}$$

Hence, we now have two equations linking the star's luminosity, temperature and radius. The first of these is in terms of the surface luminosity and effective temperature, whilst the second is in terms of the luminosity and temperature as a function of radial distance from the centre.

Step 2: *Re-scale the parameters by the contracting radius R of the star.*

Scaling the density: The mean density of a star of radius R and mass M is given by $\rho = 4M/3\pi R^3$. Under the assumption of homology, if the *mean* density scales with the contracting radius as R^{-3}, then so does the density at *any* radial distance r. Hence we can write

$$\rho(r) \propto R^{-3}.$$

Scaling the pressure: In hydrostatic equilibrium, $dP(r)/dr = -G\,m(r)\,\rho(r)/r^2$ (Equation 1.12). By the assumption of homology: a pressure interval $dP(r)$ scales in the same way as the core pressure P_{c}, a radial interval dr scales in the same way as the radius R, and the density $\rho(r)$ scales in the same way as the mean density ρ. So the hydrostatic equilibrium condition becomes

$$\frac{P_{\text{c}}}{R} \propto \frac{\rho}{R^2} \propto \frac{R^{-3}}{R^2} \propto R^{-5}$$

so $P_{\text{c}} \propto R^{-4}$. Under the assumption of homology, if the core pressure scales with the contracting radius as R^{-4}, then so does the pressure at *any* radial distance r. So,

$$P(r) \propto R^{-4}.$$

Scaling the temperature: From the ideal gas law $P(r) = \rho(r)\,kT(r)/\overline{m}$ (Equation 1.11), so $T(r) \propto P(r)/\rho(r)$ at any radius r within the star. Using the scaling relations for density ($\rho(r) \propto R^{-3}$) and pressure ($P(r) \propto R^{-4}$) derived above, we can write the scaling relation for temperature as $T(r) \propto R^{-4}/R^{-3}$, i.e.

$$T(r) \propto R^{-1}.$$

Scaling the opacity: Assuming the opacity is a Kramers opacity, $\kappa(r) \propto \rho(r)\,T(r)^{-3.5}$ (Equation 1.16), and using the relations derived above, we find $\kappa(r) \propto R^{-3}(R^{-1})^{-3.5}$, i.e.

$$\kappa(r) \propto R^{0.5}.$$

Scaling the temperature gradient: A temperature increment $\mathrm{d}T(r)$ scales with the contraction in the same way as $T(r)$, which is proportional to $1/R$. Similarly, a radial increment $\mathrm{d}r$ scales in the same way as R. Therefore the temperature gradient $\mathrm{d}T(r)/\mathrm{d}r$ responds as $(1/R)/R = 1/R^2$, i.e.

$$\frac{\mathrm{d}T(r)}{\mathrm{d}r} \propto R^{-2}.$$

Scaling the luminosity: We can now use the four scaling relations derived above:

$$\rho(r) \propto R^{-3}$$
$$T(r) \propto R^{-1}$$
$$\kappa(r) \propto R^{0.5}$$
$$\frac{\mathrm{d}T(r)}{\mathrm{d}r} \propto R^{-2}$$

to rewrite the radiative diffusion equation as

$$L(r) \propto r^2 \frac{T^3(r)}{\rho(r)\,\kappa(r)} \frac{\mathrm{d}T(r)}{\mathrm{d}r}$$

$$L(r) \propto R^2 \frac{R^{-3}}{R^{-3}\,R^{0.5}} R^{-2} \propto R^{-0.5}.$$

By the homology assumption, if the luminosity at any radius r scales as $R^{-0.5}$, then so does the surface luminosity, $L \propto R^{-0.5}$. That is, the surface luminosity increases as the reciprocal square root of the contracting radius. Equivalently,

$$R \propto L^{-2}. \tag{8.6}$$

Step 3: *Clarify the Henyey path in the H–R diagram.*

Finally, we know from Equation 1.1 that $L = 4\pi R^2 \sigma T_{\mathrm{eff}}^4$. Writing this as proportionalities, we have $L \propto R^2 T_{\mathrm{eff}}^4$. We can use the radius–luminosity relationship we have derived to eliminate R, and then write the relation between L and T_{eff} for any radius of the contracting star, giving $L \propto (L^{-2})^2 T_{\text{eff}}^4 \propto L^{-4} T_{\text{eff}}^4$ or $L^5 \propto T_{\text{eff}}^4$.

That is to say, during the Henyey contraction, the surface luminosity varies as the (4/5)-power of effective temperature:

$$L \propto T_{\text{eff}}^{4/5}. \tag{8.7}$$

Figure 8.7 shows a line of slope 4/5 (= 0.8) in the log–log version of the H–R diagram. By comparing this with the evolutionary tracks based on more detailed models shown in the figure, you can confirm that the homologous collapse approximation leads us to a good representation of the collapse process. That is, you have been able to show how a pre-main-sequence star evolves in the H–R diagram.

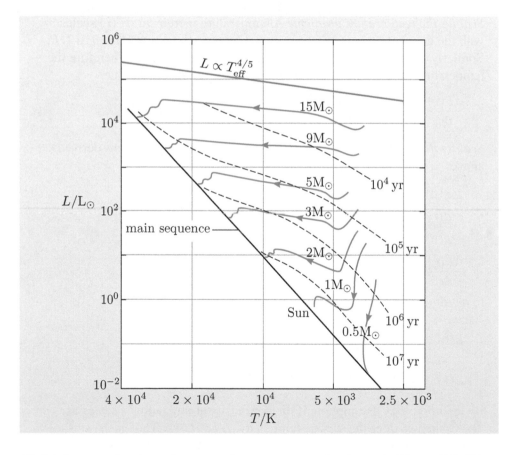

Figure 8.7 H–R diagram showing the final collapse to the main sequence – the Kelvin–Helmholtz phase – for a range of masses. Also shown is a blue line of slope 4/5 as derived under the homologous collapse assumption, which is found to be a reasonable representation of the true collapse for $M \geq M_\odot$.

● Look more closely at the pre-main-sequence tracks shown in Figure 8.7. We predicted evolutionary tracks of slope 4/5. Do both high-mass and low-mass stars actually follow such a slope, or are there differences? If there are differences, why might this be?

○ Figure 8.7 shows that the high-mass stars do evolve at close to the expected slope, while low-mass stars, e.g. the Sun, evolve slightly more steeply. This tells us that one of the assumptions in our argument that holds for high-mass stars does *not* hold so well in lower-mass stars. Either something about their structure or the way they collapse differs. The assumption that fails is that energy transport is dominated by radiative diffusion rather than convection. The envelopes of lower-mass stars experience more convection, and hence they deviate more from the expected slope of 4/5.

8.4 Observations of young pre-main-sequence objects

Having studied the *theory* of the collapse of gas clouds and the contraction of protostars towards the main sequence, it is time to turn to the *observable* Universe and see what evidence we find of **young stellar objects** (**YSOs**).

Once protostellar objects become opaque and start to heat up, we might expect them to become visible. However, for some of the high-luminosity phase of their pre-main-sequence life, stars may not be visible at optical wavelengths due to their being cloaked in dust remaining from the dense molecular cloud in which their collapse began. However, the dust may be heated by the star and radiate strongly at infrared wavelengths. Developments in **infrared astronomy** and **submillimetre astronomy** are making it possible to study young stellar objects that are still enshrouded by dust and gas from the clouds out of which they formed (e.g. see Figure 8.8). These make it possible to observe objects which are not yet visible at optical wavelengths. X-rays coming from the hot **coronae** of young **active stars** are also increasingly being used to probe these obscured systems.

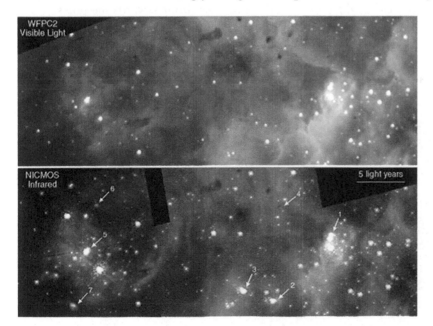

Figure 8.8 A comparison of visible light (top) and infrared (below) images of a star-forming region (30 Dor). Many sources that are not seen at visible wavelengths are revealed to be very bright in the infrared. This is due to longer wavelengths being less scattered by dust that is ubiquitous in star-forming regions.

The range of pre-main-sequence objects observed has led to an evolutionary sequence being envisaged, shown schematically in Figure 8.9 overleaf. Stars of the **T Tauri** type are believed to be very young pre-main-sequence objects in the mass range $(0.2–2.0)M_\odot$ that are making their first appearance. Radio data complement the infrared and submillimetre observations, and show many protostars to have **bipolar outflows** – gas flowing away from a central object along an axis, presumably its rotation axis, at velocities of several tens of km s^{-1}. The reason that the outflowing material streams along an axis, rather than in all directions, is believed to be due to the influence of a disc or torus (doughnut) of material which remains for a time around the protostar.

Figure 8.9   A schematic diagram of the evolution of a pre-main-sequence star from the protostar stage, where bipolar outflows can be measured, through the T Tauri stage where outflows are disappearing but the protostellar (and possibly protoplanetary) disc still remains, to the stage where the star becomes a genuine main-sequence object.

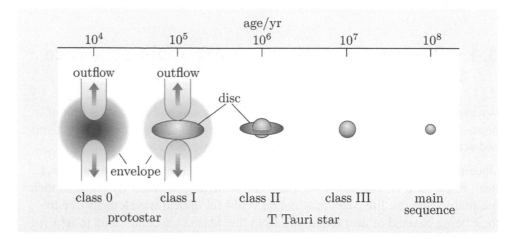

In our earlier descriptions of protostellar collapse, we ignored material which, though unstable to gravitational collapse, would possess too much angular momentum (rotation) to let it collapse down to a small sphere. The expected fate of such material is that it would collapse only partially before colliding with other high-angular-momentum material and forming a rotating disc or torus about the star. This belt of dense material would prevent outflowing gas from escaping in the orbital plane, but would allow escape in the directions towards the poles. Magnetic fields may also play a role in collimating the polar outflows. These outflows can now also be observed at optical wavelengths (see Figure 8.10). In some high-mass objects, which are hotter than T Tauri stars and hence have even stronger radiation pressure, the outflow velocities reach into the hundreds of km s^{-1}.

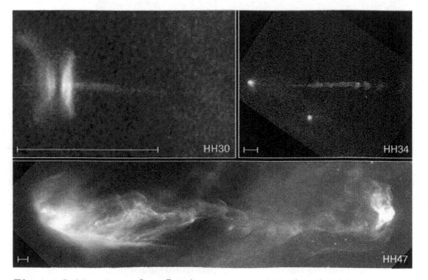

Figure 8.10   Jets of outflowing material associated with young stars. In HH30, the obscuring protostellar disc is seen almost edge-on. The scale bar in each image is 1 AU.

T Tauri stars are found in regions of space where considerable dense gas and infrared-emitting dust remains (left over from recent star formation). They show other indications of youth such as **mass loss**, especially via bipolar outflows, rapid rotation, and high abundances of the fragile element lithium, which many stars destroy during their lives. The temperatures and luminosities of T Tauri stars, i.e. their positions in the H–R diagram, are close to the Hayashi boundary at which protostars first appear (see Figure 8.11), consistent with the locations of the theoretical Hayashi tracks.

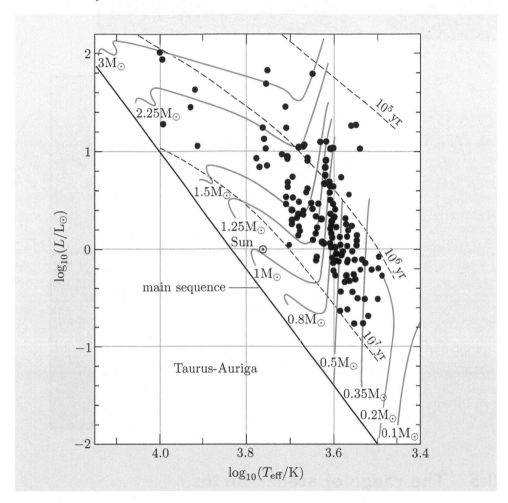

Figure 8.11   Pre-main-sequence evolutionary tracks (Hayashi tracks) for $(0.1–3)\,M_\odot$ stars, and locations of known T Tauri stars (black dots). Evolutionary times are also shown. The locations of the T Tauri stars are consistent with ages of $\sim 10^6$ years.

Some signs of youth remain visible even after stars reach the main sequence. One such class of objects is the **Vega-excess stars**, named after the first star in the class. These are very close to or sometimes on the main sequence, but show unusually large excesses of infrared radiation (compared to a bare photosphere). The radiation is thought to come from a dust shell or disc surrounding the star, which has been heated by the starlight to $T \approx 100\,K$ and which therefore emits thermal radiation at infrared wavelengths. Such circumstellar discs are believed to be the structures from which planets may form around their young host stars. A particularly well studied example is **Beta Pictoris** (β Pic), where the disc is seen almost edge-on. This enables its profile to be studied in some detail (Figure 8.12 overleaf). Circumstellar discs around young stars are regarded as likely places for the formation of planets, and are also called proto planetary discs or **proplyds**. Such discs, while primarily gaseous, contain enough dust to scatter the light and make the discs opaque. Several examples have been seen silhouetted against the bright background of the Orion nebula (Figure 8.13).

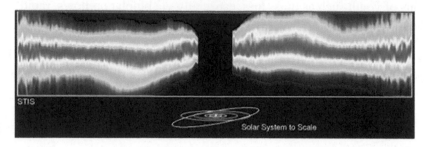

Figure 8.12   The circumstellar/protoplanetary disc around β Pic, with the central star masked to shield the telescope from its intense light. The disc is visible because it scatters light from the central star. The asymmetries in the disc are evidence that it has been influenced by the gravity of other unseen stars or giant planets. The image traces the disc to 250 AU from the star.

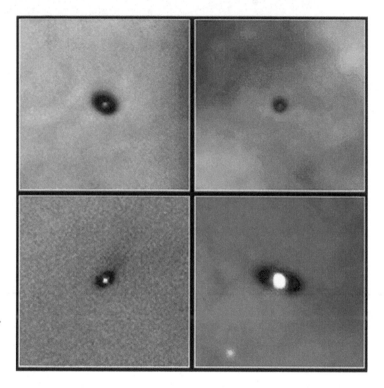

Figure 8.13   Protoplanetary discs (proplyds) and their host stars, roughly only 10^6 years old, seen silhouetted against the bright background of the Orion nebula. The discs range in size from 2–8 times the diameter of the Solar System.

8.5   The range of stellar births

Although in this Chapter we have focused on the formation of a single star, it is instructive to consider what range of masses of stars form from collapsing clouds, and what fraction of these stars are born in binary or multiple stellar systems.

8.5.1   The stellar mass function

The stellar mass function is a way of quantifying the number of stars that exist with each range of masses. A specific example of this is the initial mass function (IMF) which refers to the mass distribution of a population of stars immediately after they arrive at the main sequence.

● Why might the observed (present day) stellar mass function of a cluster of stars not be a good measure of its initial mass function?

○ Higher-mass stars have shorter lives. So, depending on the age of the cluster, more of the high-mass stars will have disappeared, meaning that the fraction of lower-mass stars will be overestimated.

For many years, the IMF was presumed to have a shape described by

$$N(M)\,\mathrm{d}M \propto M^{-2.35}\,\mathrm{d}M \qquad (8.8)$$

where $N(M)\,\mathrm{d}M$ is the number of stars within a small mass interval $\mathrm{d}M$. This form was established by Edwin Salpeter in 1955 and is known as the **Salpeter mass function**. It predicts that the number of stars within a given mass range continually increases as the mass considered becomes smaller and smaller.

Exercise 8.7 For every star born with a mass of $100\,\mathrm{M}_\odot$, how many stars with masses of $50\,\mathrm{M}_\odot$, $10\,\mathrm{M}_\odot$, $5\,\mathrm{M}_\odot$, $1\,\mathrm{M}_\odot$, $0.5\,\mathrm{M}_\odot$ and $0.1\,\mathrm{M}_\odot$ does the Salpeter mass function predict? ■

As the Exercise 8.7 shows, if the Salpeter mass function is correct, there should be over 10 million low-mass stars (just above the brown-dwarf limit) for every star formed at the upper mass limit for stars.

Checking this has proved difficult however. Low-mass stars are necessarily very faint and very difficult to detect. In recent years though, advances in infrared astronomy have enabled censuses to be carried out of low-mass stars in the solar neighbourhood. It turns out that there are rather fewer low-mass stars than predicted by the Salpeter mass function. Instead, the mass function seems to have a peak around a mass of $0.5\,\mathrm{M}_\odot$, and falls off either side of this with roughly the same slope. The most common stars in the Universe are therefore M-type stars with about half the mass of the Sun.

8.5.2 The single-star fraction

Observationally it is clear that many stars exist in binary systems, or in clusters. For instance, the nearest star system to the Sun, Alpha Centauri, is actually a triple system consisting of a close binary pair (α Cen A and α Cen B) in orbit with a more distant third star (known as Proxima Centauri).

Historically, in the 18th century it had been recognized that the fraction of visual double stars was too high to be due to chance alignments, and by 1802 William Herschel had catalogued hundreds of visual pairs and computed the first orbits of binary stars. In 1983 Helmut Abt had claimed that 70% – 80% of all solar-type stars were in binaries, and this led to a long-held belief that most stars in the galaxy were in fact formed as part of binary systems.

However, in recent years it has been realized that the binary-star fraction is a function of stellar type (and therefore stellar mass). Recent surveys have concluded that, whilst the binary-star fraction is indeed high for massive stars, it is somewhat less for lower-mass stars. Furthermore, since most stars in the Universe are low-mass ($\sim 0.5\,\mathrm{M}_\odot$) stars, the binary-star fraction is not as high as had been supposed.

It is actually easier to talk about the single-star fraction, since that avoids complications concerning binary, triple, quadruple, etc., systems containing more than one star. The fraction of stars that are single appears to be about 25% for massive stars ($M > 2\,M_\odot$), about 45% for solar-type stars, and about 75% for low-mass ($M \sim 0.5\,M_\odot$) stars. When the fraction of single stars is weighted by the initial or present day stellar mass function, it turns out that about 65%–70% of stars are actually single, and most of these are low-mass M-dwarfs.

8.6   Stellar life cycles

Having examined the process by which stars are born, we conclude the book by revisiting the idea of stellar life cycles. In Chapters 6 and 7 you read about planetary nebulae leading to white dwarfs, and supernovae leading to neutron stars and black holes. You might have thought of these as the final link in the chain of the life and death of stars. But the process does *not* end. Rather, mass-loss via stellar winds, planetary nebulae and supernovae is a vital link in a *cycle* of activity that explains the chemical evolution of entire galaxies and the Universe.

As you have just learnt in this Chapter, stars form in gravitationally contracting gas clouds, and gravitational contraction continues to drive stellar evolution throughout a star's life (Chapter 2). Stars ignite hydrogen in their cores if their masses exceed $\approx 0.08\,M_\odot$, and produce helium and possibly nitrogen (via the CNO cycle) (Chapters 1, 3 and 4). If their masses exceed $\approx 0.5\,M_\odot$, later they ignite helium to produce carbon and oxygen (Chapter 5), and if their main-sequence masses exceed $\approx 10\,M_\odot$ they produce elements up to the iron peak (Chapter 6). Carbon and nitrogen synthesized in low-mass and intermediate-mass stars ($M \leq 2\,M_\odot$ and $2\,M_\odot \leq M \leq 8\,M_\odot$ respectively) are returned to the interstellar medium via mass-loss late in the life of AGB stars (where s-process nuclei are also made) and planetary nebulae (Chapter 6). Most heavier elements are produced in supernovae or their progenitors (parent stars). Type II supernovae (SN II) eject the products of nucleosynthesis that range from helium in the outer layers of the star, down to iron-peak elements on the edge of the iron core, and even elements beyond the iron peak that are produced by the r-process during the explosion (Chapters 6 and 7). Type Ia supernovae (SN Ia), whilst being lower-mass objects not possessing an iron core *prior to* the explosion, are prolific producers of iron *during* the explosion.

The nucleosynthesis products *ejected* during (though not necessarily *produced* during) supernova explosions enrich the interstellar medium, and accompanying shock waves from the explosion are believed to compress interstellar clouds and trigger the next episode of star formation. Stars that form from gas enriched by these events have a higher metal content (i.e. elements other than hydrogen and helium) than previous generations of stars. Their stellar structure also differs, which in turn affects the ways they evolve.

The oldest stars seen in the Galaxy today are observed to have very low abundances of metals; they are referred to as **Population II** stars. These stars are found within the bulge near the centre of the Galaxy and in the halo of the Galaxy, including the globular clusters. In contrast, younger stars (such as the Sun) have higher abundances of metals; they are referred to as **Population I** stars. These are typically found in the spiral arms of the Galaxy. Of course, there is a gradual transition from metal-poor Population II stars to metal-rich Population I stars, and the disc of the Galaxy contains a population of stars with a range of metallicities.

The metallicity of a star essentially depends on how many generations of stars preceded its formation.

An important point to note is that even Population II stars contain *some* metals. It is therefore suggested that an even earlier generation of metal-free stars once existed in the Galaxy. These hypothetical **Population III** stars must have been formed from primordial hydrogen and helium only, and lived their lives before the observed Population II stars were born. Such stars are postulated to have been extremely massive (perhaps several hundred solar masses) and consequently would have had extremely short lives (less than a million years).

By observing the different chemical compositions, ages, and space motions of the stars that make up galaxies, astronomers can piece together the history of entire galaxies, and ultimately the Universe. Such investigations also have a very direct bearing on our presence, here on planet Earth. By studying extrasolar planets (planet orbiting other stars), a strong correlation has been found between the presence of planets and the metallicity of their host star – a star with a higher abundance of heavy elements is more likely to possess planets. The possibility of life in the Universe is therefore intimately linked with the processes of stellar evolution and nucleosynthesis.

The physics of extrasolar planets is discussed in the companion volume to this book: *Transiting Exoplanets* by C. A. Haswell.

Summary of Chapter 8

1. The Jeans criterion for collapse, $E_{\text{TOT}} < 0$, can be expressed in several ways. A cloud of radius R collapses if its mass M exceeds the Jeans mass:

$$M > M_{\text{J}} = \frac{3kT}{2G\overline{m}} R. \qquad \text{(Eqn 8.1)}$$

A cloud of mass M condenses if its average density exceeds the Jeans density:

$$\rho > \rho_{\text{J}} = \left(\frac{3kT}{2G\overline{m}} \right)^3 \frac{3}{4\pi M^2}. \qquad \text{(Eqn 8.3)}$$

2. The changing parameters of collapsing gas clouds may be tracked on a log(temperature)–log(density) diagram. The Jeans criterion indicates that the critical boundary for collapse has the form

$$\log_{10} T = \frac{1}{3} \log_{10} \rho + \text{constant}.$$

Adiabatic collapse for an ideal gas with 3 degrees of freedom would have a slope 2/3, and would always lead back to the Jeans line, thus *failing* to form stars. Rather, collapse is initially *non*-adiabatic. Initially clouds cool, until heating and cooling timescales become comparable. High-density clouds then collapse almost isothermally until they become optically thick, after which they contract almost adiabatically.

3. The Jeans equations show that if the cloud collapses (R decreases and ρ increases) without heating up, then progressively smaller submasses satisfy the Jeans criterion. This leads to fragmentation of the collapsing cloud.

4. A $1\,M_{\odot}$ molecular cloud fragment that has just become unstable to collapse at a temperature of 20 K would have a Jean's density $\rho_{\text{J}} \sim 4 \times 10^{-16}$ kg m^{-3} and a radius $R_{\text{J}} \sim 10^{15}$ m.

5. A cloud collapses almost isothermally and in free fall ($\tau_{\mathrm{ff}} \approx 10^5$ yr) until the liberated gravitational energy has completely dissociated molecular hydrogen (H_2) and then ionized atomic hydrogen. For $1\,M_\odot$ of hydrogen, the necessary amount of energy ($\sim 3 \times 10^{39}$ J) is liberated when the protostar collapses to $R \approx 150\,R_\odot$.

6. Ionization of hydrogen creates a plasma of protons and electrons and hence also increases the opacity. The protostar becomes opaque to its own radiation and heats up, and the pressure slows the collapse. The protostar approaches hydrostatic equilibrium. By the time hydrostatic equilibrium has been established, the internal temperature has risen to $\sim 30\,000$ K.

7. In hydrostatic equilibrium, the star undergoes the long Kelvin–Helmholtz phase in which it *slowly* collapses, remaining in hydrostatic equilibrium. The loss of gravitational potential energy due to collapse is at a rate sufficient to replenish the energy radiated from the surface.

8. The Kelvin–Helmholtz contraction comprises two parts: the Hayashi track (where the protostar is fully convective) and the Henyey phase (where the protostar is dominated increasingly by radiative transport).

9. The Hayashi convective boundary exists because cooler stellar material has higher opacity, according to the Kramers opacity law $\kappa \propto \rho T^{-3.5}$. High opacity impedes the radiative heating of layers further out and hence leads to a steep radiative temperature gradient; a steep radiative temperature gradient promotes convection.

10. The Hayashi boundary is at $T_{\mathrm{eff}} \approx 3000\text{–}5000$ K (depending on mass, composition and luminosity), and protostars descend along the boundary from high luminosity to low luminosity at roughly constant effective surface temperature. Stars below $0.5\,M_\odot$ reach the main sequence at the base of their Hayashi tracks, whilst still fully convective.

11. Stars more massive than $0.5\,M_\odot$ enter the subsequent Henyey contraction phase before reaching the main sequence. During this phase the luminosity goes as $L \propto T_{\mathrm{eff}}^{4/5}$.

12. Protostars are generally still cloaked in dust, but can be observed in radio, infrared, and X-ray wavelengths. Bipolar outflows (along the rotation axis) are often seen, along with evidence for circumstellar discs. As the dust is blown away from the luminous protostar, a T Tauri pre-main-sequence star appears in the H–R diagram near the fully convective (Hayashi) boundary.

13. The initial stellar mass function seems to peak at around $0.5\,M_\odot$ and fall off either side with a power law dependence roughly $\propto M^{\pm 2.35}$. Most stars in the Universe are therefore low-mass M-dwarfs.

14. The binary-star fraction is a function of stellar mass. Only about a quarter of high-mass stars are single, but around half of solar-type stars are single. The single-star fraction for stars around $0.5\,M_\odot$ is about 75%.

15. Stellar evolution is a cyclic process, with the nuclear-processed products of one generation seeding interstellar space to be incorporated into subsequent generations of stars.

Appendix

Table A.1 Common SI unit conversions and derived units.

Quantity	Unit	Conversion
speed	$\mathrm{m\,s^{-1}}$	
acceleration	$\mathrm{m\,s^{-2}}$	
angular speed	$\mathrm{rad\,s^{-1}}$	
angular acceleration	$\mathrm{rad\,s^{-2}}$	
linear momentum	$\mathrm{kg\,m\,s^{-1}}$	
angular momentum	$\mathrm{kg\,m^2\,s^{-1}}$	
force	newton (N)	$1\,\mathrm{N} = 1\,\mathrm{kg\,m\,s^{-2}}$
energy	joule (J)	$1\,\mathrm{J} = 1\,\mathrm{N\,m} = 1\,\mathrm{kg\,m^2\,s^{-2}}$
power	watt (W)	$1\,\mathrm{W} = 1\,\mathrm{J\,s^{-1}} = 1\,\mathrm{kg\,m^2\,s^{-3}}$
pressure	pascal (Pa)	$1\,\mathrm{Pa} = 1\,\mathrm{N\,m^{-2}} = 1\,\mathrm{kg\,m^{-1}\,s^{-2}}$
frequency	hertz (Hz)	$1\,\mathrm{Hz} = 1\,\mathrm{s^{-1}}$
charge	coulomb (C)	$1\,\mathrm{C} = 1\,\mathrm{A\,s}$
potential difference	volt (V)	$1\,\mathrm{V} = 1\,\mathrm{J\,C^{-1}} = 1\,\mathrm{kg\,m^2\,s^{-3}\,A^{-1}}$
electric field	$\mathrm{N\,C^{-1}}$	$1\,\mathrm{N\,C^{-1}} = 1\,\mathrm{V\,m^{-1}} = 1\,\mathrm{kg\,m\,s^{-3}\,A^{-1}}$
magnetic field	tesla (T)	$1\,\mathrm{T} = 1\,\mathrm{N\,s\,m^{-1}\,C^{-1}} = 1\,\mathrm{kg\,s^{-2}\,A^{-1}}$

Table A.2 Other unit conversions.

wavelength
1 nanometre (nm) $= 10\,\text{Å} = 10^{-9}\,\mathrm{m}$
1 ångstrom $= 0.1\,\mathrm{nm} = 10^{-10}\,\mathrm{m}$

mass–energy equivalence
$1\,\mathrm{kg} = 8.99 \times 10^{16}\,\mathrm{J}/c^2$ (c in $\mathrm{m\,s^{-1}}$)
$1\,\mathrm{kg} = 5.61 \times 10^{35}\,\mathrm{eV}/c^2$ (c in $\mathrm{m\,s^{-1}}$)

angular measure
$1° = 60\,\mathrm{arcmin} = 3600\,\mathrm{arcsec}$
$1° = 0.017\,45\,\mathrm{radian}$
$1\,\mathrm{radian} = 57.30°$

distance
1 astronomical unit (AU) $= 1.496 \times 10^{11}\,\mathrm{m}$
1 light-year (ly) $= 9.461 \times 10^{15}\,\mathrm{m} = 0.307\,\mathrm{pc}$
1 parsec (pc) $= 3.086 \times 10^{16}\,\mathrm{m} = 3.26\,\mathrm{ly}$

temperature
absolute zero: $0\,\mathrm{K} = -273.15\,°\mathrm{C}$
$0\,°\mathrm{C} = 273.15\,\mathrm{K}$

energy
$1\,\mathrm{eV} = 1.602 \times 10^{-19}\,\mathrm{J}$
$1\,\mathrm{J} = 6.242 \times 10^{18}\,\mathrm{eV}$

spectral flux density
1 jansky (Jy) $= 10^{-26}\,\mathrm{W\,m^{-2}\,Hz^{-1}}$
$1\,\mathrm{W\,m^{-2}\,Hz^{-1}} = 10^{26}\,\mathrm{Jy}$

cross-sectional area
$1\,\mathrm{barn} = 10^{-28}\,\mathrm{m^2}$
$1\,\mathrm{m^2} = 10^{28}\,\mathrm{barn}$

cgs units
$1\,\mathrm{erg} = 10^{-7}\,\mathrm{J}$
$1\,\mathrm{dyne} = 10^{-5}\,\mathrm{N}$
$1\,\mathrm{gauss} = 10^{-4}\,\mathrm{T}$
$1\,\mathrm{emu} = 10\,\mathrm{C}$

pressure
$1\,\mathrm{bar} = 10^5\,\mathrm{Pa}$
$1\,\mathrm{Pa} = 10^{-5}\,\mathrm{bar}$
$1\,\mathrm{atmosphere} = 1.013\,25\,\mathrm{bar}$
$1\,\mathrm{atmosphere} = 1.013\,25 \times 10^5\,\mathrm{Pa}$

Table A.3 Constants.

Name of constant	Symbol	SI value
Fundamental constants		
gravitational constant	G	$6.673 \times 10^{-11} \, \mathrm{N \, m^2 \, kg^{-2}}$
Boltzmann's constant	k	$1.381 \times 10^{-23} \, \mathrm{J \, K^{-1}}$
speed of light in vacuum	c	$2.998 \times 10^8 \, \mathrm{m \, s^{-1}}$
Planck's constant	h	$6.626 \times 10^{-34} \, \mathrm{J \, s}$
	$\hbar = h/2\pi$	$1.055 \times 10^{-34} \, \mathrm{J \, s}$
fine structure constant	$\alpha = e^2/4\pi\varepsilon_0\hbar c$	$1/137.0$
Stefan–Boltzmann constant	σ	$5.671 \times 10^{-8} \, \mathrm{J \, m^{-2} \, K^{-4} \, s^{-1}}$
Thomson cross-section	σ_{T}	$6.652 \times 10^{-29} \, \mathrm{m^2}$
permittivity of free space	ε_0	$8.854 \times 10^{-12} \, \mathrm{C^2 \, N^{-1} \, m^{-2}}$
permeability of free space	μ_0	$4\pi \times 10^{-7} \, \mathrm{T \, m \, A^{-1}}$
Particle constants		
charge of proton	e	$1.602 \times 10^{-19} \, \mathrm{C}$
charge of electron	$-e$	$-1.602 \times 10^{-19} \, \mathrm{C}$
electron rest mass	m_{e}	$9.109 \times 10^{-31} \, \mathrm{kg}$
		$= 0.511 \, \mathrm{MeV}/c^2$
proton rest mass	m_{p}	$1.673 \times 10^{-27} \, \mathrm{kg}$
		$= 938.3 \, \mathrm{MeV}/c^2$
neutron rest mass	m_{n}	$1.675 \times 10^{-27} \, \mathrm{kg}$
		$= 939.6 \, \mathrm{MeV}/c^2$
atomic mass unit	u	$1.661 \times 10^{-27} \, \mathrm{kg}$
Astronomical constants		
mass of the Sun	$\mathrm{M}_\odot$	$1.99 \times 10^{30} \, \mathrm{kg}$
radius of the Sun	$\mathrm{R}_\odot$	$6.96 \times 10^8 \, \mathrm{m}$
luminosity of the sun	$\mathrm{L}_\odot$	$3.83 \times 10^{26} \, \mathrm{W}$
mass of the Earth	$\mathrm{M}_\oplus$	$5.97 \times 10^{24} \, \mathrm{kg}$
radius of the Earth	$\mathrm{R}_\oplus$	$6.37 \times 10^6 \, \mathrm{m}$
mass of Jupiter	$\mathrm{M_J}$	$1.90 \times 10^{27} \, \mathrm{kg}$
radius of Jupiter	$\mathrm{R_J}$	$7.15 \times 10^7 \, \mathrm{m}$
astronomical unit	AU	$1.496 \times 10^{11} \, \mathrm{m}$
light-year	ly	$9.461 \times 10^{15} \, \mathrm{m}$
parsec	pc	$3.086 \times 10^{16} \, \mathrm{m}$
Hubble parameter	H_0	$(70.4 \pm 1.5) \, \mathrm{km \, s^{-1} \, Mpc^{-1}}$
		$(2.28 \pm 0.05) \times 10^{-18} \, \mathrm{s^{-1}}$
age of Universe	t_0	$(13.73 \pm 0.15) \times 10^9 \, \mathrm{years}$
current critical density	$\rho_{\mathrm{c},0}$	$(9.30 \pm 0.40) \times 10^{-27} \, \mathrm{kg \, m^{-3}}$
current dark energy density	$\Omega_{\Lambda,0}$	$(73.2 \pm 1.8)\%$
current matter density	$\Omega_{\mathrm{m},0}$	$(26.8 \pm 1.8)\%$
current baryonic matter density	$\Omega_{\mathrm{b},0}$	$(4.4 \pm 0.2)\%$
current non-baryonic matter density	$\Omega_{\mathrm{c},0}$	$(22.3 \pm 0.9)\%$
current curvature density	$\Omega_{\mathrm{k},0}$	$(-1.4 \pm 1.7)\%$
current deceleration	q_0	-0.595 ± 0.025

Solutions to exercises

Exercise 1.1 Using Equation 1.1, $L = 4\pi R^2 \sigma T_{\text{eff}}^4$, the luminosities corresponding to each combination of radius and temperature are as shown in Table S1.1. (Remember to first convert the radii from solar units to metres, in order to calculate the luminosity L in watts. Then divide by the solar luminosity $L_\odot$ to convert the answer into solar units.)

As an example, the first entry in Table S1.1 may be calculated as follows:

$$
\begin{aligned}
L &= 4\pi R^2 \sigma T_{\text{eff}}^4 \\
&= 4\pi \times (0.1 \times 6.96 \times 10^8 \text{ m})^2 \times (5.671 \times 10^{-8} \text{ J m}^{-2} \text{ K}^{-4} \text{ s}^{-1}) \times (2000 \text{ K})^4 \\
&= 5.52 \times 10^{22} \text{ J s}^{-1} = 5.52 \times 10^{22} \text{ J s}^{-1} \times \frac{L_\odot}{3.83 \times 10^{26} \text{ W}} \\
&= 1.44 \times 10^{-4} \, L_\odot.
\end{aligned}
$$

Since Figure 1.1 has logarithmic axes, in order to plot this value, note that $\log_{10}(1.44 \times 10^{-4}) = -3.84$. Hence one end of the line connecting points with $R = 0.1\,R_\odot$ lies at the point $\log_{10}(L/L_\odot) = -3.84$, $T_{\text{eff}} = 2000$ K. The set of lines of constant radii illustrating all the results in Table S1.1 is shown on Figure S1.1 overleaf.

Table S1.1 Luminosities for stars of a given temperature and radius, for use with Exercise 1.1.

T_{eff}	$R = 0.1\,R_\odot$	$R = 1\,R_\odot$
2000 K	$L = 1.44 \times 10^{-4}\,L_\odot$	$L = 1.44 \times 10^{-2}\,L_\odot$
4000 K	$L = 2.31 \times 10^{-3}\,L_\odot$	$L = 2.31 \times 10^{-1}\,L_\odot$
6000 K	$L = 1.17 \times 10^{-2}\,L_\odot$	$L = 1.17\,L_\odot$
10 000 K	$L = 9.01 \times 10^{-2}\,L_\odot$	$L = 9.01\,L_\odot$
20 000 K	$L = 1.44\,L_\odot$	$L = 144\,L_\odot$
40 000 K	$L = 23.1\,L_\odot$	$L = 2.31 \times 10^3\,L_\odot$
T_{eff}	$R = 10\,R_\odot$	$R = 100\,R_\odot$
2000 K	$L = 1.44\,L_\odot$	$L = 1.44 \times 10^2\,L_\odot$
4000 K	$L = 23.1\,L_\odot$	$L = 2.31 \times 10^3\,L_\odot$
6000 K	$L = 117\,L_\odot$	$L = 1.17 \times 10^4\,L_\odot$
10 000 K	$L = 901\,L_\odot$	$L = 9.01 \times 10^4\,L_\odot$
20 000 K	$L = 1.44 \times 10^4\,L_\odot$	$L = 1.44 \times 10^6\,L_\odot$
40 000 K	$L = 2.31 \times 10^5\,L_\odot$	$L = 2.31 \times 10^7\,L_\odot$

Exercise 1.2 Let us suppose the Sun contains N ions in total. Each hydrogen ion will be matched by one electron, whereas each helium ion will be matched by two electrons. The mean molecular mass $\mu_\odot$ is therefore

$$\mu_\odot = \frac{N_{\text{H}}(m_{\text{H}}/u) + N_{\text{He}}(m_{\text{He}}/u) + N_{\text{e}}(m_{\text{e}}/u)}{N_{\text{H}} + N_{\text{He}} + N_{\text{e}}}.$$

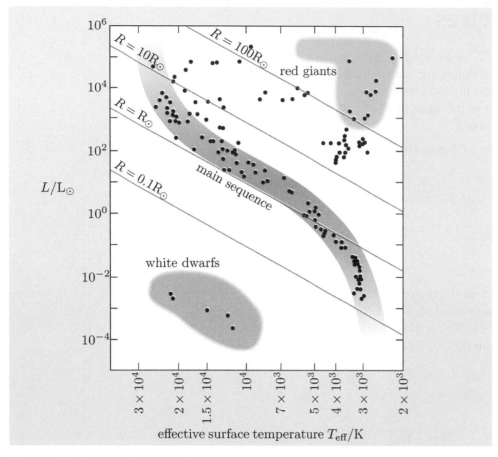

Figure S1.1 A schematic Hertzsprung–Russell diagram. The sloping lines indicate where stars would have radii $R = 0.1\,R_\odot$, $R_\odot$, $10\,R_\odot$ and $100\,R_\odot$.

Given the proportions in the question, $N_H = 0.927N$, $N_{He} = 0.073N$ and $N_e = 0.927N + (2 \times 0.073N) = 1.073N$. Assuming $m_H/u \approx 1$, $m_{He}/u \approx 4$ and $m_e/u \approx 0$, we can write

$$\mu_\odot \approx \frac{0.927N + (0.073N \times 4)}{(0.927 + 0.073 + 1.073)N} \approx 1.219/2.073 \approx 0.6.$$

So the mean molecular mass is $\mu_\odot \approx 0.6$ or $\overline{m}_\odot \approx 0.6u$.

Exercise 1.3 With a Kramers opacity, the opacity is given by $\kappa(r) \propto \rho(r)/T^{3.5}(r)$. Using the argument from the previous example, the mean opacity may be expressed as simply $\overline{\kappa} \propto \overline{\rho}/T_c^{3.5}$, where T_c is the star's core temperature and $\overline{\rho}$ is its mean density. Now, from Worked Example 1.1, we already have the relationships $\overline{\rho} \propto M/R^3$ and $T_c \propto M/R$, where M and R are the mass and radius of the star. So, the mean value of the Kramers opacity may be re-written as $\overline{\kappa} \propto R^{0.5}/M^{2.5}$.

We derive the same penultimate equation as in the previous worked example, namely $L \propto M^3/\overline{\kappa}$, where L is the star's surface luminosity. So, using the relationship above, this becomes $L \propto M^{5.5}R^{-0.5}$.

Exercise 1.4 The second branch of the proton–proton chain will include one electron–positron annihilation reaction as only one instance of the initial proton+proton reaction is involved. The overall reaction may be written as

$$2e^- + 4p \longrightarrow {}^4_2\text{He} + \nu_e(\text{pp}) + \nu_e(\text{Be}) + \gamma_{\text{pd}} + \gamma_{\text{He}} + 2\gamma_e,$$

where $\nu_e(\text{pp})$ is the electron neutrino released by the proton+proton reaction step, $\nu_e(\text{Be})$ is the electron neutrino released by the beryllium-7 electron capture reaction step, γ_{pd} is the gamma-ray released by the proton+deuterium reaction step, γ_{He} is the gamma-ray released by the helium-3 + helium-4 reaction step, and $2\gamma_e$ are the gamma-rays released by the electron–positron annihilation step.

Now, using the masses from earlier, i.e. $1.672\,623 \times 10^{-27}$ kg for the ${}^1_1\text{H}$ *nucleus* and $6.644\,656 \times 10^{-27}$ kg for the ${}^4_2\text{He}$ *nucleus*, the mass defect can be calculated as

$$\Delta m = \text{initial mass} - \text{final mass}$$
$$= 2m(e^-) + m(4p) - m({}^4_2\text{He}) - m(\nu_e(\text{pp})) - m(\nu_e(\text{Be})) - m(\gamma_{\text{pd}}) - m(\gamma_{\text{He}}) - m(2\gamma_e)$$
$$= 1.8218 \times 10^{-30}\,\text{kg} + 6.690\,492 \times 10^{-27}\,\text{kg} - 6.644\,656 \times 10^{-27}\,\text{kg} - 0 - 0 - 0 - 0 - 0$$
$$= 4.7658 \times 10^{-29}\,\text{kg}.$$

Exactly as for branch ppI, for branch ppII the energy equivalent $\Delta Q = (\Delta m)c^2$ is 4.2833×10^{-12} J or 26.74 MeV.

This includes the energy that goes into the γ-rays, which is then absorbed by the surrounding gas. As before, the two neutrinos escape the star without depositing their energy, which in this case removes 0.26 MeV for the $\nu_e(\text{pp})$ neutrino and $(0.9 \times 0.86) + (0.1 \times 0.38)$ MeV = 0.81 MeV for the $\nu_e(\text{Be})$ neutrino. This leaves $26.74\,\text{MeV} - 0.26\,\text{MeV} - 0.81\,\text{MeV} = 25.67\,\text{MeV}$ for the star.

Exercise 2.1 The free-fall time for the Sun is

$$t_{\text{ff}} = \left(\frac{3\pi}{32G\rho}\right)^{1/2} = \left(\frac{3\pi}{32 \times (6.673 \times 10^{-11}\,\text{N m}^2\,\text{kg}^{-2}) \times (1.41 \times 10^3\,\text{kg m}^{-3})}\right)^{1/2}$$
$$= 1770\,\text{s}.$$

The free-fall time of the Sun is therefore about half an hour.

Exercise 2.2 The limiting case is when $f = 0.5$. So, the first term in Equation 2.19

$$\left(\frac{36}{\pi}\frac{3c}{4\sigma}\frac{(1-f)}{f^4}\right)^{1/2} = \left(\frac{36}{\pi} \times \frac{3 \times 2.998 \times 10^8\,\text{m s}^{-1}}{4 \times 5.671 \times 10^{-8}\,\text{J m}^{-2}\,\text{K}^{-4}\,\text{s}^{-1}} \times \frac{1-0.5}{0.5^4}\right)^{1/2}$$
$$= 6.029 \times 10^8\,\text{m}^{1/2}\,\text{s K}^2\,\text{kg}^{-1/2}.$$

The second term in Equation 2.19 becomes

$$\left(\frac{k}{m}\right)^2 = \left(\frac{1.381 \times 10^{-23}\,\text{J K}^{-1}}{0.6 \times 1.661 \times 10^{-27}\,\text{kg}}\right)^2 = 1.920 \times 10^8\,\text{m}^4\,\text{s}^{-4}\,\text{K}^{-2}.$$

Finally, the third term in Equation 2.19 becomes

$$\left(\frac{1}{G}\right)^{3/2} = \left(\frac{1}{6.673 \times 10^{-11}\,\text{N m}^2\,\text{kg}^{-2}}\right)^{3/2} = 1.835 \times 10^{15}\,\text{kg}^{3/2}\,\text{m}^{-9/2}\,\text{s}^3.$$

So the upper mass limit for a star is given by

$$M \approx (6.029 \times 10^8\,\text{m}^{1/2}\,\text{s K}^2\,\text{kg}^{-1/2}) \times (1.920 \times 10^8\,\text{m}^4\,\text{s}^{-4}\,\text{K}^{-2}) \times (1.835 \times 10^{15}\,\text{kg}^{3/2}\,\text{m}^{-9/2}\,\text{s}^3)$$
$$\approx 2.12 \times 10^{32}\,\text{kg} \approx 100\,M_\odot.$$

Exercise 2.3   The Kelvin–Helmholtz timescale for the Sun is

$$\tau_{KH,\odot} = \frac{GM_\odot^2}{R_\odot L_\odot} = \frac{6.673 \times 10^{-11} \text{ N m}^2 \text{ kg}^{-2} \times (1.99 \times 10^{30} \text{ kg})^2}{6.96 \times 10^8 \text{ m} \times 3.83 \times 10^{26} \text{ J s}^{-1}} = 9.90 \times 10^{14} \text{ s}.$$

This is equivalent to 9.90×10^{14} s $/ (365.25 \times 24 \times 3600)$ s yr$^{-1} \approx 3 \times 10^7$ yr.

Exercise 2.4   For a $0.5\,M_\odot$ star, the Kelvin–Helmholtz contraction time is
$\tau_{KH,0.5} \approx 3 \times 10^7$ yr $\times\, 0.5^{-2.4} \approx 1.6 \times 10^8$ yr.

For a $2\,M_\odot$ star, the Kelvin–Helmholtz contraction time is
$\tau_{KH,2} \approx 3 \times 10^7$ yr $\times\, 2^{-2.4} \approx 5.7 \times 10^6$ yr.

For a $5\,M_\odot$ star, the Kelvin–Helmholtz contraction time is
$\tau_{KH,5} \approx 3 \times 10^7$ yr $\times\, 5^{-2.4} \approx 6.3 \times 10^5$ yr.

Exercise 3.1   The time-independent Schrödinger equation in one dimension, for a constant barrier potential V is

$$\left[-\frac{\hbar^2}{2m_r} \frac{\partial^2}{\partial r^2} + V \right] \psi_s(r) = E\psi_s(r),$$

where m_r is the reduced mass. Equation 3.4 can also be written

$$\frac{\partial^2}{\partial r^2} \psi_s(r) = \frac{2m_r}{\hbar^2} (V - E) \psi_s(r)$$

or

$$\frac{\partial^2}{\partial r^2} \psi_s(r) = \chi^2 \psi_s(r) \quad \text{where} \quad \chi^2 = \frac{2m_r}{\hbar^2} (V - E).$$

To verify that the wave function $\psi_s(r) = \exp(\chi r)$ is a solution for a constant potential (i.e. when V and hence χ do not depend on r), substitute this into the left-hand side of the Schrödinger equation:

$$\frac{\partial^2}{\partial r^2} \psi_s(r) = \frac{\partial^2}{\partial r^2} \exp(\chi r)$$

Expand the second derivative

$$\frac{\partial^2}{\partial r^2} \psi_s(r) = \frac{\partial}{\partial r} \frac{\partial}{\partial r} \exp(\chi r).$$

Evaluate the first derivative

$$\frac{\partial^2}{\partial r^2} \psi_s(r) = \frac{\partial}{\partial r} \chi \exp(\chi r)$$

and then the second, but note that $\exp(\chi r) = \psi_s(r)$

$$\frac{\partial^2}{\partial r^2} \psi_s(r) = \chi^2 \exp(\chi r) = \chi^2 \psi_s(r),$$

which equals the right-hand side of Schrödinger equation, as required. Note that if the barrier potential had not been constant, then χ would depend on r, and the differentiation would not be so straightforward.

Exercise 3.2   The Gamow energy is $E_G = 2m_r c^2 (\pi \alpha Z_A Z_B)^2$, where m_r is the reduced mass of the two-body system, given by $m_r = m_A m_B / (m_A + m_B)$.

(a) Begin by calculating the reduced mass:

$$m_{\mathrm{r}} = \frac{m_{\mathrm{p}}m_{\mathrm{p}}}{m_{\mathrm{p}} + m_{\mathrm{p}}} = \frac{m_{\mathrm{p}}^2}{2m_{\mathrm{p}}} = \frac{m_{\mathrm{p}}}{2} = \frac{1.673 \times 10^{-27} \text{ kg}}{2} = 8.365 \times 10^{-28} \text{ kg.}$$

Then

$$E_{\mathrm{G}} = 2m_{\mathrm{r}}c^2(\pi\alpha Z_{\mathrm{p}}Z_{\mathrm{p}})^2$$

$$= 2 \times 8.365 \times 10^{-28} \text{ kg} \times (2.998 \times 10^8 \text{ m s}^{-1})^2 \times (\pi \times \frac{1}{137.0} \times 1 \times 1)^2$$

$$= 7.907 \times 10^{-14} \text{ kg m}^2 \text{ s}^{-2} = 7.907 \times 10^{-14} \text{ J.}$$

Since 1 eV = 1.602×10^{-19} J, $E_{\mathrm{G}} = 7.907 \times 10^{-14}$ J/1.602×10^{-19} J eV^{-1} = 493.6 keV.

(b) In this case, the reduced mass is

$$m_{\mathrm{r}} = \frac{m_3 m_3}{m_3 + m_3} = \frac{m_3^2}{2m_3} = \frac{m_3}{2} = \frac{3m_{\mathrm{p}}}{2} = \frac{3 \times 1.673 \times 10^{-27} \text{ kg}}{2} = 2.510 \times 10^{-27} \text{ kg.}$$

Then

$$E_{\mathrm{G}} = 2m_{\mathrm{r}}c^2(\pi\alpha Z_{3\mathrm{He}}Z_{3\mathrm{He}})^2$$

$$= 2 \times 2.510 \times 10^{-27} \text{ kg} \times (2.998 \times 10^8 \text{ m s}^{-1})^2 \times (\pi \times \frac{1}{137.0} \times 2 \times 2)^2$$

$$= 3.796 \times 10^{-12} \text{ kg m}^2 \text{ s}^{-2} = 3.796 \times 10^{-12} \text{ J.}$$

Since 1 cV = 1.602×10^{-19} J, $E_{\mathrm{G}} = 3.796 \times 10^{-12}$ J/1.602×10^{-19} J eV^{-1} = 23.70 MeV.

Exercise 3.3 The probability of barrier penetration is

$$P_{\mathrm{pen}} \approx \exp\left[-\left(\frac{E_{\mathrm{G}}}{E}\right)^{1/2}\right] \approx \exp\left[-\left(\frac{E_{\mathrm{G}}}{kT_{\mathrm{c}}}\right)^{1/2}\right].$$

(a) proton–proton:

$$P_{\mathrm{pen}} \approx \exp\left[-\left(\frac{493.6 \text{ keV}}{1.3 \text{ keV}}\right)^{1/2}\right] = 3.4 \times 10^{-9}.$$

(b) $_2^3$He–$_2^3$He:

$$P_{\mathrm{pen}} \approx \exp\left[-\left(\frac{23\,700 \text{ keV}}{1.3 \text{ keV}}\right)^{1/2}\right] = 2.3 \times 10^{-59}.$$

Note that the answers have been given to only 2 significant figures, rather than the 3 s.f. available, because the approximation that the energy is given by $E \approx kT_{\mathrm{c}}$ degrades the accuracy further.

Exercise 3.4 Since $S(E)$ is being treated as a constant, the integrand can be written

$$f(E) = S\exp\left[-\left(\frac{E_{\mathrm{G}}}{E}\right)^{1/2} - \frac{E}{kT}\right].$$

Differentiating gives

$$\frac{\mathrm{d}f(E)}{\mathrm{d}E} = S\frac{\mathrm{d}}{\mathrm{d}E}\exp\left[-\left(\frac{E_G}{E}\right)^{1/2} - \frac{E}{kT}\right]$$

but using Hint 2 gives

$$\frac{\mathrm{d}f(E)}{\mathrm{d}E} = S\exp\left[-\left(\frac{E_G}{E}\right)^{1/2} - \frac{E}{kT}\right] \times \frac{\mathrm{d}}{\mathrm{d}E}\left[-\left(\frac{E_G}{E}\right)^{1/2} - \frac{E}{kT}\right]$$

$$= S\exp\left[-\left(\frac{E_G}{E}\right)^{1/2} - \frac{E}{kT}\right] \times \left[-E_G^{1/2}\left(-\frac{1}{2}E^{-3/2}\right) - \frac{1}{kT}\right]$$

$$= S\exp\left[-\left(\frac{E_G}{E}\right)^{1/2}\right]\exp\left(-\frac{E}{kT}\right) \times \left[\frac{E_G^{1/2}}{2}(E^{-3/2}) - \frac{1}{kT}\right].$$

The integrand $f(E)$ is either a minimum or maximum when $\mathrm{d}f(E)/\mathrm{d}E = 0$, which is when one of the following terms is zero:

(i) $S = 0$; this is a trivial, uninteresting case.

(ii) $\exp[-(E_G/E)^{1/2}] \to 0$; this occurs when E becomes very small.

(iii) $\exp[-E/kT] \to 0$; this occurs when E becomes very large.

(iv)

$$\left[\frac{E_G^{1/2}}{2}(E^{-3/2}) - \frac{1}{kT}\right] = 0.$$

We rearrange this to get an expression for E

$$\frac{E_G^{1/2}}{2}(E^{-3/2}) = \frac{1}{kT}, \quad \text{and simplifying, we get} \quad \frac{E_G^{1/2}}{E^{1/2}} = \frac{2E}{kT}.$$

Comment: This result will be useful later.

Collecting terms in E, we have

$$E^{3/2} = \frac{kT}{2}E_G^{1/2},$$

then take the $(2/3)$-power of all terms

$$E = \left(\frac{kT}{2}\right)^{2/3}E_G^{1/3} = \left(E_G\left(\frac{kT}{2}\right)^2\right)^{1/3}.$$

This is the interesting case, and the energy $E_0 = [E_G(kT/2)^2]^{1/3}$ is called the Gamow peak.

(Do not confuse this with the Gamow energy E_G.)

Exercise 3.5 (a) First calculate the energy ratio:

$$\frac{E_G}{4kT} = \frac{2m_p m_{13C}}{m_p + m_{13C}} \times \frac{c^2(\pi \alpha Z_p Z_{13C})^2}{4kT}$$

$$= \frac{2 \times 1u \times 13u}{1u + 13u} \times \frac{(2.998 \times 10^8 \text{ m s}^{-1})^2 \times (\pi \times \frac{1}{137} \times 1 \times 6)^2}{4 \times 1.381 \times 10^{-23} \text{ J K}^{-1} \times 15.6 \times 10^6 \text{ K}}$$

$$= 3.667 \times 10^{30} u \frac{(\text{m s}^{-1})^2}{\text{J}}$$

$$= 3.667 \times 10^{30} \times 1.661 \times 10^{-27} \text{ kg} \frac{\text{m}^2 \text{ s}^{-2}}{\text{kg m}^2 \text{ s}^{-2}} = 6091.$$

Next compute the fusion rate per unit mass fraction:

$$\frac{R_{p13C}}{X_{13C}} = 6.48 \times 10^{-24} \times \frac{(A_p + A_{13C})\rho_c^2 X_p}{(A_p A_{13C} u)^2 [\text{m}^{-6}] Z_p Z_{13C}} \times \frac{S(E_0)}{[\text{keV barns}]} \left(\frac{E_G}{4kT}\right)^{2/3} \exp\left[-3\left(\frac{E_G}{4kT}\right)^{1/3}\right] \text{ m}^{-3} \text{ s}^{-1}$$

$$= 6.48 \times 10^{-24} \times \frac{(1 + 13) \times (1.48 \times 10^5 \text{ kg m}^{-3})^2 \times 0.5}{(1 \times 13 \times 1.661 \times 10^{-27} \text{ kg})^2 \times [\text{m}^{-6}] \times 1 \times 6}$$

$$\times \frac{5.5 \text{ keV barns}}{[\text{keV barns}]} \times (6091)^{2/3} \times \exp\left[-3 \times (6091)^{1/3}\right] \text{ m}^{-3} \text{ s}^{-1}$$

$$= 1.0 \times 10^{18} \text{ m}^{-3} \text{ s}^{-1}.$$

(b) First calculate the energy ratio:

$$\frac{E_G}{4kT} = \frac{2m_p m_{14N}}{m_p + m_{14N}} \times \frac{c^2(\pi \alpha Z_p Z_{14N})^2}{4kT}$$

$$= \frac{2 \times 1u \times 14u}{1u + 14u} \times \frac{(2.998 \times 10^8 \text{ m s}^{-1})^2 \times (\pi \times \frac{1}{137} \times 1 \times 7)^2}{4 \times 1.381 \times 10^{-23} \text{ J K}^{-1} \times 15.6 \times 10^6 \text{ K}}$$

$$= 5.017 \times 10^{30} u \frac{(\text{m s}^{-1})^2}{\text{J}}$$

$$= 5.017 \times 10^{30} \times 1.661 \times 10^{-27} \text{ kg} \frac{\text{m}^2 \text{ s}^{-2}}{\text{kg m}^2 \text{ s}^{-2}} = 8333.$$

Next compute the fusion rate per unit mass fraction:

$$\frac{R_{p14N}}{X_{14N}} = 6.48 \times 10^{-24} \times \frac{(A_p + A_{14N})\rho_c^2 X_p}{(A_p A_{14N} u)^2 [\text{m}^{-6}] Z_p Z_{14N}} \times \frac{S(E_0)}{[\text{keV barns}]} \left(\frac{E_G}{4kT}\right)^{2/3} \exp\left[-3\left(\frac{E_G}{4kT}\right)^{1/3}\right] \text{ m}^{-3} \text{ s}^{-1}$$

$$= 6.48 \times 10^{-24} \times \frac{(1 + 14) \times (1.48 \times 10^5 \text{ kg m}^{-3})^2 \times 0.5}{(1 \times 14 \times 1.661 \times 10^{-27} \text{ kg})^2 \times [\text{m}^{-6}] \times 1 \times 7}$$

$$\times \frac{3.3 \text{ keV barns}}{[\text{keV barns}]} \times (8333)^{2/3} \times \exp\left[-3 \times (8333)^{1/3}\right] \text{ m}^{-3} \text{ s}^{-1}$$

$$= 1.5 \times 10^{15} \text{ m}^{-3} \text{ s}^{-1}.$$

(c) First calculate the energy ratio:

$$\frac{E_G}{4kT} = \frac{2m_p m_{15N}}{m_p + m_{15N}} \times \frac{c^2 (\pi \alpha Z_p Z_{15N})^2}{4kT}$$

$$= \frac{2 \times 1u \times 15u}{1u + 15u} \times \frac{(2.998 \times 10^8 \text{ m s}^{-1})^2 \times (\pi \times \frac{1}{137} \times 1 \times 7)^2}{4 \times 1.381 \times 10^{-23} \text{ J K}^{-1} \times 15.6 \times 10^6 \text{ K}}$$

$$= 5.039 \times 10^{30} u \frac{(\text{m s}^{-1})^2}{\text{J}}$$

$$= 5.039 \times 10^{30} \times 1.661 \times 10^{-27} \text{ kg} \frac{\text{m}^2 \text{ s}^{-2}}{\text{kg m}^2 \text{ s}^{-2}} = 8370.$$

Next compute the fusion rate per unit mass fraction:

$$\frac{R_{p15N}}{X_{15N}} = 6.48 \times 10^{-24} \times \frac{(A_p + A_{15N})\rho_c^2 X_p}{(A_p A_{15N} u)^2 [\text{m}^{-6}] Z_p Z_{15N}} \times \frac{S(E_0)}{[\text{keV barns}]} \left(\frac{E_G}{4kT}\right)^{2/3} \exp\left[-3\left(\frac{E_G}{4kT}\right)^{1/3}\right] \text{ m}^{-3} \text{ s}^{-1}$$

$$= 6.48 \times 10^{-24} \times \frac{(1+15) \times (1.48 \times 10^5 \text{ kg m}^{-3})^2 \times 0.5}{(1 \times 15 \times 1.661 \times 10^{-27} \text{ kg})^2 \times [\text{m}^{-6}] \times 1 \times 7}$$

$$\times \frac{78 \text{ keV barns}}{[\text{keV barns}]} \times (8370)^{2/3} \times \exp\left[-3 \times (8370)^{1/3}\right] \text{ m}^{-3} \text{ s}^{-1}$$

$$= 3.0 \times 10^{16} \text{ m}^{-3} \text{ s}^{-1}.$$

Exercise 3.6   (a) In equilibrium $R_{p12C} = R_{p14N}$, so $3.5 \times 10^{17} X_{12C} = 0.015 \times 10^{17} X_{14N}$. This means

$$X_{14N}/X_{12C} = 3.5/0.015 = 230 \quad \text{and therefore} \quad {}^{14}_{7}\text{N}/{}^{12}_{6}\text{C} = 12/14 \times 230 = 200.$$

(b) In equilibrium $R_{p14N} = R_{p15N}$, so $0.015 \times 10^{17} X_{14N} = 0.30 \times 10^{17} X_{15N}$. This means

$$X_{14N}/X_{15N} = 0.30/0.015 = 20 \quad \text{and therefore} \quad {}^{14}_{7}\text{N}/{}^{15}_{7}\text{N} = 15/14 \times 20 = 21.$$

Exercise 3.7   (a) We begin with

$$R_{AB} = \frac{6.48 \times 10^{-24}}{A_r Z_A Z_B} \times \frac{n_A n_B}{[\text{m}^{-6}]} \times \frac{S(E_0)}{[\text{keV barns}]} \times \left(\frac{E_G}{4kT}\right)^{2/3} \exp\left[-3\left(\frac{E_G}{4kT}\right)^{1/3}\right] \text{ m}^{-3} \text{ s}^{-1}.$$

Following Hint 1, define

$$a = \frac{6.48 \times 10^{-24}}{A_r Z_A Z_B} \times \frac{n_A n_B}{[\text{m}^{-6}]} \times \frac{S(E_0)}{[\text{keV barns}]}$$

so

$$R_{AB} = a \left(\frac{E_G}{4kT}\right)^{2/3} \exp\left[-3\left(\frac{E_G}{4kT}\right)^{1/3}\right] \text{ m}^{-3} \text{ s}^{-1}$$

and in preparation for using Hint 2, write

$$u = a \left(\frac{E_G}{4kT}\right)^{2/3} \quad \text{and} \quad v = \exp\left[-3\left(\frac{E_G}{4kT}\right)^{1/3}\right]$$

so $R_{AB} = uv$ and then $\dfrac{dR_{AB}}{dT} = u\dfrac{dv}{dT} + v\dfrac{du}{dT}$.

For the sake of clarity, calculate these two parts separately.

Step 1: Calculate dv/dT.

$$\frac{dv}{dT} = \frac{d}{dT}\exp\left[-3\left(\frac{E_G}{4kT}\right)^{1/3}\right]$$

but using Hint 3

$$\frac{d\exp(y)}{dx} = \frac{d\exp(y)}{dy} \times \frac{dy}{dx} = \exp(y)\frac{dy}{dx}.$$

Hence

$$\frac{dv}{dT} = \exp\left[-3\left(\frac{E_G}{4kT}\right)^{1/3}\right]\frac{d}{dT}\left[-3\left(\frac{E_G}{4kT}\right)^{1/3}\right].$$

Note the the first exponential is just v again. Taking constants out of the differentiation gives

$$\frac{dv}{dT} = v\left[-3\left(\frac{E_G}{4k}\right)^{1/3}\right]\frac{d}{dT}T^{-1/3}.$$

So, differentiating the $T^{-1/3}$ part

$$\frac{dv}{dT} = v\left[-3\left(\frac{E_G}{4k}\right)^{1/3}\right]\left(-\frac{1}{3}\right)T^{-4/3} = v\left[\left(\frac{E_G}{4kT}\right)^{1/3}\right]\frac{1}{T}.$$

Step 2: Calculate du/dT.

$$\frac{du}{dT} = \frac{d}{dT}a\left(\frac{E_G}{4kT}\right)^{2/3}$$

taking the constants out of the differentiation gives

$$\frac{du}{dT} = a\left(\frac{E_G}{4k}\right)^{2/3}\frac{d}{dT}T^{-2/3} = a\left(\frac{E_G}{4k}\right)^{2/3}\left(-\frac{2}{3}\right)T^{-5/3} = a\left(\frac{E_G}{4kT}\right)^{2/3}\left(-\frac{2}{3}\right)\frac{1}{T}$$

but the first term on the right-hand side of the equation is just u again, so

$$\frac{du}{dT} = u\left(-\frac{2}{3}\right)\frac{1}{T}.$$

Step 3: Calculate $dR_{AB}/dT = u\,dv/dT + v\,du/dT$.

Substitute the results from Steps 1 and 2:

$$\frac{dR_{AB}}{dT} = uv\left[\left(\frac{E_G}{4kT}\right)^{1/3}\right]\frac{1}{T} + vu\left(-\frac{2}{3}\right)\frac{1}{T}$$

take out the common factor uv/T

$$\frac{dR_{AB}}{dT} = \frac{uv}{T}\left[\left(\frac{E_G}{4kT}\right)^{1/3} - \frac{2}{3}\right]$$

and note that uv is simply R_{AB}

$$\frac{dR_{AB}}{dT} = \frac{R_{AB}}{T}\left[\left(\frac{E_G}{4kT}\right)^{1/3} - \frac{2}{3}\right].$$

(b) From the chain rule

$$\frac{d\log_e R_{AB}}{d\log_e T} = \frac{d\log_e R_{AB}}{dR_{AB}}\frac{dR_{AB}}{dT}\frac{dT}{d\log_e T} = \frac{1}{R_{AB}}\frac{dR_{AB}}{dT}\left(\frac{d\log_e T}{dT}\right)^{-1} = \frac{T}{R_{AB}}\frac{dR_{AB}}{dT}.$$

So finally

$$\frac{d\log_e R_{AB}}{d\log_e T} = \left[\left(\frac{E_G}{4kT}\right)^{1/3} - \frac{2}{3}\right].$$

Exercise 3.8 $R_{AB} \propto T^{[(E_G/4kT)^{1/3} - \frac{2}{3}]}$, so you need to evaluate the value

$$\nu = \left(\frac{E_G}{4kT}\right)^{1/3} - \frac{2}{3}$$

for each reaction. Recall that $E_G = 2m_r c^2 (\pi\alpha Z_A Z_B)^2$ so

$$\nu = \left(\frac{2m_r c^2 (\pi\alpha Z_A Z_B)^2}{4kT}\right)^{1/3} - \frac{2}{3}.$$

(a) For p + p, begin by calculating the reduced mass:

$$m_r = \frac{m_p m_p}{m_p + m_p} = \frac{m_p^2}{2m_p} = \frac{m_p}{2} = \frac{1.673 \times 10^{-27}\,\text{kg}}{2} = 8.365 \times 10^{-28}\,\text{kg}.$$

Then, using $T_{\odot,c} = 15.6 \times 10^6$ K we obtain

$$\nu = \left(\frac{2m_r c^2 (\pi\alpha Z_p Z_p)^2}{4kT}\right)^{1/3} - \frac{2}{3}$$

$$= \left(\frac{2 \times 8.365 \times 10^{-28}\,\text{kg} \times (2.998 \times 10^8\,\text{m s}^{-1})^2 \times (\pi \times \frac{1}{137.0} \times 1 \times 1)^2}{4 \times 1.381 \times 10^{-23}\,\text{J K}^{-1} \times 15.6 \times 10^6\,\text{K}}\right)^{1/3} - \frac{2}{3}$$

$$= 3.84$$

i.e. $R_{pp} \propto T^{3.8}$.

(b) For p + $^{14}_{7}$N begin by calculating the reduced mass (where the reduced mass of $^{14}_{7}$N is given as m_{14}):

$$m_r = \frac{m_p m_{14}}{m_p + m_{14}} = \frac{1u \times 14u}{1u + 14u} = \frac{14u^2}{15u} = \frac{14}{15}u = \frac{14}{15} \times 1.673 \times 10^{-27}\,\text{kg} = 1.550 \times 10^{-27}\,\text{kg}.$$

Then, using $T_{\odot,c} = 15.6 \times 10^6$ K we obtain

$$\nu = \left(\frac{2m_r c^2 (\pi\alpha Z_p Z_{14N})^2}{4kT}\right)^{1/3} - \frac{2}{3}$$

$$= \left(\frac{2 \times 1.550 \times 10^{-27}\,\text{kg} \times (2.998 \times 10^8\,\text{m s}^{-1})^2 \times (\pi \times \frac{1}{137.0} \times 1 \times 7)^2}{4 \times 1.381 \times 10^{-23}\,\text{J K}^{-1} \times 15.6 \times 10^6\,\text{K}}\right)^{1/3} - \frac{2}{3}$$

$$= 19.6. \quad \text{Therefore } R_{p14N} \propto T^{19.6}.$$

Exercise 3.9 Your amended Figure 3.6 should resemble Figure S3.1.

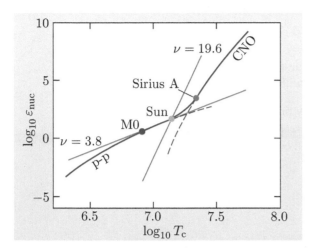

Figure S3.1 Copy of Figure 3.6, but with lines for $\varepsilon \propto T^\nu$ added, for $\nu = 3.8$ (p–p chain) and $\nu = 19.6$ (CNO cycle).

Exercise 4.1 The initial hydrogen content of the Sun is $0.70\,M_\odot$.

If it converted all of this into helium via the proton–proton chain, 0.0066 (i.e. $\approx 0.7\%$) of the hydrogen mass would be converted into energy. The total mass consumed would be

$$m = M_\odot \times 0.70 \times 0.0066 = 1.99 \times 10^{30}\ \text{kg} \times 0.70 \times 0.0066 = 9.2 \times 10^{27}\ \text{kg}.$$

This corresponds to an energy

$$E_\text{fusion} = mc^2 = 9.2 \times 10^{27}\ \text{kg} \times (2.998 \times 10^8\ \text{m s}^{-1})^2 = 8.3 \times 10^{44}\ \text{J}$$

over its lifetime.

The Sun's current luminosity is $L_\odot = 3.83 \times 10^{26}\ \text{J s}^{-1}$, so it could radiate at this rate for a lifetime given by

$$\tau_\text{nuc} = E_\text{fusion}/L_\odot = 8.3 \times 10^{44}\ \text{J}/3.83 \times 10^{26}\ \text{J s}^{-1} = 2.2 \times 10^{18}\ \text{s} \approx 70 \times 10^9\ \text{yr}$$

if it could indeed burn all of its hydrogen to helium.

Exercise 4.2 Lifetime $\propto M/L \propto M/M^{3.5} = 1/M^{2.5}$. If the solar lifetime is 10×10^9 yr, then the lifetime of a $0.5\,M_\odot$ star will be $1/0.5^{2.5} = 5.7$ times longer, i.e. 57×10^9 yr, and the lifetime of a $10\,M_\odot$ star will be $1/10^{2.5} = 0.0032$ times as long, i.e. 32×10^6 yr.

Exercise 4.3 For high-mass stars, the CNO cycle dominates energy production and the opacity is due entirely to electron scattering. For this case, $\nu \approx 17$. For stars of uniform chemical composition, μ is constant, so the μ-term can be absorbed into the unknown constant of proportionality.

(a) Equation 4.6 becomes $L \propto M^3 \mu^4$, i.e. $L \propto M^3$.

(b) Equation 4.8 becomes

$$T_\text{c} \propto M^{4/(\nu+3)}\, \mu^{7/(\nu+3)} \propto M^{4/(17+3)} \propto M^{4/20}$$

i.e. $T_\text{c} \propto M^{0.2}$.

(c) Equation 4.7 becomes

$$R \propto M^{(\nu-1)/(\nu+3)}\,\mu^{(\nu-4)/(\nu+3)} \propto M^{(17-1)/(17+3)} \propto M^{16/20}$$

i.e. $R \propto M^{0.8}$.

That is, the luminosity increases strongly with mass, the core temperature increases weakly with mass, and the radius increases almost linearly with mass.

Exercise 4.4   For high-mass stars, the CNO cycle dominates energy production and the opacity is due entirely to electron scattering. For this case, $\nu \approx 17$. For stars of constant mass, The M-term can be absorbed into the unknown constant of proportionality.

(a) Equation 4.6 becomes $L \propto M^3\mu^4$, i.e. $L \propto \mu^4$.

(b) Equation 4.8 becomes

$$T_c \propto M^{4/(\nu+3)}\mu^{7/(\nu+3)} \propto \mu^{7/(17+3)} \propto \mu^{7/20}$$

i.e. $T_c \propto \mu^{0.4}$.

(c) Equation 4.7 becomes

$$R \propto M^{\frac{\nu-1}{\nu+3}}\mu^{\frac{\nu-4}{\nu+3}} \propto \mu^{\frac{17-4}{17+3}} \propto \mu^{\frac{13}{20}}$$

i.e. $R \propto \mu^{0.7}$.

That is, the luminosity increases steeply with mean molecular mass, the radius increases moderately and the core temperature increases very moderately with mean molecular mass.

Exercise 4.5   (a) (i) For fully ionized hydrogen and helium in Big Bang proportions, the mean molecular mass is

$$\mu = \frac{\sum_i n_i \frac{m_i}{u}}{\sum_i n_i} = \frac{N_H m_p + N_{He} m_{He} + N_e m_e}{u(N_H + N_{He} + N_e)}.$$

Let N_{nuc} be the (unknown) total number of nuclei, so $N_H = 0.93 N_{nuc}$, $N_{He} = 0.07 N_{nuc}$, $N_e = N_H + 2N_{He} = 0.93 N_{nuc} + 2 \times 0.07 N_{nuc} = 1.07 N_{nuc}$, $m_{He} \approx 4 m_p$, and $m_e/m_p = 9.109 \times 10^{-31}$ kg/1.673×10^{-27} kg $= 1/1837$, so $m_e = m_p/1837$. Note also that $m_p/u = 1.673 \times 10^{-27}$ kg/1.661×10^{-27} kg $= 1.007$.

Substituting these into the expression for μ gives:

$$\mu = \frac{0.93 N_{nuc} m_p + 0.07 N_{nuc} 4 m_p + 1.07 N_{nuc}(m_p/1837)}{u(0.93 N_{nuc} + 0.07 N_{nuc} + 1.07 N_{nuc})}$$

but the N_{nuc} terms cancel, and m_p is a common factor on the top line, so

$$\mu = \frac{(0.93 + 0.28 + 0.00058)}{(0.93 + 0.07 + 1.07)}\frac{m_p}{u} = 0.58 \times 1.007 = 0.58.$$

Note that the electrons make a negligible contribution to the mass of the material (the numerator of the equation), but account for more than half of the *number* of particles, and hence greatly affect the denominator.

(ii) For fully ionized helium, the mean molecular mass is

$$\mu = \frac{\sum_i n_i \frac{m_i}{u}}{\sum_i n_i} = \frac{N_{\text{He}} m_{\text{He}} + N_e m_e}{u(N_{\text{He}} + N_e)}.$$

Now, $N_e = 2N_{\text{He}}$, $m_{\text{He}} = 4m_p$ and $m_e = m_p/1837$, so

$$\mu = \frac{N_{\text{He}} 4m_p + 2N_{\text{He}}(m_p/1837)}{u(N_{\text{He}} + 2N_{\text{He}})} = \frac{(4 + 0.00109)}{1 + 2} \frac{m_p}{u} = 1.33 \times 1.007 = 1.34.$$

The ratio of μ in case (ii) to μ in case (i) is $1.34/0.58 = 2.3$.

(b) (i) For stars burning hydrogen by the p–p chain: $R \propto \mu^0$, so $R_{\text{final}}/R_{\text{initial}} \propto (\mu_{\text{final}}/\mu_{\text{initial}})^0 \propto 2.3^0 \approx 1$. So the star would be the same size.

$L \propto \mu^4$, so $L_{\text{final}}/L_{\text{initial}} \propto (\mu_{\text{final}}/\mu_{\text{initial}})^4 \propto 2.3^4 \approx 28$. So the star would be a lot brighter!

$T_c \propto \mu^1$, so $T_{c,\text{final}}/T_{c,\text{initial}} \propto (\mu_{\text{final}}/\mu_{\text{initial}})^1 \propto 2.3^1 \approx 2.3$. So the star would be hotter.

(ii) For stars burning hydrogen by the CNO-cycle: $R \propto \mu^{0.7}$, so $R_{\text{final}}/R_{\text{initial}} \propto (\mu_{\text{final}}/\mu_{\text{initial}})^{0.7} \propto 2.3^{0.7} \approx 1.8$. So the star would expand.

$L \propto \mu^4$, so $L_{\text{final}}/L_{\text{initial}} \propto (\mu_{\text{final}}/\mu_{\text{initial}})^4 \propto 2.3^4 \approx 28$. So the star would be a lot brighter!

$T_c \propto \mu^{0.4}$, so $T_{c,\text{final}}/T_{c,\text{initial}} \propto (\mu_{\text{final}}/\mu_{\text{initial}})^{0.4} \propto 2.3^{0.4} \approx 1.4$. So the star would be hotter.

Exercise 4.6 (a) For $s = 3$, $\gamma = (1 + s/2) \div (s/2) = (5/2) \div (3/2) = 5/3$, so the coefficient $(\gamma - 1)/\gamma = ((5/3) - 1) \div (5/3) = 2/5$.

(b) As $s \to \infty$, $\gamma \to (1/s + 1/2) \div (1/2) = (0 + 1) \div 1 = 1$, so the coefficient $(\gamma - 1)/\gamma = (1 - 1) \div 1 = 0$.

The critical temperature gradient for convection is

$$\frac{dT}{dr} < \frac{(\gamma - 1)}{\gamma} \frac{T}{P} \frac{dP}{dr}.$$

Since, in part (b), $dT/dr < 0$ and $(\gamma - 1)/\gamma = 0$, this material is *always* unstable to convection.

Exercise 5.1 (a) $\mu'(^{8}_{4}\text{Be}) = \mu'(^{4}_{2}\text{He}) + \mu'(^{4}_{2}\text{He}) = 2\mu'(^{4}_{2}\text{He})$.

(b) Since the chemical potential is $\mu' = mc^2 - kT \log_e(g_s n_{\text{QNR}}/n)$, in this case the result of part (a) and Equation 5.1 give

$$m_8 c^2 - kT \log_e\left(\frac{g_8\, n_{Q8}}{n_8}\right) = 2\left[m_4 c^2 - kT \log_e\left(\frac{g_4\, n_{Q4}}{n_4}\right)\right].$$

We need to find an expression for n_8/n_4, so rearrange the equation to work towards that goal. As a first step, collect the logarithms on one side and the mc^2 terms on the other:

$$m_8 c^2 - 2m_4 c^2 = kT\left(\log_e\left(\frac{g_8\, n_{Q8}}{n_8}\right) - 2\log_e\left(\frac{g_4\, n_{Q4}}{n_4}\right)\right)$$

$$\frac{(m_8 - 2m_4)c^2}{kT} = \log_e\left[\frac{(g_8\, n_{Q8}/n_8)}{(g_4\, n_{Q4}/n_4)^2}\right]$$

$$\exp\left[\frac{(m_8 - 2m_4)c^2}{kT}\right] = \frac{(g_8\, n_{Q8}/n_8)}{(g_4\, n_{Q4}/n_4)^2}$$

$$\frac{n_8}{n_4^2} = \exp\left[\frac{-(m_8 - 2m_4)c^2}{kT}\right]\frac{g_8\, n_{Q8}}{(g_4\, n_{Q4})^2}. \tag{S5.1}$$

Since $n_{QA} = (2\pi m_A kT/h^2)^{3/2}$, the n_Q-term at the end of Equation S5.1 is

$$\frac{n_{Q8}}{(n_{Q4})^2} = \frac{\left(2\pi m_8 kT/h^2\right)^{3/2}}{\left[\left(2\pi m_4 kT/h^2\right)^{3/2}\right]^2} = \frac{\left(2\pi m_8 kT/h^2\right)^{3/2}}{\left(2\pi m_4 kT/h^2\right)^3} = \left(\frac{2\pi kT}{h^2}\right)^{-3/2}\left(\frac{m_8}{m_4^2}\right)^{3/2}.$$

Substituting this into Equation S5.1 gives

$$\frac{n_8}{n_4^2} = \exp\left[\frac{-(m_8 - 2m_4)c^2}{kT}\right]\frac{g_8}{g_4^2}\left(\frac{m_8}{m_4^2}\right)^{3/2}\left(\frac{h^2}{2\pi kT}\right)^{3/2}.$$

As we want to find the relative abundances of the nuclei, n_8/n_4, we must multiply both sides by n_4. Doing this, and using $n_4 = \rho X_4/m_4$, we obtain the final expression

$$\frac{n_8}{n_4} = \exp\left[\frac{-(m_8 - 2m_4)c^2}{kT}\right]\frac{g_8}{g_4^2}\rho X_4 \frac{m_8^{3/2}}{m_4^4}\left(\frac{h^2}{2\pi kT}\right)^{3/2}. \tag{S5.2}$$

We assume that the material is primarily $^4_2$He, so $X_4 = 1$. In evaluating the ratio $m_8^{3/2}/m_4^4$ we can use the approximation $m_4 = 4u$ and $m_8 = 8u$, but the term $m_8 - 2m_4$ involves the subtraction of nearly equal numbers, and for that we cannot use this approximation. However, we are given that $\Delta Q = (2m_4 - m_8)c^2 = -91.8\,\text{keV}$ so $(m_8 - 2m_4)c^2 = 91.8\,\text{keV}$. We calculated the polarizations g_4 and g_8 in the bulleted question at the end of Section 5.2.

So, evaluating Equation S5.2 at $T = 2 \times 10^8$ K and $\rho = 10^8$ kg m^{-3} gives:

$$\frac{n_8}{n_4} = \exp\left[-\frac{91.8\,\text{keV} \times 1.602 \times 10^{-16}\,\text{J keV}^{-1}}{1.381 \times 10^{-23}\,\text{J K}^{-1} \times 2 \times 10^8\,\text{K}}\right] \times \frac{1}{1^2} \times 10^8\,\text{kg m}^{-3} \times 1$$

$$\times \frac{(8u)^{3/2}}{(4u)^4} \times \left(\frac{(6.626 \times 10^{-34}\,\text{J s})^2}{2\pi \times 1.381 \times 10^{-23}\,\text{J K}^{-1} \times 2 \times 10^8\,\text{K}}\right)^{3/2}$$

$$= 4.87 \times 10^{-3} \times 10^8\,\text{kg m}^{-3} \times 0.0884u^{-5/2} \times 1.27 \times 10^{-79}\,\text{J}^{3/2}\,\text{s}^3$$

$$= 5.47 \times 10^{-75} \times (1.661 \times 10^{-27})^{-5/2}\,\text{kg}^{-3/2}\,\text{m}^{-3}\,\text{J}^{3/2}\,\text{s}^3$$

$$= 4.86 \times 10^{-8}.$$

Equivalently, $n_4/n_8 = 1/4.86 \times 10^{-8} = 2.1 \times 10^7$. That is, there is roughly one $^8_4$Be nucleus for every 21 million $^4_2$He nuclei!

Exercise 5.2 (a) The Gamow energy is $E_G = 2m_r c^2 (\pi\alpha Z_A Z_B)^2$, where m_r is the reduced mass of the two-body system, given by $m_r = m_A m_B/(m_A + m_B)$.

Begin by calculating the reduced mass:

$$m_r = \frac{m_4 m_4}{m_4 + m_4} = \frac{m_4^2}{2m_4} = \frac{m_4}{2} = 2u = 2 \times 1.661 \times 10^{-27}\,\text{kg} = 3.322 \times 10^{-27}\,\text{kg}.$$

Then

$$E_G = 2m_r c^2 (\pi \alpha Z_4 Z_4)^2$$

$$= 2 \times 3.322 \times 10^{-27} \text{ kg} \times (2.998 \times 10^8 \text{ m s}^{-1})^2 \times \left(\pi \times \frac{1}{137.0} \times 2 \times 2 \right)^2$$

$$= 5.024 \times 10^{-12} \text{ J}.$$

Since $1 \text{ eV} = 1.602 \times 10^{-19}$ J, the Gamow energy,
$E_G = 5.024 \times 10^{-12}$ J$/1.602 \times 10^{-19}$ J eV$^{-1} = 31.4$ MeV.

(b) The energy of the Gamow peak is $E_0 = (E_G (kT)^2/4)^{1/3}$. So, cubing both
sides and multiplying by $4/E_G$ gives $(kT)^2 = 4E_0^3/E_G$. Taking the square root
and dividing by k gives

$$T = \frac{2}{k} \sqrt{\frac{E_0^3}{E_G}}.$$

Now $\Delta Q = 91.8$ keV can be written in SI units as
$\Delta Q = 91.8 \text{ keV} \times 1.602 \times 10^{-16}$ J keV$^{-1} = 1.47 \times 10^{-14}$ J. So, if the energy of
the Gamow peak coincides with this value of ΔQ, then

$$T = \frac{2}{k} \sqrt{\frac{E_0^3}{E_G}} = \frac{2}{1.381 \times 10^{-23} \text{ J K}^{-1}} \sqrt{\frac{(1.47 \times 10^{-14} \text{ J})^3}{5.024 \times 10^{-12} \text{ J}}}$$

$$= 1.448 \times 10^{23} \text{J}^{-1} \text{ K} \times 7.95 \times 10^{-16} \text{ J} = 1.15 \times 10^8 \text{ K}.$$

Exercise 5.3 (a) The Gamow energy is $E_G = 2m_r c^2 (\pi \alpha Z_A Z_B)^2$, where m_r
is the reduced mass of the two-body system, given by $m_r = m_A m_B/(m_A + m_B)$.

Begin by calculating the reduced mass:

$$m_r = \frac{m_4 m_8}{m_4 + m_8} = \frac{4u \times 8u}{4u + 8u} = \frac{32u^2}{12u} = \frac{8u}{3} = 8 \times 1.661 \times 10^{-27} \text{ kg}/3 = 4.429 \times 10^{-27} \text{ kg}.$$

Then

$$E_G = 2m_r c^2 (\pi \alpha Z_4 Z_8)^2$$

$$= 2 \times 4.429 \times 10^{-27} \text{ kg} \times (2.998 \times 10^8 \text{ m s}^{-1})^2 \times \left(\pi \times \frac{1}{137.0} \times 2 \times 4 \right)^2$$

$$= 2.679 \times 10^{-11} \text{ J}.$$

Since $1 \text{ eV} = 1.602 \times 10^{-19}$ J, the Gamow energy,
$E_G = 2.679 \times 10^{-11}$ J$/1.602 \times 10^{-19}$ J eV$^{-1} = 167.2$ MeV.

(b) The energy of the Gamow peak is $E_0 = (E_G (kT)^2/4)^{1/3}$. So, cubing both
sides and multiplying by $4/E_G$ gives $(kT)^2 = 4E_0^3/E_G$. Taking the square root
and dividing by k gives

$$T = \frac{2}{k} \sqrt{\frac{E_0^3}{E_G}}.$$

Now $\Delta Q = 287.7$ keV can be written in SI units as
$\Delta Q = 287.7 \text{ keV} \times 1.602 \times 10^{-16}$ J keV$^{-1} = 4.611 \times 10^{-14}$ J. So, if the energy

of the Gamow peak coincides with this value of ΔQ, then

$$T = \frac{2}{k}\sqrt{\frac{E_0^3}{E_G}} = \frac{2}{1.381 \times 10^{-23}\text{ J K}^{-1}}\sqrt{\frac{(4.611 \times 10^{-14}\text{ J})^3}{2.679 \times 10^{-11}\text{ J}}}$$

$$= 1.448 \times 10^{23}\text{J}^{-1}\text{ K} \times 1.913 \times 10^{-15}\text{ J} = 2.77 \times 10^8\text{ K}.$$

Exercise 5.4   (a) Ignoring the electronic parts, we find the mass defect is
$m_{\text{initial}} - m_{\text{final}} = 3 \times 4.00260\text{ amu} - 12\text{ amu} = 0.007800\text{ amu}.$

(b) Using $E = \Delta m\,c^2$ we have $E = (0.007800 \times 1.661 \times 10^{-27}\text{ kg})$
$\times(2.998 \times 10^8\text{ m s}^{-1})^2 = 1.164 \times 10^{-12}\text{ J}$. Since $1\text{ eV} = 1.602 \times 10^{-19}\text{ J}$, this is
equivalent to $E = 1.164 \times 10^{-12}\text{ J} / 1.602 \times 10^{-19}\text{ J eV}^{-1} = 7.269 \times 10^6\text{ eV}$ or
about 7.27 MeV.

In the earlier subsections, Step 1 is said to require 91.8 keV and Step 2 is said to
require 287.7 keV, whilst Step 3 releases 7.65 MeV. The net energy released is
therefore $(7.65 - 0.2877 - 0.0918)\text{ MeV} = 7.27\text{ MeV}$, in agreement with the
above.

(c) As a fraction of the initial mass this is
$0.007800\text{ amu}/(3 \times 4.00260\text{ amu}) = 0.00065$. Recall that for hydrogen burning,
the mass defect corresponds to 0.0066 of the initial mass, a factor of ten larger.

Exercise 5.5   (a) (i) For the proton

$$\lambda_{\text{dB}}(\text{p}) = h/(3m_\text{p}kT)^{1/2}$$

$$= \frac{6.626 \times 10^{-34}\text{ J s}}{\sqrt{3 \times 1.673 \times 10^{-27}\text{ kg} \times 1.381 \times 10^{-23}\text{ J K}^{-1} \times 15.6 \times 10^6\text{ K}}}$$

$$= 6.372 \times 10^{-13}\text{ m}.$$

(ii) For the electron

$$\lambda_{\text{dB}}(\text{e}) = h/(3m_\text{e}kT)^{1/2}$$

$$= \frac{6.626 \times 10^{-34}\text{ J s}}{\sqrt{3 \times 9.109 \times 10^{-31}\text{ kg} \times 1.381 \times 10^{-23}\text{ J K}^{-1} \times 15.6 \times 10^6\text{ K}}}$$

$$= 2.731 \times 10^{-11}\text{ m}.$$

(b) The ratio of de Broglie wavelengths is therefore

$$\frac{\lambda_{\text{dB}}(\text{e})}{\lambda_{\text{dB}}(\text{p})} = \frac{h}{(3m_\text{e}kT)^{1/2}} \times \frac{(3m_\text{p}kT)^{1/2}}{h} = \left(\frac{m_\text{p}}{m_\text{e}}\right)^{1/2} = \left(\frac{1.673 \times 10^{-27}\text{ kg}}{9.109 \times 10^{-31}\text{ kg}}\right)^{1/2} = 42.85.$$

The de Broglie wavelength of an electron is greater than that of a proton, by a
factor of ≈ 43.

(c) From part (b), this ratio depends only on the mass of the particles, and
hence is independent of the environment and hence of the temperature. The
electron's wavelength is about 40 times longer than the proton's in any star. Since
a proton and a neutron have nearly the same mass, we can say that the de Broglie
wavelength of an electron is about 40 times greater than that of a nucleon in any
star.

Exercise 5.6   The second condition is $n \gg n_Q$, i.e. $n \gg (2\pi mkT/h^2)^{3/2}$.

Since $n = 1/l^3$, substituting for n gives $1/l^3 \gg (2\pi mkT/h^2)^{3/2}$.

Taking the (1/3)-power gives $1/l \gg (2\pi mkT/h^2)^{1/2}$, and multiplying through by $l/(2\pi mkT/h^2)^{1/2}$ gives

$$l \ll h/(2\pi mkT)^{1/2} = (3/2\pi)^{1/2} \times h/(3mkT)^{1/2} = (3/2\pi)^{1/2} \times \lambda_{dB} < 0.7\lambda_{dB}.$$

That is, from the second degeneracy condition $n \gg n_Q$ we obtain the first degeneracy condition $l \ll \lambda_{dB}$.

Exercise 5.7   (a) $n_Q = (2\pi mkT/h^2)^{3/2}$, so the degeneracy condition $n \gg n_Q$ implies that $n \gg (2\pi mkT/h^2)^{3/2}$.

Taking the (2/3)-power and multiplying both sides by $h^2/2\pi m$ gives $n^{2/3}h^2/(2\pi m) \gg kT$, i.e. $kT \ll n^{2/3}h^2/(2\pi m)$.

The third equivalent condition for degeneracy is: the gas is degenerate if its temperature $T \ll n^{2/3}h^2/(2\pi mk)$.

(b) (i) $n_p = \rho_{\odot,c}X_H/m_p = 1.48 \times 10^5$ kg m^{-3} $\times 0.5/1.673 \times 10^{-27}$ kg $= 4.42 \times 10^{31}$ m^{-3} (i.e. hydrogen nuclei per m^3).

Therefore

$$\frac{h^2 n_p^{2/3}}{2\pi m_p k} = \frac{(6.626 \times 10^{-34} \text{ J s})^2 \times (4.42 \times 10^{31} \text{ m}^{-3})^{2/3}}{(2 \times \pi \times 1.673 \times 10^{-27} \text{ kg} \times 1.381 \times 10^{-23} \text{ J K}^{-1})}$$
$$= 3.78 \times 10^3 \text{ J s}^2 \text{ m}^{-2} \text{ kg}^{-1} \text{ K} \approx 3780 \text{ K}.$$

(ii) In the solar core, all atoms are ionized. The electrons are provided by the hydrogen and helium which each account for 0.5 of the composition by mass.

$$n_e = n_p + 2n_{He} = \frac{\rho_{\odot,c}X_H}{m_p} + \frac{2\rho_{\odot,c}X_{He}}{m_{He}} = \rho_{\odot,c}\left(\frac{X_H}{m_p} + \frac{2X_{He}}{4u}\right)$$
$$= 1.48 \times 10^5 \text{ kg m}^{-3} \left(\frac{0.5}{1.673 \times 10^{-27} \text{ kg}} + \frac{2 \times 0.5}{4 \times 1.661 \times 10^{-27} \text{ kg}}\right)$$
$$= 6.65 \times 10^{31} \text{ m}^{-3}.$$

Therefore

$$\frac{h^2 n_e^{2/3}}{2\pi m_e k} = \frac{(6.626 \times 10^{-34} \text{ J s})^2 \times (6.65 \times 10^{31} \text{ m}^{-3})^{2/3}}{(2 \times \pi \times 9.109 \times 10^{-31} \text{ kg} \times 1.381 \times 10^{-23} \text{ J K}^{-1})}$$
$$= 9.12 \times 10^6 \text{ J s}^2 \text{ m}^{-2} \text{ kg}^{-1} \text{ K} \approx 9.12 \times 10^6 \text{ K}.$$

(c) The temperature condition for degeneracy is $T \ll n^{2/3}h^2/(2\pi mk)$. $T_{\odot,c} = 15.6 \times 10^6$ K, so the temperature in the core of the Sun is (i) much too high for proton degeneracy to have set in, and (ii) marginally too high for electron degeneracy to have set in.

Exercise 6.1   (a) Equating the core pressure $P_c = (\pi/36)^{1/3}GM^{2/3}\rho_c^{4/3}$ to the pressure of non-relativistic degenerate electrons $P_{NR} = K_{NR}(\rho_c Y_e/m_H)^{5/3}$, where $K_{NR} = (h^2/5m_e)(3/8\pi)^{2/3}$, we have $K_{NR}(\rho_c Y_e/m_H)^{5/3} = (\pi/36)^{1/3}GM^{2/3}\rho_c^{4/3}$.

Collecting terms in ρ_c on the left-hand side, and all others on the right, we get

$$\rho_c^{1/3} = \left(\frac{\pi}{36}\right)^{1/3} \frac{G}{K_{NR}} \frac{m_H^{5/3}}{Y_e^{5/3}} M^{2/3}$$

cubing this gives

$$\rho_c = \left(\frac{\pi}{36}\right) \left(\frac{G}{K_{NR}}\right)^3 \frac{m_H^5}{Y_e^5} M^2$$

and substituting for K_{NR} gives

$$\rho_c = \left(\frac{\pi}{36}\right) \left(\frac{5m_e}{h^2}\right)^3 \left(\frac{8\pi}{3}\right)^2 G^3 \frac{m_H^5}{Y_e^5} M^2.$$

Consolidating the numerical factors, we have

$$\rho_c = \left(\frac{16\pi^3}{81}\right) \left(\frac{5m_e}{h^2}\right)^3 G^3 \frac{m_H^5}{Y_e^5} M^2.$$

(b) The electron number density in the core of the star is $n_e = \rho_c Y_e / m_H$, so substituting the core density from above gives

$$n_e = \left(\frac{16\pi^3}{81}\right) \left(\frac{5m_e}{h^2}\right)^3 G^3 \frac{m_H^4}{Y_e^4} M^2.$$

Exercise 6.2   The Fermi energy is

$$E_F \approx \frac{25}{2} \left(\frac{2\pi^2}{27}\right)^{2/3} \left(\frac{G}{h}\right)^2 \left(\frac{m_H}{Y_e}\right)^{8/3} M^{4/3} m_e$$

which as a fraction of the electron rest-mass energy is

$$\frac{E_F}{m_e c^2} \approx \frac{25}{2} \left(\frac{2\pi^2}{27}\right)^{2/3} \left(\frac{G}{h}\right)^2 \left(\frac{m_H}{Y_e}\right)^{8/3} \frac{M^{4/3}}{c^2}.$$

So in this case for a $0.4\,M_\odot$ white dwarf, the Fermi energy as a fraction of the electron rest-mass energy is

$$\frac{E_F}{m_e c^2} \approx \frac{25}{2} \left(\frac{2\pi^2}{27}\right)^{2/3} \left(\frac{6.673 \times 10^{-11}\ \text{N m}^2\ \text{kg}^{-2}}{6.626 \times 10^{-34}\ \text{J s}}\right)^2 \left(\frac{1.673 \times 10^{-27}\ \text{kg}}{0.5}\right)^{8/3} \frac{(0.4 \times 1.99 \times 10^{30}\ \text{kg})^{4/3}}{(2.998 \times 10^8\ \text{m s}^{-1})^2}$$

$$\approx 0.212.$$

Exercise 6.3   (a) For the ultra-relativistic case, equating P_{UR} to P_c gives $P_{UR} = K_{UR}(\rho_c Y_e/m_H)^{4/3} = (\pi/36)^{1/3} G M^{2/3} \rho_c^{4/3}$ where $K_{UR} = (hc/4)(3/8\pi)^{1/3}$.

Note that in contrast to the non-relativistic case, the density term is the same on both sides ($\rho_c^{4/3}$), so cancels, leaving

$$K_{UR} \left(\frac{Y_e}{m_H}\right)^{4/3} = \left(\frac{\pi}{36}\right)^{1/3} G M^{2/3}.$$

Collecting M on one side and swapping left and right sides, gives

$$M^{2/3} = \left(\frac{36}{\pi}\right)^{1/3} \left(\frac{K_{UR}}{G}\right) \left(\frac{Y_e}{m_H}\right)^{4/3}.$$

Taking the (3/2)-power and substituting for K_{UR} gives

$$M = \left(\frac{36}{\pi}\right)^{1/2} \left(\frac{(hc/4)(3/8\pi)^{1/3}}{G}\right)^{3/2} \left(\frac{Y_e}{m_H}\right)^2 = \left(\frac{3 \times 36}{4^3 \times 8\pi^2}\right)^{1/2} \left(\frac{hc}{G}\right)^{3/2} \left(\frac{Y_e}{m_H}\right)^2$$

$$= \frac{1}{\pi} \left(\frac{27}{128}\right)^{1/2} \left(\frac{hc}{G}\right)^{3/2} \left(\frac{Y_e}{m_H}\right)^2 .$$

(b) Putting in the numbers we have

$$M = \frac{1}{\pi} \left(\frac{27}{128}\right)^{1/2} \left(\frac{hc}{G}\right)^{3/2} \left(\frac{Y_e}{m_H}\right)^2$$

$$= \frac{1}{\pi} \left(\frac{27}{128}\right)^{1/2} \left(\frac{6.626 \times 10^{-34}\,\mathrm{J\,s} \times 2.998 \times 10^8\,\mathrm{m\,s^{-1}}}{6.673 \times 10^{-11}\,\mathrm{N\,m^2\,kg^{-2}}}\right)^{3/2} \left(\frac{0.5}{1.673 \times 10^{-27}\,\mathrm{kg}}\right)^2$$

$$= 2.121 \times 10^{30}\,\mathrm{kg}.$$

Since $M_\odot = 1.99 \times 10^{30}\,\mathrm{kg}$, $M = 2.121 \times 10^{30}\,\mathrm{kg}/1.99 \times 10^{30}\,\mathrm{kg}\,M_\odot^{-1} = 1.07\,M_\odot$.

Exercise 6.4   (a) The radius as a function of mass and average density is $R = (3M/4\pi\langle\rho\rangle)^{1/3}$.

(b) If the average density is 1/6 of the core density, then $R = (9M/2\pi\rho_c)^{1/3}$.

(c) Substituting the core density gives:

$$R_{\mathrm{WD}} = \left(\frac{9M}{2\pi}\right)^{1/3} \left(\frac{16\pi^3}{81}\right)^{-1/3} \left(\frac{5m_e}{h^2}\right)^{-1} G^{-1} \frac{m_H^{-5/3}}{Y_e^{-5/3}} M^{-2/3}.$$

Rearranging the terms with negative powers

$$R_{\mathrm{WD}} = \left(\frac{9M}{2\pi}\right)^{1/3} \left(\frac{81}{16\pi^3}\right)^{1/3} \left(\frac{h^2}{5m_e}\right) \frac{Y_e^{5/3}}{Gm_H^{5/3}} M^{-2/3}.$$

Now consolidate the mass terms and the numerical factors

$$R_{\mathrm{WD}} = \left(\frac{729}{32\pi^4}\right)^{1/3} \left(\frac{h^2}{5m_e}\right) \frac{Y_e^{5/3}}{Gm_H^{5/3}} M^{-1/3}.$$

Now for convenience, express the mass in solar units

$$R_{\mathrm{WD}} = \left(\frac{729}{32\pi^4}\right)^{1/3} \left(\frac{h^2}{5m_e}\right) \frac{Y_e^{5/3}}{Gm_H^{5/3}} \left(\frac{M}{M_\odot}\right)^{-1/3} \times (1.99 \times 10^{30}\,\mathrm{kg})^{-1/3}.$$

Then putting in the numbers

$$R_{\mathrm{WD}} = \left(\frac{729}{32\pi^4}\right)^{1/3} \frac{(6.626 \times 10^{-34}\,\mathrm{J\,s})^2}{(5 \times 9.109 \times 10^{-31}\,\mathrm{kg})} \frac{(0.5)^{5/3} \times (1.99 \times 10^{30}\,\mathrm{kg})^{-1/3}}{(6.673 \times 10^{-11}\,\mathrm{N\,m^2\,kg^{-2}}) \times (1.673 \times 10^{-27}\,\mathrm{kg})^{5/3}} \left(\frac{M}{M_\odot}\right)^{-1/3}.$$

And so

$$R_{\mathrm{WD}} = 9.45 \times 10^6 \left(\frac{M}{M_\odot}\right)^{-1/3}\,\mathrm{m}.$$

Since $R_{\odot} = 6.96 \times 10^8$ m,

$$R_{\mathrm{WD}} = \left(\frac{9.45 \times 10^6 \text{ m}}{6.96 \times 10^8 \text{ m R}_{\odot}^{-1}}\right) \times \left(\frac{M}{\mathrm{M}_{\odot}}\right)^{-1/3}$$

$$= \frac{\mathrm{R}_{\odot}}{74} \times \left(\frac{M}{\mathrm{M}_{\odot}}\right)^{-1/3}.$$

(d) Since $R_{\mathrm{Earth}} = 6.37 \times 10^6$ m,

$$R_{\mathrm{WD}} = \left(\frac{9.45 \times 10^6 \text{ m}}{6.37 \times 10^6 \text{ m R}_{\mathrm{Earth}}^{-1}}\right) \times \left(\frac{M}{\mathrm{M}_{\odot}}\right)^{-1/3}$$

$$= 1.5 \, \mathrm{R}_{\mathrm{Earth}} \times \left(\frac{M}{\mathrm{M}_{\odot}}\right)^{-1/3}.$$

Exercise 6.5

Table S6.1   Nucleosynthesis processes

Process	Major reactions	Products	Mass range of stars	Ignition temp/K	Timescale
Big Bang	(not studied)	$^1_1$H, $^2_1$H, $^3_2$He $^4_2$He, $^7_4$Li	(not applicable)	(not studied)	~ 15 minutes
H-burning	p–p chain (3 branches)	$^4_2$He	$M_{\mathrm{ms}} \geq 0.08\,\mathrm{M}_{\odot}$	$(2\text{--}10) \times 10^6$	$\sim 10^7$ to 10^{10} yr
	CNO cycle	$^4_2$He, $^{13}_6$C, $^{14}_7$N			
He-burning	triple-alpha process	$^{12}_6$C, $^{16}_8$O	$M_{\mathrm{ms}} \geq 0.5\,\mathrm{M}_{\odot}$	$(1\text{--}2) \times 10^8$	$\sim 10^6$ yr
C-burning	$^{12}_6$C + $^{12}_6$C	$^{20}_{10}$Ne + $^4_2$He	$M_{\mathrm{ms}} \geq 8\,\mathrm{M}_{\odot}$	$(5\text{--}9) \times 10^8$	~ 500 yr
	$^{12}_6$C + $^{12}_6$C	$^{23}_{11}$Na + p			
	$^{12}_6$C + $^{12}_6$C	$^{23}_{12}$Mg + n			
Ne-burning	$^{20}_{10}$Ne + γ	$^{16}_8$O + $^4_2$He	$M_{\mathrm{ms}} \geq 10\,\mathrm{M}_{\odot}$	$(1\text{--}2) \times 10^9$	~ 1 yr
	$^{20}_{10}$Ne + $^4_2$He	$^{24}_{12}$Mg			
O-burning	$^{16}_8$O + $^{16}_8$O	$^{28}_{14}$Si + $^4_2$He	$M_{\mathrm{ms}} \geq 10\,\mathrm{M}_{\odot}$	$(2\text{--}3) \times 10^9$	~ 6 months
Si-burning	$^{28}_{14}$Si + γ	$^{24}_{12}$Mg + $^4_2$He	$M_{\mathrm{ms}} \geq 11\,\mathrm{M}_{\odot}$	$(3\text{--}4) \times 10^9$	~ 1 day
	$^{28}_{14}$Si + $n\,^4_2$He	$^{32}_{16}$S, $^{36}_{18}$Ar,			
	(successive captures	$^{40}_{20}$Ca, $^{44}_{20}$Ca,			
	of α-particles)	$^{48}_{22}$Ti, $^{52}_{24}$Cr, $^{56}_{26}$Fe			
neutron-capture	^A_ZX + n	$^{A+1}_Z$X			
β-decay	$^{A+1}_Z$X	$^{A+1}_{Z+1}$(X + 1) + e$^-$ + $\bar{\nu}_e$			
	s-process	Zr, Mo, Ba, Ce, Pb, Bi	$M_{\mathrm{ms}} \geq 1\,\mathrm{M}_{\odot}$		$\sim 10^4$ yr
	r-process	Kr, Sr, Te, Xe, Cs, Os, Pt, Au, Hg, Th, U	$M_{\mathrm{ms}} \geq 10\,\mathrm{M}_{\odot}$		~ 1 second

Exercise 7.1   We begin by balancing the chemical potentials:

$$\mu_4' = 2\mu_p' + 2\mu_n'$$

and since $\mu' = mc^2 - kT \log_e(g_s n_{QNR}/n)$ we have

$$m_4c^2 - kT \log_e\left(\frac{g_4\, n_{Q4}}{n_4}\right) = 2m_pc^2 - 2kT \log_e\left(\frac{g_p\, n_{Qp}}{n_p}\right) + 2m_nc^2 - 2kT \log_e\left(\frac{g_n\, n_{Qn}}{n_n}\right)$$

where m_4, m_p and m_n are the masses of a helium-4 nucleus, a proton and a neutron respectively; n_{Q4}, n_{Qp} and n_{Qn} are the non-relativistic quantum concentrations of a helium-4 nucleus, a proton and a neutron respectively; g_4, g_p and g_n are the number of polarizations of helium-4 nuclei, protons and neutrons respectively; and n_4, n_p and n_n are the number densities of helium-4 nuclei, protons and neutrons respectively. This may be rearranged as

$$2m_pc^2 + 2m_nc^2 - m_4c^2 = kT \log_e\left[\left(\frac{g_p\, n_{Qp}}{n_p}\right)^2 \left(\frac{g_n\, n_{Qn}}{n_n}\right)^2 \left(\frac{g_4\, n_{Q4}}{n_4}\right)^{-1}\right].$$

The left-hand side is simply ΔQ, so taking the exponential of both sides and rearranging slightly, we have:

$$\exp\left(\frac{\Delta Q}{kT}\right) = \frac{(g_p\, n_{Qp}/n_p)^2(g_n\, n_{Qn}/n_n)^2}{g_4\, n_{Q4}/n_4}.$$

Since we are interested in the proportion of helium-4 nuclei that are dissociated, we take this fraction onto the left-hand side to get:

$$\frac{n_p^2\, n_n^2}{n_4} = \frac{g_p^2\, g_n^2}{g_4} \frac{n_{Qp}^2\, n_{Qn}^2}{n_{Q4}} \exp\left(-\frac{\Delta Q}{kT}\right).$$

Exercise 7.2   The number of particles is $N = M/m$. For a total core mass of $M = (1/2) \times 1.4\, M_\odot$, the number of helium-4 nuclei is $N = (1/2) \times 1.4\, M_\odot / 4u = (0.7 \times 1.99 \times 10^{30}\ \text{kg}) / (4 \times 1.661 \times 10^{-27}\ \text{kg}) = 2.10 \times 10^{56}$ nuclei.

Each nucleus absorbs 28.3 MeV = 28.3×10^6 eV $\times\, 1.602 \times 10^{-19}$ J eV^{-1} = 4.534×10^{-12} J.

So the core absorbs $2.10 \times 10^{56} \times 4.534 \times 10^{-12}$ J = 9.5×10^{44} J by the photodisintegration of helium-4 nuclei.

Exercise 7.3

Table S7.1 Properties of white dwarfs and neutron stars

White dwarf	Neutron star
(a) Pressure of non-relativistic degenerate matter $$P_{\mathrm{NR}} = \frac{h^2}{5m_{\mathrm{e}}} \left(\frac{3}{8\pi}\right)^{2/3} n_{\mathrm{e}}^{5/3}$$	$$P_{\mathrm{NR}} = \frac{h^2}{5m_{\mathrm{n}}} \left(\frac{3}{8\pi}\right)^{2/3} n_{\mathrm{n}}^{5/3}$$
(b) In terms of mass density $$n_{\mathrm{e}} = Y_{\mathrm{e}}\rho_{\mathrm{c}}/m_{\mathrm{H}}$$ $$P_{\mathrm{NR}} = \frac{h^2}{5m_{\mathrm{e}}} \left(\frac{3}{8\pi}\right)^{2/3} \left(\frac{Y_{\mathrm{e}}\rho_{\mathrm{c}}}{m_{\mathrm{H}}}\right)^{5/3}$$	$$n_{\mathrm{n}} = \rho_{\mathrm{c}}/m_{\mathrm{n}}$$ $$P_{\mathrm{NR}} = \frac{h^2}{5m_{\mathrm{n}}} \left(\frac{3}{8\pi}\right)^{2/3} \left(\frac{\rho_{\mathrm{c}}}{m_{\mathrm{n}}}\right)^{5/3}$$
(c) Clayton model $$P_{\mathrm{c}} = (\pi/36)^{1/3} G M^{2/3} \rho_{\mathrm{c}}^{4/3}$$ $$\rho_{\mathrm{c}} = (36/\pi)^{1/4} G^{-3/4} M^{-1/2} P_{\mathrm{c}}^{3/4}$$	$$P_{\mathrm{c}} = (\pi/36)^{1/3} G M^{2/3} \rho_{\mathrm{c}}^{4/3}$$ $$\rho_{\mathrm{c}} = (36/\pi)^{1/4} G^{-3/4} M^{-1/2} P_{\mathrm{c}}^{3/4}$$
(d) Put (b) into (c) $$\rho_{\mathrm{c}} = \left(\frac{36}{\pi}\right)^{1/4} G^{-3/4} M^{-1/2}$$ $$\times \left[\frac{h^2}{5m_{\mathrm{e}}} \left(\frac{3}{8\pi}\right)^{2/3} \left(\frac{Y_{\mathrm{e}}\rho_{\mathrm{c}}}{m_{\mathrm{H}}}\right)^{5/3} \right]^{3/4}$$ $$= \left(\frac{36}{\pi}\right)^{1/4} G^{-3/4} M^{-1/2}$$ $$\times \left(\frac{h^2}{5m_{\mathrm{e}}}\right)^{3/4} \left(\frac{3}{8\pi}\right)^{1/2} \left(\frac{Y_{\mathrm{e}}\rho_{\mathrm{c}}}{m_{\mathrm{H}}}\right)^{5/4}$$	$$\rho_{\mathrm{c}} = \left(\frac{36}{\pi}\right)^{1/4} G^{-3/4} M^{-1/2}$$ $$\times \left[\frac{h^2}{5m_{\mathrm{n}}} \left(\frac{3}{8\pi}\right)^{2/3} \left(\frac{\rho_{\mathrm{c}}}{m_{\mathrm{n}}}\right)^{5/3} \right]^{3/4}$$ $$= \left(\frac{36}{\pi}\right)^{1/4} G^{-3/4} M^{-1/2}$$ $$\times \left(\frac{h^2}{5m_{\mathrm{n}}}\right)^{3/4} \left(\frac{3}{8\pi}\right)^{1/2} \left(\frac{\rho_{\mathrm{c}}}{m_{\mathrm{n}}}\right)^{5/4}$$
Collect powers of ρ_{c} $$\rho_{\mathrm{c}}^{-1/4} = \left(\frac{36}{\pi}\right)^{1/4} G^{-3/4} M^{-1/2}$$ $$\times \left(\frac{h^2}{5m_{\mathrm{e}}}\right)^{3/4} \left(\frac{3}{8\pi}\right)^{1/2} \left(\frac{Y_{\mathrm{e}}}{m_{\mathrm{H}}}\right)^{5/4}$$	$$\rho_{\mathrm{c}}^{-1/4} = \left(\frac{36}{\pi}\right)^{1/4} G^{-3/4} M^{-1/2}$$ $$\times \left(\frac{h^2}{5m_{\mathrm{n}}}\right)^{3/4} \left(\frac{3}{8\pi}\right)^{1/2} \left(\frac{1}{m_{\mathrm{n}}}\right)^{5/4}$$
Raise both sides to the power 4/3 $$\rho_{\mathrm{c}}^{-1/3} = \left(\frac{36}{\pi}\right)^{1/3} G^{-1} M^{-2/3}$$ $$\times \left(\frac{h^2}{5m_{\mathrm{e}}}\right) \left(\frac{3}{8\pi}\right)^{2/3} \left(\frac{Y_{\mathrm{e}}}{m_{\mathrm{H}}}\right)^{5/3}$$ $$= \frac{324^{1/3}}{4\pi G} \left(\frac{h^2}{5m_{\mathrm{e}}}\right) \left(\frac{Y_{\mathrm{e}}}{m_{\mathrm{H}}}\right)^{5/3} M^{-2/3}$$	$$\rho_{\mathrm{c}}^{-1/3} = \left(\frac{36}{\pi}\right)^{1/3} G^{-1} M^{-2/3}$$ $$\times \left(\frac{h^2}{5m_{\mathrm{n}}}\right) \left(\frac{3}{8\pi}\right)^{2/3} \left(\frac{1}{m_{\mathrm{n}}}\right)^{5/3}$$ $$= \frac{324^{1/3}}{4\pi G} \left(\frac{h^2}{5m_{\mathrm{n}}}\right) \left(\frac{1}{m_{\mathrm{n}}}\right)^{5/3} M^{-2/3}$$
(e) Adopt $\langle\rho\rangle = \rho_{\mathrm{c}}/6$ in density expression $$R = (3M/4\pi\langle\rho\rangle)^{1/3} = (9M/2\pi\rho_{\mathrm{c}})^{1/3}$$ $$R = (9/2\pi)^{1/3} \rho_{\mathrm{c}}^{-1/3} M^{1/3}$$	$$R = (3M/4\pi\langle\rho\rangle)^{1/3} = (9M/2\pi\rho_{\mathrm{c}})^{1/3}$$ $$R = (9/2\pi)^{1/3} \rho_{\mathrm{c}}^{-1/3} M^{1/3}$$
(f) Substitute for ρ_{c} using (d) $$R_{\mathrm{WD}} = \left(\frac{729}{32\pi^4}\right)^{1/3} \frac{1}{G} \left(\frac{h^2}{5m_{\mathrm{e}}}\right) \left(\frac{Y_{\mathrm{e}}}{m_{\mathrm{H}}}\right)^{5/3} M^{-1/3}$$	$$R_{\mathrm{NS}} = \left(\frac{729}{32\pi^4}\right)^{1/3} \frac{1}{G} \left(\frac{h^2}{5m_{\mathrm{n}}}\right) \left(\frac{1}{m_{\mathrm{n}}}\right)^{5/3} M^{-1/3}$$

Exercise 7.4   (a) Equation 7.5 may be re-arranged as

$$R_{\mathrm{max}} = \left[GM \left(\frac{P}{2\pi}\right)^2 \right]^{1/3}.$$

So in this case,

$$R_{\text{max}} = \left[6.673 \times 10^{-11} \text{ N m}^2 \text{ kg}^{-2} \times 1.99 \times 10^{30} \text{ kg} \times \left(\frac{1 \text{ s}}{2\pi} \right)^2 \right]^{1/3}$$

$$= 1.5 \times 10^6 \text{ m} = 1500 \text{ km}.$$

So any object larger than 1500 km radius cannot rotate more quickly than once a second. This effectively rules out rotating white dwarfs as the origin of pulsars.

(b) Equation 7.5 is

$$P_{\text{min}} = 2\pi \left(\frac{R^3}{GM} \right)^{1/2}.$$

So in this case

$$P_{\text{min}} = 2\pi \left(\frac{(10^4 \text{ m})^3}{6.673 \times 10^{-11} \text{ N m}^2 \text{ kg}^{-2} \times 1.99 \times 10^{30} \text{ kg}} \right)^{1/2}$$

$$= 0.55 \times 10^{-3} \text{ s} = 0.55 \text{ ms}.$$

So a $1 \, M_\odot$ neutron star can rotate as fast as 2000 times per second. Rapidly rotating neutron stars clearly can provide an explanation for pulsars.

Exercise 7.5   The magnetic field strength is

$$B = \frac{\mu_0 m}{4\pi R^3}.$$

Substituting for the magnetic dipole moment,

$$m = \left[-\frac{\dot{E}_{\text{rot}}}{\omega^4 \sin^2 \theta} \frac{3c^3}{2} \frac{4\pi}{\mu_0} \right]^{1/2}$$

we have

$$B = \frac{\mu_0}{4\pi R^3} \left[-\frac{\dot{E}_{\text{rot}}}{\omega^4 \sin^2 \theta} \frac{3c^3}{2} \frac{4\pi}{\mu_0} \right]^{1/2}$$

and then substituting for the rate of loss of energy $\dot{E}_{\text{rot}} = I\omega\dot{\omega} = 2MR^2\omega\dot{\omega}/5$ we have

$$B = \frac{\mu_0}{4\pi R^3} \left[-\frac{2MR^2\omega\dot{\omega}}{5\omega^4 \sin^2 \theta} \frac{3c^3}{2} \frac{4\pi}{\mu_0} \right]^{1/2}.$$

Then collecting together some terms we get

$$B = \left(\frac{\mu_0}{4\pi} \right)^{1/2} \frac{1}{R^2} \left(\frac{2M}{5} \right)^{1/2} \left(\frac{3c^3}{2} \right)^{1/2} \left(\frac{-\dot{\omega}}{\omega^3} \right)^{1/2} \frac{1}{\sin \theta}.$$

Now we note that, since $P = 2\pi/\omega$, then $\dot{P} = \dot{\omega} \times (-2\pi/\omega^2)$, so $-\dot{\omega}/\omega^3 = \dot{P}P/4\pi^2$, hence

$$B = \left(\frac{\mu_0}{4\pi} \right)^{1/2} \frac{1}{R^2} \left(\frac{2M}{5} \right)^{1/2} \left(\frac{3c^3}{2} \right)^{1/2} \frac{(\dot{P}P)^{1/2}}{2\pi} \frac{1}{\sin \theta}.$$

As required, this is an expression for the magnetic field strength of a pulsar in terms of its rotation period, rate of change of rotation period, mass, radius and other physical constants.

Exercise 7.6 Equation 7.10 may be rearranged to give

$$1 + \frac{2\dot{\omega}\tau}{\omega} = \frac{\omega^2}{\omega_0^2}$$

or

$$\omega_0 = \left(\frac{\omega^3}{\omega + 2\dot{\omega}\tau} \right)^{1/2}.$$

So for the Crab pulsar we have an initial angular frequency of

$$\omega_0 = \left(\frac{(190 \text{ s}^{-1})^3}{(190 \text{ s}^{-1}) + (2 \times -2.4 \times 10^{-9} \text{ s}^{-2} \times 950 \times 365 \times 24 \times 3600 \text{ s})} \right)^{1/2}$$

$$\omega_0 \sim 400 \text{ s}^{-1}.$$

Exercise 8.1 (a) The Jeans mass is $M_J = 3kTR/2G\overline{m}$. Rearranging this, setting the mass equal to that of the Sun and setting $\overline{m} = 2u$, gives

$$R = M_J \frac{2G\overline{m}}{3kT} = M_\odot \frac{2G\,2u}{3kT}.$$

Substituting in values gives

$$R = 1.99 \times 10^{30} \text{ kg} \times \frac{2 \times 6.673 \times^{-11} \text{ N m}^2 \text{ kg}^{-2} \times 2 \times 1.661 \times 10^{-27} \text{ kg}}{3 \times 1.381 \times 10^{-23} \text{ J K}^{-1} \times 20 \text{ K}}$$

$$= 1.07 \times 10^{15} \text{ m}.$$

This is the Jeans length for a solar mass of molecular hydrogen at a temperature of 20 K.

Since 1 pc = 3.086×10^{16} m, 1 AU = 1.496×10^{11} m, and $R_\odot = 6.96 \times 10^8$ m, we can also express this distance as

$$R = 1.07 \times 10^{15} \text{ m}/3.086 \times 10^{16} \text{ m pc}^{-1} = 0.0347 \text{ pc} \quad (\approx 0.03 \text{ pc})$$

$$R = 1.07 \times 10^{15} \text{ m}/1.496 \times 10^{11} \text{ m AU}^{-1} = 7.15 \times 10^3 \text{ AU} \quad (\approx 7000 \text{ AU})$$

$$R = 1.07 \times 10^{15} \text{ m}/6.96 \times 10^8 \text{ m R}_\odot^{-1} = 1.54 \times 10^6 \text{ R}_\odot \quad (\approx 1.5 \text{ million R}_\odot).$$

So, a cloud of molecular hydrogen at 20 K with a mass equal to that of the Sun will collapse if its radius is less than about 7000 AU.

(b) The Jeans density is

$$\rho_J = \frac{3}{4\pi M^2} \left(\frac{3kT}{2G\overline{m}} \right)^3$$

so in this case

$$\rho_J = \frac{3}{4\pi(1.99 \times 10^{30} \text{ kg})^2} \left(\frac{3 \times 1.381 \times 10^{-23} \text{ J K}^{-1} \times 20 \text{ K}}{2 \times 6.673 \times 10^{-11} \text{ N m}^2 \text{ kg}^{-2} \times 2 \times 1.661 \times 10^{-27} \text{ kg}} \right)^3$$

$$\sim 4 \times 10^{-16} \text{ kg m}^{-3}.$$

The Jeans density for a $1 \text{ M}_\odot$ cloud of molecular hydrogen at 20 K is 4×10^{-16} kg m^{-3} or about 120 billion molecules per cubic metre.

Exercise 8.2 The density is

$$\rho = \rho_J = \frac{3}{4\pi M^2} \left(\frac{3kT}{2G\overline{m}} \right)^3$$

so the free-fall time is found by substituting this in Equation 2.5:

$$\tau_{\text{ff}} = \left(\frac{3\pi}{32G}\frac{1}{\rho}\right)^{1/2} = \left(\frac{3\pi}{32G}\frac{4\pi M^2}{3}\left(\frac{2G\overline{m}}{3kT}\right)^3\right)^{1/2} = \left((\pi G)^2 M^2\left(\frac{\overline{m}}{3kT}\right)^3\right)^{1/2}.$$

The mass of the molecules of H_2 is $\overline{m} \approx 2u$, so putting $M = M_\odot$ we have

$$\tau_{\text{ff}} = \left((\pi G)^2 M_\odot^2\left(\frac{2u}{3kT}\right)^3\right)^{1/2} = \pi G M_\odot\left(\frac{2u}{3kT}\right)^{3/2}$$

$$= \pi \times 6.673 \times 10^{-11}\,\text{N m}^2\,\text{kg}^{-2} \times 1.99 \times 10^{30}\,\text{kg} \times \left(\frac{2 \times 1.661 \times 10^{-27}\,\text{kg}}{3 \times 1.381 \times 10^{-23}\,\text{J K}^{-1} \times 20\,\text{K}}\right)^{3/2}$$

$$= 3.35 \times 10^{12}\,\text{s} \approx 1.1 \times 10^5\,\text{yr}.$$

So, a $1\,M_\odot$ cloud of molecular hydrogen at $20\,\text{K}$ with the Jeans density will collapse (if unopposed by internal pressure) within about one hundred thousand years.

Exercise 8.3   The Jeans density is

$$\rho_J = \frac{3}{4\pi M^2}\left(\frac{3kT}{2G\overline{m}}\right)^3$$

which can be rearranged to give

$$M_J = \sqrt{\frac{3}{4\pi\rho_J}\left(\frac{3kT}{2G\overline{m}}\right)^3}.$$

We now substitute into this $\rho_J = n\overline{m}$ to give

$$M_J = \left(\frac{3}{4\pi n\overline{m}}\right)^{1/2}\left(\frac{3kT}{2G\overline{m}}\right)^{3/2}.$$

We can now evaluate the Jeans masses.

(a) Using $\overline{m} \approx 1u$ (i.e. 1 amu) for neutral atomic hydrogen,

$$M_J = \left(\frac{3 \times 1.381 \times 10^{-23}\,\text{J K}^{-1} \times 100\,\text{K}}{2 \times 6.673 \times 10^{-11}\,\text{N m}^2\,\text{kg}^{-2} \times 1.661 \times 10^{-27}\,\text{kg}}\right)^{3/2} \times \left(\frac{3}{4\pi \times 10^6\,\text{m}^{-3} \times 1.661 \times 10^{-27}\,\text{kg}}\right)^{1/2}$$

$$= 3.06 \times 10^{34}\,\text{kg}.$$

Since $1\,M_\odot = 1.99 \times 10^{30}\,\text{kg}$, we have $M_J = 3.06 \times 10^{34}\,\text{kg}/1.99 \times 10^{30}\,\text{kg}\,M_\odot^{-1}$
$= 1.54 \times 10^4\,M_\odot$.

(b) Using $\overline{m} \approx 2u$ (i.e. 2 amu) for molecular hydrogen,

$$M_J = \left(\frac{3 \times 1.381 \times 10^{-23}\,\text{J K}^{-1} \times 10\,\text{K}}{2 \times 6.673 \times 10^{-11}\,\text{N m}^2\,\text{kg}^{-2} \times 2 \times 1.661 \times 10^{-27}\,\text{kg}}\right)^{3/2}\left(\frac{3}{4\pi \times 10^9\,\text{m}^{-3} \times 2 \times 1.661 \times 10^{-27}\,\text{kg}}\right)^{1/2}$$

$$= 7.66 \times 10^{30}\,\text{kg}.$$

Since $1\,M_\odot = 1.99 \times 10^{30}\,\text{kg}$, we have $M_J = 7.66 \times 10^{30}\,\text{kg}/1.99 \times 10^{30}\,\text{kg}\,M_\odot^{-1}$
$= 3.85\,M_\odot$.

Exercise 8.4 (a) Beginning with the Jeans mass (written in terms of density and temperature)

$$M_J = \sqrt{\frac{3}{4\pi\rho_J}\left(\frac{3kT}{2G\overline{m}}\right)^3}$$

we re-arrange this to get

$$\left(\frac{4\pi\rho_J}{3}\right)^{1/2}\left(\frac{2G\overline{m}}{3k}\right)^{3/2}M_J = T^{3/2}.$$

Raising each side to the (2/3)-power, and re-ordering the terms gives

$$T = \frac{2G\overline{m}}{3k}\left(\frac{4\pi}{3}\right)^{1/3}M_J^{2/3}\rho_J^{1/3}.$$

(b) Since $\log_{10} AB = \log_{10} A + \log_{10} B$, and $\log_{10} A^k = k\log_{10} A$, we get

$$\log_{10} T = \log_{10}\left(\frac{2G\overline{m}}{3k}\left(\frac{4\pi}{3}\right)^{1/3}\right) + \frac{2}{3}\log_{10} M_J + \frac{1}{3}\log_{10}\rho_J.$$

The first term on the right-hand side of the equation is merely a constant, whereas the second and third terms depend on mass and density. Drawn in the log temperature versus log density plane, the curve of T against ρ_J for a *given* protostellar mass M_J is a straight line of slope 1/3. The Jeans line for a protostellar mass 10 times higher (or lower) is also a straight line of slope 1/3, but offset vertically from the first by +2/3 (or –2/3) logarithmic units.

Exercise 8.5 (a) We have two expressions for pressure, $P \propto \rho^\gamma$ for an adiabatic process, and $P = \rho kT/\overline{m}$ for an ideal gas.

Substituting for P in the first gives $\rho kT/\overline{m} \propto \rho^\gamma$.

Dividing both sides by $\rho k/\overline{m}$ gives $T \propto (\overline{m}/k)\rho^{\gamma-1}$, but $\overline{m}$ and k are both constants, so we can write $T \propto \rho^{\gamma-1}$.

(b) Since $T \propto \rho^{\gamma-1}$, we can also write $T = \text{constant} \times \rho^{\gamma-1}$.
Taking logarithms of both sides of the equation gives
$\log_{10} T = \log_{10}(\text{constant}) + (\gamma - 1)\log_{10}\rho$.
Therefore the adiabats are straight lines of slope $\gamma - 1$.

(c) Since

$$\gamma = \frac{1 + (s/2)}{(s/2)},$$

so for $s = 3$, then $\gamma = 5/3$, and the slope $\gamma - 1$ of the adiabat of an ideal gas with three degrees of freedom is 5/3 – 3/3 = 2/3.

(d) s takes values from 3 to ∞, so γ takes values from 5/3 to 1, and the slope is in the range from 2/3 to 0.

(e) If $\gamma = 4/3$, then the adiabat has slope $\gamma - 1 = 1/3$.

Exercise 8.6 (a) Using Equation 8.5,

$$E_{DI} = \frac{M}{2m_H}E_D + \frac{M}{m_H}E_I = \frac{1.99 \times 10^{30}\text{ kg}}{1.673 \times 10^{-27}\text{ kg}}\left(\frac{4.5\text{ eV}}{2} + 13.6\text{ eV}\right) = 1.885 \times 10^{58}\text{ eV}.$$

Since 1 eV $= 1.602 \times 10^{-19}$ J, this is equivalent to
$(1.885 \times 10^{58}$ eV$) \times (1.602 \times 10^{-19}$ J eV$^{-1}) \sim 3 \times 10^{39}$ J.

(b) Equating this energy to the change in gravitational potential energy, we have

$$\frac{GM^2}{R_2} - \frac{GM^2}{R_1} \sim 3 \times 10^{39} \text{ J}$$

so

$$R_2 = GM^2 \left(\frac{GM^2}{R_1} + 3 \times 10^{39} \text{ J} \right)^{-1}$$

$$= 6.673 \times 10^{-11} \text{ N m}^2 \text{ kg}^{-2} \times (1.99 \times 10^{30} \text{ kg})^2$$

$$\times \left(\frac{6.673 \times 10^{-11} \text{ N m}^2 \text{ kg}^{-2} \times (1.99 \times 10^{30} \text{ kg})^2}{10^{15} \text{ m}} + 3 \times 10^{39} \text{ J} \right)^{-1}$$

$$= 8.8 \times 10^{10} \text{ m}.$$

Hence the cloud collapses to a radius of about 10^{11} m or $\approx 150\,R_\odot$.

Exercise 8.7   Equation 8.8 can be re-written as

$$\frac{N(M_1)}{N(M_2)} = \left(\frac{M_1}{M_2} \right)^{-2.35}$$

where M_1 and M_2 are two particular masses. So the number of stars with each mass are:

$$N(50) = \left(\frac{50}{100} \right)^{-2.35} \sim 5 \qquad \text{and} \qquad N(10) = \left(\frac{10}{100} \right)^{-2.35} \sim 200,$$

$$N(5) = \left(\frac{5}{100} \right)^{-2.35} \sim 1000 \qquad \text{and} \qquad N(1) = \left(\frac{1}{100} \right)^{-2.35} \sim 50\,000,$$

$$N(0.5) = \left(\frac{0.5}{100} \right)^{-2.35} \sim 250\,000 \qquad \text{and} \qquad N(0.1) = \left(\frac{0.1}{100} \right)^{-2.35} \sim 11 \text{ million}.$$

Acknowledgements

Grateful acknowledgement is made to the following sources:

Figures

Cover image courtesy of NASA and The Hubble Heritage Team (AURA/STScl); Figures 2.2 and 3.5: Adapted from Phillips, A. C. (1999) *The Physics of Stars*, 2nd edition, John Wiley and Sons Limited; Figure 3.2: Wikipedia.com; Figure 3.6: Adapted from Schwarzschild, M. (1958) *Structure and Evolution of the Stars*, Dover Publications Inc; Figure 4.1: Adapted from Arp, H. C. (1958) *Handbook of Physics*, Vol. 51. Springer Verlag; Figure 4.2: Adapted from Pont, F. et al. (1998) 'Hipparcos subdwarfs and globular cluster ages: M92', *Astronomy & Astrophysics*, Vol. 329, pp. 87–100, Kluwer Academic Publishers; Figure 4.3a: Royal Observatory Edinburgh/Anglo–Australian Telescope Board/David Malin; Figure 4.3b: Nigel Sharp, Mark Hanna/AURA/NOAO/NSF; Figure 4.3c: Hubble Heritage Team (AURA/ST ScI/NASA); Figure 4.6: Böhm-Vitense, E. (1992) *Introduction to Stellar Astrophysics*, Cambridge University Press; Figure 5.1: Adapted from Mateo, M., (1987) Ph.D. Thesis, University of Washington; Figure 5.2: Adapted from Stetson, P. B. et al. (1999) 'Ages for globular clusters in the outer galactic halo', *The Astronomical Journal*, Vol. 117, January 1999, University of Chicago Press, ©Stetson et al.; Figure 5.3: Strohmeier, W. (1972) *Variable Stars*, Pergamon Press. ©1972 W. Strohmeier; Figure 5.6: Hesser, Harris and Vandenberg (1987) *Publications of the Astronomical Society of the Pacific*, Vol. 99, 1987, The Astronomical Society of the Pacific; Figure 6.1: S. Kwok (University of Calgary)/R. Rubin (NASA Ames Research Center)/ H. Bond (ST ScI) and NASA; Figure 6.2a: H. Bond (ST ScI) and NASA; Figure 6.2b: NASA, A. Fruchter and the ERO team (ST ScI); Figure 6.4: Kitt Peak National Observatory 0.9-meter telescope, National Optical Astronomy Observatories, courtesy of M. Bolte, University of California, Santa Cruz; Figures 7.1a and b: Anglo–Australian Observatory; Figure 7.1c: Hubble Heritage Team (AURA/ST ScI/NASA); Figure 7.2a: MDM Observatory; Figures 7.2b, 7.2c and 7.3: NASA/CXC/SAO; Figure 7.2d: NASA/CXC/SAO/Rutgers/J Hughes; Figure 7.4: Adapted from Lorimer, D. R. and Kramer, M. (2004) *Handbook of Pulsar Astronomy*, Cambridge University Press; Figure 8.1: Jeff Hester and Paul Scowen (AZ State Univ)/NASA; Figure 8.8: N. Walborn (ST ScI), R. Barbva (La Plata Observatory) and NASA; Figure 8.9: Yohko Tsuboi, Pennsylvania State University; Figure 8.10: C. Burrows (ST ScI)/J. Hester (AZ State University)/ J. Morse (ST ScI)/NASA; Figure 8.11: Adapted from Strahler, S. (1983), Astrophysical Journal, vol. 274, University of Chicago Press; Figure 8.12: A. Schultz (Computer Sciences Corp), S. Head (NASA Goddard Space Flight Center and NASA); Figure 8.13: M. J. McCaughrean (MPIA)/ C. R. O'Dell (Rice University)/NASA.

Every effort has been made to contact copyright holders. If any have been inadvertently overlooked the publishers will be pleased to make the necessary arrangements at the first opportunity.

Index

Items that appear in the Glossary have page numbers in **bold type**. Ordinary index items have page numbers in Roman type.